The Politics of Immunity

The Politics of Immunity

Security and the Policing of Bodies

Mark Neocleous

VERSO
London • New York

First published by Verso 2022

1 3 5 7 9 10 8 6 4 2

Verso
UK: 6 Meard Street, London W1F 0EG
US: 20 Jay Street, Suite 1010, Brooklyn, NY 11201
versobooks.com

Verso is the imprint of New Left Books

ISBN-13: 978-1-83976-483-7
ISBN-13: 978-1-83976-486-8 (US EBK)
ISBN-13: 978-1-83976-485-1 (UK EBK)

British Library Cataloguing in Publication Data
A catalogue record for this book is available from the British Library
Library of Congress Cataloging-in-Publication Data
A catalog record for this book is available from the Library of Congress

Typeset in Minion by Hewer Text UK Ltd, Edinburgh
Printed and bound by CPI Group (UK) Ltd, Croydon, CR0 4YY

We know the system doesn't work because we're living in its ruins.

'Leaders', poem in *The Seed* (1967)

Contents

Acknowledgements

For listening to me wonder aloud and ad nauseam about bodies, selves, and systems, and their movement towards death and destruction, as well as for being there through mourning and melancholia, I am grateful to Debbie Broadhurst and Lola Broadhurst.

For offering me information, links, thoughts, and provocations, I am grateful to Andrea Bardin, Travis Linnemann, Brendan McQuade, Andrea Miller, Colin Perrin, Guillermina Seri, Deniz Türker, Manuela Trindade Viana, and Tyler Wall.

For providing me with a whole gamut of other things, I am grateful to Poli Neocleous (2001–2018), Sheila Neocleous (1936–2018), and Renos Neocleous (1935–2019). Their 'life death', to use a term explained below, undergirds this book in ways that will not be understood by the reader, since they are not yet understood by the author.

Introduction

> *Though I have spent my working life almost wholly at the bench, immunology has always seemed to me more a problem in philosophy than a practical science.*
>
> Sir Frank Macfarlane Burnet,
> 'The Darwinian Approach to Immunity' (1965)

In an age of HIV/AIDS, virulent viruses such as Ebola, common colds, influenza that kills thousands every year, antibiotic-resistant bacteria, parasites that deplete the body, and, most recently, the shock to our bodies and minds produced by Covid, we dream a dream of immunity.

In an age when we are told that washing our hands will save the life of the state as well as the lives of people, that immune systems of individual bodies hold the key to the security of the body politic, that viruses pose a threat to national security, and that the future of global capitalism depends on fighting wars against viruses through police measures such as quarantine, lockdown, border closures, immunity passports, and immune surveillance systems, the dream we dream of immunity turns out to be decidedly political.

This is a book about the politics of immunity. It is a book about the ways in which immunity permeates how we think about the body and the body politic. It is a book about *cells*, *selves*, *systems*, and *sovereignty*.

'Immunity' has a rather odd history. In its ancient origins, one finds nothing remotely 'biological'. In Roman law, immunity conferred *exemption* from various kinds of state obligations. A compound of *in-* (not), *munus* (a gift as well as a service, but also the root of our term 'municipal'), and *-tas* (denoting an abstraction), the Latin word *immunitas* has a range of meanings and implications concerned with 'exemption' or 'freedom' from public burdens such as taxes, duties, services, and participation. An immunity might be granted by the state to an individual, group or community, it could be for a limited duration, and it could always be revoked. Thus, what was originally at stake in immunity was a political decision about *privilege*, in the sense of a law that applied only to certain classes of persons or individuals. Only with the discovery of what came to be called the body's 'immune response' in the second half of the nineteenth century did 'immunity' take a biological turn. In so doing, it became a concept with which we think about bodies, being used to describe the process by which an organism maintains continuity of life in the face of biological threats and diseases. In other words, in the late nineteenth century, 'immunity' was transported from the juridico-political world into the biological world.[1] Ed Cohen has pointed out that it is more than a little strange that a term that had for two millennia served as a political and legal concept should have so quickly and easily been assimilated into the conceptual framework of biomedical knowledge.[2] We shall have cause to discuss this assimilation at various points in this book, as we take up the complicated conceptual history of immunity and its various political uses; in chapter 6 in particular we take up some of the shifts in the concept of immunity that take place between the Roman usage and the adoption of the idea by biology in the nineteenth century. Here, we can note that the strangeness of biology's adoption of a legal term is compounded by three further issues.

First, there is the fact that being transported into the world of biology did not mean that immunity gave up its foothold in the legal and political world. In fact, the law sought to retain custody of the concept, or at least one meaning of the concept, finding in 'immunity' a perfect foil for sovereign violence. Such violence will be a key theme in this book. For

1 Anne Marie Moulin, *Le dernier langage de la médecine: Histoire de l'immunologie de Pasteur au sida* (Paris: Presses Universitaires de France, 1991).

2 Ed Cohen, 'Figuring Immunity: Towards the Genealogy of a Metaphor', in Anne-Marie Moulin and Alberto Cambrosio (eds.), *Singular Selves: Historical Issues and Contemporary Debates in Immunology* (Paris: Elsevier, 2001), 179–80.

example, we read again and again of officials of the state being granted immunity from prosecution for acts of violence that would see the rest of us imprisoned for an awfully long time. When we think about immunity in these terms, we can see the dust being shaken off some incredibly old law books. Such 'grants' or 'privileges' attached to certain offices of state have their roots in the application of Roman law notions of immunity to the early modern diplomatic Embassy and the figure of the Ambassador or Diplomat, and when we find them applied to more modern offices of state, such as the police officer or soldier, we see them being put to some very contemporary political uses. The second issue is that as immunity came to be understood as a biological category, so its biological connotations filtered back into legal language, and legal scholars started using the term's biological connotations as a means of explaining some legal ideas. For example, writing in the late nineteenth century about the reception of Roman law in England, legal historian Heinrich Brunner commented that the early reception of Roman ideas 'operated as a sort of prophylactic inoculation, and had rendered the national law immune against destructive infection'.[3] Here we find a lawyer discussing a term from legal history, using the language of a newly emerging biological discourse surrounding that same term, to explain an aspect of legal history. By the end of the nineteenth century, then, as the biology surrounding immunity began to be more widely understood, ideas about immunity started oscillating back and forth between the legal and biological worlds. The third issue compounding immunity's strangeness is that the biological notion of immunity has itself been reinterpreted in all sorts of political ways. 'Our nation has been put on notice: we are not immune from attack.' This comment from a presidential authority could be from early 2020 when it became clear that Covid-19 was spreading across the globe. In fact, the comment is from 20 September 2001, to justify the creation of a new Office of Homeland Security to defend the American body politic from the 'virus' of terrorism said to be spreading across the globe.

A book on the politics of immunity, then, is a book on a concept that spent two millennia being used by jurists and political theorists to

3 Heinrich Brunner, 'The Sources of English Law', in Association of American Law Schools (ed.), *Select Essays in Anglo-American Legal History*, vol. II (Boston: Little, Brown, 1908), 42, citing one of his own earlier essays from 1896.

understand the sovereign grant of privilege or exemption, which was then being picked up by biologists in the late nineteenth century to capture something about organismic survival, and which then spent the following 150 years tacking back and forth between biology and law. This tacking back and forth between biology and law was decidedly political, and remains so. The more immunity tacked back and forth, so the more political uses to which it was put, so the more it came to occupy a significant space in our cultural and intellectual worlds. We encounter lines such as 'How to Build a Corporate Immune System for the Crisis' run by major international organizations (that one from the European Commission's Microfinance Centre in April 2020) and are expected to simply treat the idea as entirely unproblematic. A corporate immune system? Well, if the corporation is a body, which it is (a *corpus*), then why not?

Commenting on immunity's iconic status, Donna Haraway highlights the ways in which immunity moves easily across an incredibly diverse range of practices and powers. Immunity touches on and appears in

> global and local politics; Nobel Prize-winning research; heteroglossic cultural productions, from popular dietary practices, feminist science fiction, religious imagery, and children's games, to photographic techniques and military strategic theory; clinical medical practice; venture capital investment strategies; world-changing developments in business and technology; and the deepest personal and collective experiences of embodiment [and] vulnerability.[4]

Immunity's iconic status enables it to generate a rich assortment of tropes, images, and assumptions that underpin the concept's vibrancy and political complexity and which will run through this book: Self and non-Self, friend and enemy, identity and foreignness, body and machine, recognition and toleration, system and survival, defence and destruction, war and protection, nature and nation, inoculation and incorporation. The popularity of even the most simplistic of views about immunity bears witness to a profound resonance between ideas that dominate

4 Donna J. Haraway, *Simians, Cyborgs, and Women: The Reinvention of Nature* (London: Free Association, 1991), 204–05.

immunology as a specialist discipline within the biological sciences, ideas about selfhood and the defence of the body that permeate popular culture, and ideas about what is or is not permitted in the name of sovereignty. It is this depth and complexity that explains why immunology as a biological science is 'a rich philosophical mine', as Alfred Tauber puts it, to the extent that 'philosophical and scientific inquiries are inseparable'. It is the same depth and complexity that explains why Frank Macfarlane Burnet, one of the twentieth century's leading immunologists, Nobel Prize winner, and pioneer of the clonal selection theory that became a major immunological paradigm, came to offer the observation in the epigraph to this chapter: that immunology seems to be more a question of philosophy than science.[5] This book aims to show how it is also a question of politics.

Biological categories often perform crucial ideological functions, and what appears as an objective category about natural processes often turns out to be a political claim about the social world. Nowhere is this clearer than in the concepts used to describe the body. But the image of the body is in turn always already inscribed in the image of the body politic. The idea of the body politic goes back to the ancient world, being found in the work of both Plato and Aristotle. It was heavily pursued in early modern political texts such as John of Salisbury's *Policraticus* (1159), Marsilius of Padua's *Defensor pacis* (1324), Christine de Pizan's *Book of the Body Politic* (1406), Sir John Fortescue's *In Praise of the Laws of England* (1468–1471), and Thomas Starkey's *Dialogue between Reginald Pole and Thomas Lupset* (circa 1533–1536). But many of these texts did little more than propose that, for example, the position of the head is occupied by the prince, the hands are like state officials and soldiers, and the feet are the peasants perpetually bound to the soil. More inventive accounts of the body politic begin to appear in modern political thought in the seventeenth century. In Thomas Hobbes's *Leviathan*, for example, the Commonwealth 'is but an Artificiall Man; though of greater stature and strength than the Naturall, for whose protection and defence it was intended'.

5 Alfred I. Tauber, *Immunity: The Evolution of an Idea* (Oxford: Oxford University Press, 2017), ix, xi, 228; Sir F. M. Burnet, 'The Darwinian Approach to Immunity', in J. Sterzl et al., *Molecular and Cellular Basis of Antibody Formation: Proceedings of a Symposium Held in Prague on June 1–5, 1964* (New York: Academic Press, 1965), 17.

> The *Soveraignty* is an Artificiall *Soul*, as giving life and motion to the whole body; The *Magistrates*, and other *Officers* of Judicature and Execution, artificiall *Joynts*; *Reward* and *Punishment* . . . are the *Nerves*, that do the same in the Body Naturall; The *Wealth* and *Riches* of all the particular members, are the *Strength* . . . *Counsellors*, by whom all things needfull for it to know, are suggested unto it, are the *Memory*; *Equity* and *Lawes*, an artificiall *Reason* and *Will*; *Concord*, *Health*; *Sedition*, *Sicknesse*; and *Civill war*, *Death*. Lastly, the *Pacts* and *Covenants*, by which the parts of this Body Politique were at first made, set together, and united, resemble that *Fiat*, or the *Let us make man*, pronounced by God in the Creation.

This 'man' is, like all automata, an 'artificial life'.

> For what is the *Heart*, but a *Spring*; and the *Nerves*, but so many *Strings*; and the *Joynts*, but so many *Wheeles*, giving motion to the whole Body, such as was intended by the Artificer? *Art* goes yet further, imitating that Rationall and most excellent work of Nature, *Man*. For by Art is created that great LEVIATHAN called a COMMON-WEALTH, OR STATE.[6]

The state is a product of the work of man and in that sense an artifice, yet, despite this 'artificiality', the state is a body because 'every part of the Universe, is Body'. That which is not body does not exist. The state is a *body politic* designed to exert authority over individual *bodies-in-motion*. Hence the body politic is a *body* – to the extent that we can speak 'Of the Nutrition, and the Procreation of a Common-wealth', as chapter 24 of *Leviathan* is titled – yet is also *artificial*. At the same time, man's *natural body* is also part of this *artificial body* of the state.[7] These ideas remind us that Hobbes was one of the most important mechanical philosophers of the mid-seventeenth century, transferring accounts of the body's *mechanics* onto the much older image of the body politic. Conceptualized as a machine, sovereignty is truly impersonal, while as a body it retains

6 Thomas Hobbes, *Leviathan* (1651), ed. Richard Tuck (Cambridge: Cambridge University Press, 1991), 9.

7 Thomas Hobbes, *De Homine* (1658), in Hobbes, *Man and Citizen (De Homine and De Cive)*, ed. Bernard Gert (Indianapolis: Hackett, 1991), 35.

some of the 'human' features characteristic of an age of personal power. This latter image is reinforced by the frontispiece of *Leviathan*, which has the form of a huge human body made up of the individual bodies of its subjects.

Now, far from disappearing from political discourse with the rise of liberal democracies, as many claim, the idea of the body politic shifts and develops along with new scientific discoveries, new biological concepts and new images of man.[8] The idea of a mechanical body politic, for example, gets transformed with developments in technology, as Anson Rabinbach has shown. The mimetic body of the eighteenth century is exemplified by the clockwork machine and the automata, whereas the digital body of the twentieth century derives its inspiration from information-processing and computing technology. Between them was a period represented by the development of motors that convert energy into motion and hence ideas about energetics.[9] At the same time, such images of the mechanical body have often involved a conception of nerves. This is evident in the passages from Hobbes ('*Reward* and *Punishment* . . . are the *Nerves*, that do the same in the Body Naturall'), and the nervous system has long been considered the substratum of life and a way of thinking about the social order, such as we find in the eighteenth-century Scottish Enlightenment (about which we shall say more in chapter 5). Much later, there emerges the *neural body*, the *genetic body* and the *immune body*. As each of these conceptions of the body emerges and gets refined, so conceptions of the body politic change. Is the body politic a body of nerves (and perhaps 'nervous' too)? Is the body politic a body of genetic information, an image that dominates the twentieth century and the 'information age'? Is the body politic geared to protect itself through its immune system? The images are far from mutually exclusive, of course, yet they privilege different notions of identity corresponding to different types of body politic, and they project alternative ways of understanding the relations between subject and sovereign power. To put that another way: each image makes a different kind of claim about how we imagine the *Self*,

8 I have shown the extent to which liberalism transforms rather than rejects the image of the body politic in *Imagining the State* (2003).

9 Anson Rabinbach, *The Human Motor: Energy, Fatigue, and the Origins of Modernity* (Berkeley: University of California Press, 1990); *The Eclipse of the Utopias of Labor* (New York: Fordham University Press, 2018).

how we imagine the *System*, and how we imagine *Sovereignty*. Each image conjures up different ways of thinking politically. In chapters to follow, we will discuss in more detail various ideas concerning nervous bodies (chapter 5), bodies of energy (chapter 4) and bodies of information (chapter 3), but the point is to consider their relation to the idea of an immune body and hence the politics of immunity, because as medical science develops its research on the immune system through the twentieth century, creating immunology as a discipline and forcing ideas about immunity into the popular culture, so the body politic is increasingly imagined through the idea of immunity.

The richness and ideological complexity of immunity and the idea of immune bodies have recently led many thinkers from a diverse range of philosophical traditions to dig around in immunity's philosophical mine. Haraway uses the way in which the immune system guides 'recognition and misrecognition of self and other in the dialectics of Western biopolitics' to explore the relation between postmodern bodies and strategic military culture.[10] Niklas Luhmann appropriates the idea of immunity for a social systems theory developed over three decades. His work on immunity begins in the heyday of immunological research and systems theory, out of which he develops a socio-legal metabiology which develops autopoiesis as a sociological as well as biological concept. In books such as *A Sociological Theory of Law* (1972), *Social Systems* (1984) and *Law as a Social System* (1993), as well as many essays, Luhmann imagines a society immunizing itself against social 'infections' via a legal subsystem which maintains the balance of the social system overall. Conflict and crisis function as an alarm in the society's immune system and generate the problem of system integration. Society's immune system corrects 'deviations' while accepting and adapting to change. In that sense, the society's immune system protects and maintains its autopoietic self-reproduction. Because Luhmann regards the legal system as a crucial immunizing process, he eventually comes to develop a 'juridical immunology' in which the legal system's immune processes are said to prevent the annulment of the social system.

Building on the work of Haraway and Luhmann, Roberto Esposito has, more recently, sought to establish a philosophical paradigm of

10 Haraway, *Simians*, 204–05, 222.

immunization with the kind of conceptual weight previously attached to 'rationalization', 'legitimization' and 'secularization'. In books such as *Immunitas* (2002), *Bíos* (2004), and *Terms of the Political* (2008), Esposito treats immunization as the coagulating point of contemporary existence and thus the most fruitful interpretative key for understanding current politics. Why? Because *immunization* is a negative form of the *protection* of life. Thinking about *immunitas* is a question of thinking about *communitas*, merging the fashionable term 'biopolitics' with 'immunization' as a means of grasping the protection of community. If *communitas* is the relation which binds us in reciprocal obligations, but which can also jeopardize individual identity, then *immunitas* is a dispensation from such obligations and a defence against the dangers that might lie in *communitas*. *Immunitas* slips here from the dispensing of an obligation (the Roman legal principle) into defence (in the account of the biologists), allowing Esposito to claim that community cannot exist without some form of immunitary apparatus. In other words, *immunitas* protects the one who bears it from risky contact with those who lack it, thereby reinforcing or restoring borders. For Esposito, modern communities need a series of immunitary apparatuses, and the task for immunization theorists – for Esposito, philosophers and social theorists need to become theorists of immunization – is to think with *immunitas* but without succumbing to its destructive tendencies: to create an immunizing democracy that sustains rather than destroys community.

Esposito's work overlaps somewhat with the work of Peter Sloterdijk, who also characterizes modernity through the technical production of its immunities in such a way that demands a general theory of immune systems. Through a series of works written roughly during the same decade as Esposito's publications, including *Spheres* (in three volumes, 1998–2004), *Neither Sun Nor Death* (2001), *Not Saved* (2001), *In the World Interior of Capital* (2005), and *You Must Change Your Life* (2009), Sloterdijk traces the history and being of what he calls *homo immunologicus*, a creature who lives within and utilizes three types of immune system: biological, the first system in evolutionary terms but the most recent to have been 'discovered'; socio-immunological, which incorporates the legal and military systems; and symbolic- or psycho-immunological. This is an argument that runs close to Esposito's: as humanity seeks the creation of a new immune constitution, what is needed is a more mature immunity of power necessitating an expansion of the concept of immunity. Such an expansive concept imagines immunity as possible only through 'co-immunity'. In the current state of the world, the 'co-immunitary units' are formatted tribally, nationally, and imperially, but our aim should be a genuine 'global immunization' that would be nothing less than 'co-immunism'. This would improve the global immune situation through new and more liveable immune relationships. Such a co-immunism would consist of a post-theological and post-metaphysical configuration of human immunities that would constitute a protective shield for the earth and humankind. Esposito's image of philosophers as theorists of immunization is paralleled by Sloterdijk's idea of the philosopher as an immunologist of culture. This is co-immunism as a form of *political immunology*.

We shall have reason to question and criticize some of the claims made by Luhmann, Esposito, and Sloterdijk. Here we can note that, set in the wider context of the body taking centre stage in social and political thought, there has been a plethora of studies of aspects of our contemporary condition conducted through a similar paradigm of social immunization or political immunology. These include Jean Baudrillard's essay 'AIDS: Virulence or Prophylaxis' (1987) and his book *The Transparency of Evil* (1990); Emily Martin, *Flexible Bodies* (1994); John McMurtry, *The Cancer Stage of Capitalism* (1999); A. David Napier, *The Age of Immunology* (2003); James M. Wilce's edited volume *Social and Cultural Lives of Immune Systems* (2003); Priscilla Wald, *Contagious* (2008); Ed Cohen, *A Body Worth Defending* (2009); Eula Biss, *On Immunity* (2014); Inge Mutsaers, *Immunological Discourse in Political Philosophy* (2016); and Nik Brown, *Immunitary Life* (2019). The extent of immunity's traction in contemporary thought is also reflected in a number of journals running special issues on the subject, including *Cultural Anthropology* in 2012 (edited by Napier), *Angelaki* in 2013 (edited by Greg Bird and Jonathan Short), and *Parallax* in 2017 (edited by Stefan Herbrechter and Michelle Jamieson).

The language of immunity in social and political thought has become so commonplace that one encounters it time and again in all sorts of places. To give just one example: in their book *Assembly* (2017), a consideration of the current situation of social movements as a follow-up volume to their massively influential work on dominant powers called *Empire* (2000), Michael Hardt and Antonio Negri point to new forms of movement with broader democratic bases than traditional movements focused on leaders. One way such movements are stronger, they claim, is because of their 'highly developed immune systems'. 'The immune systems of the movements have become so developed that every emergence of the leadership virus is immediately attacked by antibodies.' Black Lives Matter, for example, is one example 'of how developed the immune systems of the movements against leadership have become'.[11] It is not entirely clear how this use of immunity helps their argument, but it is surely a good example of the extent to which 'recent times have witnessed the proliferation of discourses about society that explicitly employ immunological models of explanation'.[12]

What might be the basis for this growth of immunological models of thought? One answer lies in the growth of immunological understandings of the body and hence the body politic, as we have noted and about which much more will be said throughout the book. Yet might there be another reason for the growth of this field, beyond the mere fact that we now have the image of an immune body? Might it be that this rising interest in *immunity* is connected to the seemingly ever-increasing political and cultural obsession with *security*?

In March 2011, the US Department of Homeland Security published a report called *Enabling Distributed Security in Cyberspace: Building a Healthy and Resilient Cyber Ecosystem with Automated Collective Action.* The report articulates the idea that 'in cyberspace, intelligent adversaries exploit vulnerabilities and create incidents that propagate at machine speeds to steal identities, resources, and advantage'. To tackle this, the report posits a cyber ecosystem consisting of private firms, non-profits, governments, individuals, processes, and cyber devices such as computers, software, and communications technologies, which together form 'a healthy, resilient – and fundamentally more secure – cyber ecosystem'. The cyber ecosystem is thus about national security, but also the security of the cyber ecosystem itself. To get us to imagine security, the report draws inspiration from another 'ecosystem':

11 Michael Hardt and Antonio Negri, *Assembly* (Oxford: Oxford University Press, 2017), 6, 11, 44.

12 Byung-Chul Han, *The Burnout Society* (2010), trans. Erik Butler (Stanford, CA: Stanford University Press, 2015), 2.

> To paint a picture that mirrors the body's ability to defend itself is complex. It might include layered defenses and countermeasures that work in tandem; specialized roles; powerful methods for rapidly identifying attackers; surge capabilities; and the ability to learn and rapidly adapt . . .
>
> A healthy cyber ecosystem might employ an automation strategy of fixed local defenses supported by mobile and global defenses at multiple levels. Such a strategy could enable the cyber ecosystem to sustain itself and supported missions while fighting through attacks.

The report spells out the point: 'We draw inspiration from the human body's immune system.' To this end, the report offers us a diagram of how a cyber security ecosystem might be imagined.

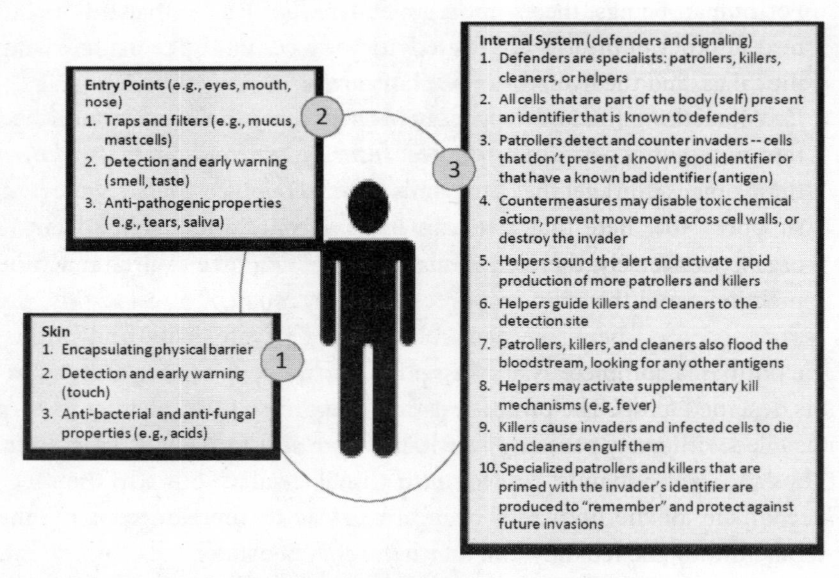

Figure A.1: Department of Homeland Security, Healthy and Resilient Cyber Ecosystem (March 2011).[13]

13 Department of Homeland Security, *Enabling Distributed Security in Cyberspace: Building a Healthy and Resilient Cyber Ecosystem with Automated Collective Action* (11 March 2011), 2, 9.

Now, such use of 'immune system' in security circles was far from original. Colonel John Warden III had developed a similar idea some years previously. Warden was perhaps the leading US air power strategist of the latter decades of the twentieth century, widely credited for being the brains behind Operation Desert Storm. His book *The Air Campaign* (1989) emerged from US war strategy of the previous forty years but was in turn influential on that strategy. The book outlines an approach based on identifying and then attacking the enemy's centres of gravity, later presented in terms of a 'five ring model', with leadership the inner ring and the military forces the outer ring, with population, infrastructure and 'organic essentials' between them; each ring constitutes a 'centre of gravity' which might be targeted. In the 1990s, Warden developed these arguments into a 'universal system model'.

> All systems seem to require certain organic essentials, normally some form of input energy and the facilities to convert it to another form. For human beings, the essential inputs are food and oxygen. Thus, next in order of priority are those organs we call vital, like the heart, the lungs, and the liver, the ones that convert or convey food and air into something the body can use. Without these organic essentials, the brain cannot perform its strategic function, and without the brain, these organs don't get the commands they need to provide integrated support. Note here that a machine can substitute for all the vital organs; conversely, there is no machine that can take over strategic functions from the brain.

The body is 'a complete system', in other words, that 'can do everything it is designed to do'. The problem is that 'the world is not perfect; rather, it is filled with nasty parasites and viruses that attack the body whenever they can'. The body must protect itself with specialist cells. The 'fighting mechanism' of the body is the immune system (represented in Table A.1, below, by the leukocytes), which parallel the state's security system (in the conjoined power of *both* police and military; this conjunction will become important in chapter 1).

A few years prior to Warden's article, the journal *Military Review* published an essay called 'Future Warriors'. The author, Colonel Frederick Timmerman of the US Army Command, suggests that, since war is a human activity, perhaps the most appropriate model to use to

	Body	State	Drug Cartel
Leadership	Brain eyes nerves	Government communication security	Leader communication security
Organic Essentials	Food and oxygen (conversion via vital organs)	Energy (electricity, oil, food) and money	Coca source plus conversion
Infrastructure	Vessels, bones, muscles	Roads, airfields, factories	Roads, airways, sea lanes
Population	Cells	People	Growers, distributors, processors
Fighting Mechanism	Leukocytes	Military, police, firemen	Street soldiers

Table A.1: Section from 'Systems', in Colonel John Warden III, 'The Enemy as a System', *Airpower Journal*, 9:1 (1995).[14]

understand it is a biological one. More specifically, the model to use is 'the most complex biological model we know – the body's immune system'. Building on an important and influential article by Peter Jaret in *National Geographic* in 1986, called 'Our Immune System: The Wars Within', about which we shall say more in chapter 1, Timmerman makes the following point:

> Within the body there exists a remarkably complex corps of internal bodyguards. In absolute numbers they are small – only about one percent of the body's cells. Yet they consist of reconnaissance specialists, killers, reconstitution specialists and communicators that can seek out invaders, sound the alarm, reproduce rapidly, and swarm to the attack to repel the enemy. They do this while preserving the other important cells of the body and operating in limited numbers when on watch.[15]

14 Colonel John Warden III, 'The Enemy as a System', *Airpower Journal*, vol. 9, no. 1 (1995), 40–56; John A. Warden III, 'Air Theory for the Twenty-First Century', in Karl P. Magyar (ed.), *Challenge and Response: Anticipating US Military Security Concerns* (Maxwell Air Force Base, AL: Air University Press, 1994).

15 Colonel Frederick W. Timmerman Jr., 'Future Warriors', *Military Review* (Sept. 1987), 46–55 (52).

We find similar arguments in publications from other departments of state, such as articles by officials in the US Department of Health on the immune system as a Clausewitzian security lesson, to dissertations produced within the Air University at Maxwell Air Force Base, Alabama, on the need to develop an immunity against the metastasizing force of terrorism that lives within the body politic.[16]

Such ideas and images were also developed in the work of leading security intellectuals during the war on terror. David Kilcullen, for example, identifies four phases of guerrilla warfare: infection, contagion, intervention, and rejection. In the infection phase, insurgent groups establish a presence 'just as a virus or bacterium is more easily able to affect a host whose immune system is compromised'. Intervention leads to a societal immune response against the guerrillas, and the rejection phase can be understood as 'a social version of an immune response in which the body rejects the intrusion of a foreign object'. In his more recent work, arguing that counter-insurgency practice needs to come out of the mountains and into the cities, Kilcullen deals with the problem of how cities are to be defended. One way is to imagine them as biological entities with metabolisms: 'If cities have metabolisms, they also have immune systems – ways to deal with internal challenges, absorb toxins, and neutralize threats.'[17]

Clearly, then, there is something about 'immunity' that has captured the imagination of security intellectuals as well as social theorists and philosophers (although, interestingly, neither group has much to say about the work of those in the other group). It has also captured thinking about security in general. In 2011, one of the world's leading corporations took the idea of a corporate immune system to its logical conclusion and announced that security and immunity were now as one: welcome to the Facebook Immune System. 'We call it the Facebook Immune System (FIS) because it learns, adapts, and protects in much

16 Eugene G. Hayunga, 'Parasites and Immunity: Tactical Considerations in the War Against Disease – Or, How Did the Worms Learn about Clausewitz?', *Perspectives in Biology and Medicine*, vol. 32, no. 3 (1989), 349–70; Douglas R. Stickle, *Malignants in the Body Politic: Redefining War through Metaphor* (Maxwell Air Force Base, AL: Air University Press, 2002).

17 David Kilcullen, *The Accidental Guerrilla: Fighting Small Wars in the Midst of a Big One* (London: Hurst, 2009), 35–38, 244; *Out of the Mountains: The Coming Age of the Urban Guerrilla* (London: Hurst, 2013), 248.

the same way as a biological immune system.'[18] To explain the FIS, three Facebook engineers adopt the post-9/11 security state's concept of the 'adversarial cycle':

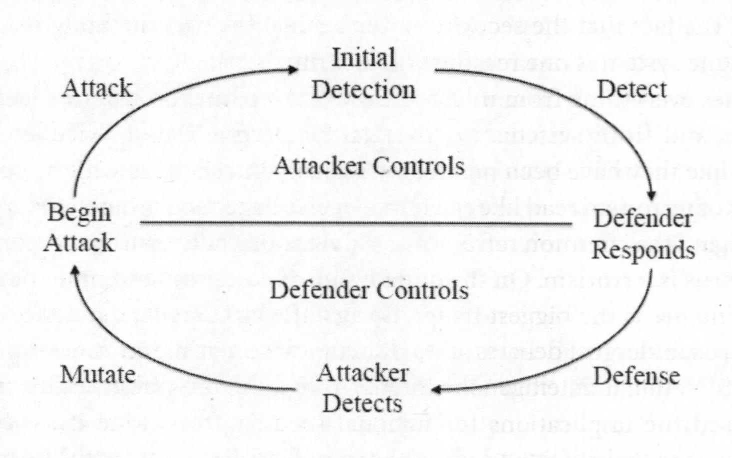

Figure A.2: Facebook Immune System, from Stein, Chen, and Mangla, 'Facebook Immune System' (2011).[19]

The corporation's choice of title is part of a wider process in which large 'social media' companies have adopted ideas around the 'organic' to present as natural the ways in which subjects and objects are ordered in their algorithms, and it also plays to a logic of information flows and data communication that dominates our assumptions about 'system'; both issues are taken up in chapters to follow. But it is also evidence of the close cultural as well as political assumptions that now exist connecting immunity and security.

To put it bluntly, and to capture the main thrust of the argument in this book: it is security that lies at the heart of the political appropriation and application of immunity. Conversely, security also lies at the heart of the immunological appropriation and application of political ideas.

18 Facebook, *National Cybersecurity Awareness Month Recap and the Facebook Immune System*, 10 November 2011, facebook.com.

19 Tao Stein, Erdong Chen, and Karan Mangla, 'Facebook Immune System', *EuroSys Social Network Systems*, 10 April 2011; facebook.com.

As is well known, the contemporary security agenda has been recently expanded to incorporate a politics of all that exists. As the passages just cited indicate, we are encouraged to believe that a body politic without a *system of security* is as defenceless as a body without a *system of immunity*. The fact that the security system is imagined as the body politic's immune system is one reason why security's strategic vision now incorporates everything from microbiology to the global order: from cells to selves and from systems to sovereignty. Descriptions of viruses now read like they have been penned by security intellectuals while descriptions of terrorism read like they have been penned by virologists. On the one hand the common refrain that 'a new and deadly virus has emerged: the virus is terrorism'. On the other hand, and equally common, the idea that 'nature is the biggest bioterrorist'.[20] The virus-as-terror and terror-as-virus undergird debates about 'biological security' and 'bioterrorism'. A US National Intelligence Council (NIC) report published in 2000 stressed the implications for national security from what it saw as a growing global infectious disease threat and suggested that new and mutating infectious diseases 'will pose a rising global health threat and will complicate US and global security over the next 20 years'. The same argument concerning infectious diseases as security threats continued in further NIC documents. Likewise, a report by the Rockefeller Foundation argues that 'emerging infectious disease may provide the strongest example of an environmental challenge that poses a clear threat to national security'.[21] The Covid-19 outbreak of 2020 merely reinforced this trend, with headlines such as 'In Times of Coronavirus and Climate Change, We Must Rethink National Security' (*Guardian*, 20 April 2020) far from uncommon. Hence, organizations created to collect and analyse data about Covid infection rates, identify local spikes, and recommend appropriate responses have been established under the rubric of 'security', such as the UK's Joint Biosecurity Centre

20 Tony Blair, July 2003 speech to the US Congress; Ron A. M. Fouchier, leading researcher on viruses at Erasmus Medical Centre, Rotterdam, cited in Doreen Carvajal, 'Security in Flu Study Was Paramount, Scientist Says', *New York Times*, 21 Dec. 2011.

21 National Intelligence Council, *The Global Infectious Disease Threat and Its Implications for the United States* (NIE 99-17D, Jan. 2000), 5; National Intelligence Council, *Strategic Implications of Global Health* (ICA 2008-10, Dec. 2008), 7; Christopher F. Chyba, *Biological Terrorism, Emerging Diseases, and National Security* (New York: Rockefeller Fund, 1998), 5, 14.

(JBC). And to reinforce the point of 'immunity-as-security', the JBC is to be headed by a senior counterterrorism official, modelled on the Joint Terrorism Analysis Centre, and is to use a 'levels of threat' model adopted from the same 'levels of threat' used to assess terrorism. 'National security' and 'natural security' coincide, held together by the idea of immunity. Health security, reinforced over and again by health terrors, becomes a politics of life itself.

A politics of immunity must therefore address security's desire to cover the whole realm of human experience, from cells to selves, from systems to sovereignty, and from the health and welfare of the body to the health and welfare of the body politic. By the same token, the politics of security must address the integration of ideas about biological immunity into the political field. The invocation to imagine the body as a security system is equally an invocation to imagine the body politic as an immunity system. Immunity is articulated as security, security articulated as immunity; immunity imagined as security, security as immunity; round and round we go, as security draws on immunity to reinforce its power just as immunity draws on security to insist on its importance. This book, configured as part of an ongoing *critique of security*,[22] will therefore tack back and forth between immunity and security to the point where the moves become faster and faster until they coincide: 'immunity-security'.

If security is articulated as immunity and immunity as security, this begs an important question: What then of the autoimmune disease? If the body politic is imagined as an immune system, then how do we think politically with the autoimmune disease?

An autoimmune disease is a disorder caused by the immune system attacking the cells of the body it is meant to be protecting. In patients with an autoimmune disease, the immune system appears unable to tell the difference between healthy body tissue and antigens such as viruses and bacteria, resulting in an immune response that destroys normal and healthy body tissue. In multiple sclerosis (MS), for example, the immune system mistakes myelin (a substance which protects the nerve fibres in the central nervous system) for a dangerous body and attacks it; in

22 Mark Neocleous, *Critique of Security* (Edinburgh: Edinburgh University Press, 2008); *War Power, Police Power* (Edinburgh: Edinburgh University Press, 2014); *The Universal Adversary: Security, Capital and 'The Enemies of All Mankind'* (Abingdon: Routledge, 2016); *A Critical Theory of Police Power* (London: Verso, 2021).

rheumatoid arthritis, the immune system treats the linings of joints as a threat and starts to destroy them; in Hashimoto's thyroiditis, the immune system attacks the thyroid gland. The range of autoimmune diseases is vast, with some eighty or more different diseases falling under the umbrella term 'autoimmune', which is why leading researchers in the field such as Noel Rose consider autoimmune diseases, taken together, as one of the 'big three' along with cancer and heart disease.[23] There is also an increasing amount of research that shows such diseases to be increasing, which may well be one reason why immunity has come to resonate so widely through our culture.

> If it feels like you're hearing about autoimmune diseases like rheumatoid arthritis more and more these days, it's not just coincidence. According to a new study . . . the number of people with autoimmune diseases – basically, when your immune system attacks your own body by mistake – is on the rise.[24]

This comment from early 2020 foregrounded the findings of a major research project conducted at the US National Institutes for Health, which tested for the prevalence of anti-nuclear antibodies, markers of a body's immune responses against its own cells, and found that these had increased significantly throughout the population since the 1980s.[25] As the comment points out, there is now a wealth of evidence that the number of people diagnosed with individual autoimmune diseases has risen over the past five decades in industrialized countries. The incidence of lupus, for example, has nearly tripled in the US, while rates of MS over the same period have tripled in a number of countries including Finland, the UK, the Netherlands, Denmark, and Sweden. As well as being on the rise, autoimmune diseases are known to affect life expectancy, especially for women, who suffer in far greater numbers than men from virtually all the diseases.[26]

---

23 Noel Rose, speaking to the *Washington Post* in 1995, and widely cited again in obituaries following his death in August 2020.

24 Lara DeSanto, 'Have You Noticed? Autoimmune Diseases Are on the Rise', *HealthCentral*, 27 April 2020.

25 Frederick W. Miller et al., 'Increasing Prevalence of Antinuclear Antibodies in the United States', *Arthritis Rheumatology* (April 2020).

26 Donna Jackson Nakazawa, *The Autoimmune Epidemic* (New York: Simon and Schuster, 2008).

All illness and disease remind us of the nature of embodiment, of bodies going wrong and letting us down. If that is true of disease in general, then it is truer still of an autoimmune disease, in which our body does not simply let us down or go wrong but also appears intent on destroying itself. This is a situation which is impossible to avoid: immunity creates the possibility of the autoimmune disease. The autoimmune disease reveals a large part of immunity's confounding nature: you think you are getting self-defence and instead you get self-destruction; you think your body is being well policed by the immune system, but then the police power turns against you; you find security in immunity, and then immunity starts to destroy you. Immunity confounds us at every turn. If we still do not know what the body can do, as Spinoza put it in a frequently cited passage in the *Ethics* (1677), then the autoimmune disease reminds us that we still do not know what the body can do *to destroy itself*. And what the autoimmune disease reveals above all else is just how self-destructive a body can be. As accounts from those suffering from an autoimmune disease testify (and which we will explore further in chapter 2), discovering that your immune system thinks of you as an enemy to be destroyed, that your body's security system is slowly killing you, is a truly discomforting experience.[27] This is perhaps why Burnet, just a few years after commenting that immunology is really a problem in philosophy rather than a practical science, went on to add that 'one cannot discuss autoimmune disease without getting into deep water philosophically'; medical thought is, after all, always engaged in one way or another with the philosophical status of man, as Foucault sought to show.[28] As we shall see, we must likewise get into some deep and very murky water politically and, it turns out, also psychoanalytically.

Remarkably, however, the autoimmune disease does not feature especially highly in the work of many of the thinkers we have cited who have

27 To use that discovery to write a book about the politics of immunity and how it shapes our image of cells, selves, systems, and sovereigns is to probably raise that level of discomfort higher still. Nonetheless, one learns more from discomfort than comfort. A splinter in the eye is often the best magnifying glass, as Adorno comments in *Minima Moralia* (1951).

28 Macfarlane Burnet, *Walter and Eliza Hall Institute, 1915–1965* (Sydney: Melbourne University Press, 1971), 146; Michel Foucault, *The Birth of the Clinic: An Archaeology of Medical Perception* (1963), trans. Alan Sheridan (London: Routledge, 1973), 198.

wanted to place immunity at the heart of social and political thought. Haraway has little to say about autoimmune disease, other than a comment to the effect that such a disease carries an 'awful significance'. Luhmann is so keen for social systems to continue their merry autopoietic way that the autoimmune disease appears to never cross his mind: social systems just carry on, protected by their immune mechanisms. Others who articulate a 'social immune system model' from a more radical position than Luhmann, such as Baudrillard in 'AIDS: Virulence or Prophylaxis' and *The Transparency of Evil*, or McMurtry in *The Cancer Stage of Capitalism*, also forgo any engagement with what this might mean when it comes to autoimmune disease, no doubt because of the focus of their central concepts: for Baudrillard, AIDS, and for McMurtry, cancer. Esposito does at least occasionally address the autoimmune disease, but his interest in immunization and borders leads him into some rather odd formulations. For Esposito, in the process of protection from infectious diseases or terrorist attacks, 'the risk has to do with trespassing or violating borders'. His interest in the disease or threat tends to be on when 'someone or something *penetrates* a body', with 'a violent *intrusion* into the body politic', a 'deviant message *entering* the body'. The threat is located 'always on the *border* between the inside and the outside'.[29] The question of borders reinforces the idea that what really animates Esposito is less immunity and more *immunization*. Moreover, the prevalence of the border in his arguments means that he ends up thinking of the closing down of the border as 'the very essence of immunization'. Aside from the fact that this gets nowhere close to capturing what is at stake in 'immunization', which requires at least a partial opening of the border to the other, the more telling point is that it adds to what is clearly a confusion about autoimmune disease in his work, in that he more or less conflates immunization with autoimmune disease. Trying to explain the process of immunization, Esposito outlines the process of introducing a controlled and tolerable portion of the disease into an organism and plays on the idea of the *pharmakon*: as disease and

29 Roberto Esposito, *Immunitas: The Protection and Negation of Life* (2002), trans. Zakiya Hanafi (Cambridge: Polity, 2011), 2. The same problematic focus on borders permeates appropriations of Esposito's work, as in Nick Vaughan-Williams, *Europe's Border Crisis: Biopolitical Security and Beyond* (Oxford: Oxford University Press, 2015) and Alexej Ulbricht in *Multicultural Immunisation: Liberalism and Esposito* (Edinburgh: Edinburgh University Press, 2015).

antidote, poison and cure, potion and counter-potion, the *pharmakon* incorporates the other. This can also be understood socio-politically: 'If we bring this immunological practice to bear on the social body, we see the same antinomy, the same counterfactual paradox,' which 'risks spiralling out of control'. He explains this as follows:

> To have a nonmetaphorical idea of what I mean, consider what happens in autoimmune diseases when the immune system becomes so strong that it turns against the very mechanism that should be defended and winds up destroying it. Certainly, we need immune systems. No individual or body could do without them, but when they grow out of proportion they end up forcing the entire organism to explode or implode.[30]

The best we can say about this is that it is a bizarre and somewhat confusing way of explaining immunization, which appears to equate immunization and autoimmune disease.

This might explain why, in a series of books and articles on *immunitas*, Esposito, in fact, has surprisingly little to say about the autoimmune disease, despite recognizing that it emerges as 'the natural impulse of every immune system', and what he does have to say is often misleading. For example, after noting that autoimmune diseases express, by their very name, the immune system's 'most acute contradiction', he suggests that such a disease is a 'reversal against itself', and earlier in the same book he writes that in the autoimmune disease 'the warring potential of the immune system is so great that at a certain point it turns against itself'. He repeats the point elsewhere. Asked in an interview about whether 'there is something radically different in the autoimmunity crisis . . . that marks a break with the standard periodizations of modernity', Esposito suggests that an autoimmunitary crisis is a moment when 'the immunitary mechanism . . . begins to turn on itself', creating 'an immunity that is destined to destroy itself'. But, in an autoimmune disease, the immune system does not turn against *itself* at all; it turns against the organism's tissues, against the 'Self' (as Esposito does occasionally note and about which we shall say more in chapter 2). As Inge

30 Roberto Esposito, *Terms of the Political: Community, Immunity, Biopolitics* (2008), trans. Rhiannon Noel Welch (New York: Fordham University Press, 2013), 62.

Mutsaers points out, Esposito sometimes conflates autoimmune disease with immunization and sometimes with autoimmunity in general, but also, and worse, sometimes seems to think that in an autoimmune disease the immune system attacks itself as an immune system.[31]

In similar fashion, Sloterdijk's interest in architectures of immunity and spatialized immune systems means that he also has little to say about anything that might be conceived of as an autoimmune disorder. He occasionally notes that through the immune system's complexity and its owner's longing for security, the immune system can become 'perverted to become their own kind of reasons for illness', leading to the 'growing universe of auto-immune pathologies'.[32] But this argument cannot be developed at any length for fear of undermining the overall affirmation of his own immunological spheres of protection. By imagining his work to be a speculative extension of the concept of immunity as a defensive basis for co-immunism, he is forced to remain with immunity as a protective and defensive system. As such, autoimmune disease cannot be discussed in any meaningful way.[33] When it does arise as an issue, it is quickly passed over.

Disease and illness come with a constellation of meanings, and that constellation is always cultural and often political. But some diseases also come to possess a kind of 'abstract universality', as Foucault comments in his history of madness. In a world dominated by ideas of immunity-security, the autoimmune disease appears to be one such disease. As we shall see in chapters 1 and 2, immunology is replete with comments along the lines that 'autoimmune diseases happen when our guardian [the immune system] becomes our antagonist'.[34] But what happens when our supposed *political* guardian, the security system, becomes our antagonist? The critique of security that underpins this book asks us to formulate a series of questions about immunity that are

31 Esposito, *Immunitas*, 17, 162, 165; *Terms*, 86; Roberto Esposito, Timothy Campbell, and Anna Paparcone, 'Interview', *Diacritics*, vol. 36, no. 2 (2006), 49–56 (55); Inge Mutsaers, *Immunological Discourse in Political Philosophy: Immunisation and Its Discontents* (Abingdon, Oxon: Routledge, 2016), 114.

32 Sloterdijk, *Spheres*, vol. III, 185–86.

33 Mutsaers, *Immunological Discourse*, 80, 91; Timothy C. Campbell, *Improper Life: Technology and Biopolitics from Heidegger to Agamben* (Minneapolis: University of Minnesota Press, 2011), 100–01.

34 Irun R. Cohen, *Tending Adam's Garden: Evolving the Cognitive Immune Self* (San Diego: Academic Press, 2000), 6.

the very same questions to be asked of security: How is it that the power that goes by the name immunity (security) comes to kill the very thing being protected? How does immunity (security) instigate the death of the very thing it is meant to immunize (secure)? How is it that immunity and thus security are said to *make life live*, and yet also turn out to *make that same life die*?[35]

Herein lies the significance of Jacques Derrida's interest in the politics of autoimmunity. Derrida writes that one finds in autoimmunity a 'source of absolute terror'. In being a terror that comes from within, turning the body's protective resources against the body, autoimmune disease is 'terrible, terrifying, terrorizing'. Hence our vulnerability to such a threat is, by definition, without limit, since it is a threat from the interior, from an 'enemy' that is lodged on the inside of the system that it terrorizes.[36] We will, in chapter 2, explore this terror through the lens of immunology's key category of the Self, and we will read it through Freud's concept of the death drive in order to grapple with some of the self-destructive things the body *can* do, developing in the process an account of the death drive in the context of early immunological research and using the death drive to make headways into understanding autoimmune disease. In chapter 5, we will take this argument in a different political direction by connecting it to the idea of the nervous state. Before we go any further, however, we need to note an issue regarding terminology.

Too many of the commentaries within this 'immunitarian turn' in social and political thought assume that *autoimmunity* is the problem, rather than *autoimmune disease*. Derrida tends to conflate autoimmunity and autoimmune disease. The conflation leads him to make some rather bizarre claims, such as 'autoimmunity is not an absolute ill or

35 Readers will note these questions as a reformulation of a series of questions asked by Foucault of disciplinary power, in Michel Foucault, *'Society Must Be Defended': Lectures at the College de France, 1975–76*, trans. David Macey (London: Allen Lane, 2003), 254.

36 Jacques Derrida, 'Autoimmunity: Real and Symbolic Suicides – A Dialogue with Jacques Derrida', in Giovanna Borradori, *Philosophy in a Time of Terror: Dialogues with Jürgen Habermas and Jacques Derrida* (Chicago: University of Chicago Press, 2003), 99, 124, 187; Jacques Derrida, 'Faith and Knowledge: The Two Sources of "Religion" at the Limits of Reason Alone' (1996), trans. Samuel Weber, in Jacques Derrida, *Acts of Religion*, ed. Gil Anidjar (New York: Routledge, 2010), 80.

evil'.[37] Virtually all the writers who have followed Derrida's train of thought have tended to repeat the claim: '"Autoimmunity" . . . cannot be all bad', claims Geoffrey Bennington.[38] But autoimmunity is not 'bad' at all. Autoimmunity is a normal immune function that is part and parcel of a stable homeostasis, involving autoreactive antibodies, B-cells, and T-cells, maintaining the body's optimal state, 'engulfing' and 'cleaning up' dead cells. This makes autoimmunity essential and perfectly normal. As Tauber and others have noted, if low-grade autoimmunity is abnormal, then we are all suffering from autoimmune disease.[39] Autoimmunity is a *feature* of the immune process; without it, one dies. Autoimmune disease is a *failure* of the immune process; with it, one dies (or rather: one is killed . . . by one's own immune system; hence the terror). Immune reactivity against an organism's own constituents is intrinsic to the immune process, which is, in Tauber's terms, *bidirectional*, in that 'the immune system becomes Janus-like by facing inward and outward simultaneously'.[40] (This is a view opposed to the 'one-way' definition which sees a Self whose constitutive agents 'see' and respond to the non-self or foreign, about which we will have much more to say in chapters 1 and 2.) Hence the real issue is the problem of *regulating autoimmunity*, which is, in effect, a problem of regulating the immune system.

At the same time, Derrida sometimes treats autoimmunity as though it is the Self that finds itself infected.[41] Yet there is no 'infected Self' in autoimmune disorders, but rather an aggressively attacked and damaged Self. He also sometimes sees the autoimmune disease as an attack on the body's immune system, as we have noted with Esposito, in which the body conducts 'a terrible war against *that which protects it*' and so

37 Jacques Derrida, *Rogues: Two Essays on Reason* (2002), trans. Pascale-Anne Brault and Michael Naas (Stanford, CA: Stanford University Press, 2005), 152.

38 Geoffrey Bennington, *Not Half No End: Militantly Melancholic Essays in Memory of Jacques Derrida* (Edinburgh: Edinburgh University Press, 2010), 59. Also Timothy Campbell, '"Bios", Immunity, Life: The Thought of Roberto Esposito', *Diacritics*, vol. 36, no. 2 (2006), 2–22 (8).

39 Alfred I. Tauber, 'Moving beyond the Immune Self?', *Immunology*, vol. 12 (2000), 241–48 (244); Tauber, *Immunity*, 14, 112–14; Cohen, *Tending Adam's Garden*, 236–37; Thomas Pradeu, *The Limits of the Self: Immunology and Biological Identity* (Oxford: Oxford University Press, 2012), 85–86.

40 Alfred I. Tauber, 'Reconceiving Immunity: An Overview', *Journal of Theoretical Biology*, vol. 375 (2015), 52–60 (53–54).

41 Derrida, *Rogues*, 109.

destroys 'its *own* immune system'.[42] This confuses autoimmune disorders with immunodeficiency disorders such as HIV/AIDS; again, this has rather misled many of those who cite him approvingly on this score.[43] Some have suggested that Derrida's 'stretching' of the concept of autoimmunity, and the weak correspondence or disconnection between how Derrida uses the term 'autoimmunity' and how it is employed in biological science, is largely irrelevant, because the term is employed by Derrida as part of a deconstructive logic.[44] True as this might be, it can only go so far before stumbling into a biological and conceptual quagmire. To refer to 'the immune-system of democracy', for example, as Martin Hägglund does, two pages after making the comment about deconstructive logic, is to ask something of the reader's knowledge of biology as much as it is to play on their willingness to concede everything to deconstruction.

Immunologists have themselves recognized that there is a difficulty with the terminology. (As there is with many other immunological concepts: to give just one example, the term 'antibody' seems counterintuitive, since it seems to suggest that they go against the body when in fact the opposite is the case.) Tauber suggests that we might be better off using *autoimmunity* to refer to the actual autoimmune diseases, and using *concinnity* (from the Latin verb *concinno*, meaning to join fitly together, to order, arrange or adjust appropriately) to describe what has hitherto been understood as normal autoimmunity. Michal Schwartz suggests 'protective autoimmunity' to capture normal autoimmunity, while leading Russian immunologists have suggested using the term

42 Derrida, 'Faith', 82, emphases added.

43 During the early period of HIV/AIDS research, some researchers thought it might be an autoimmune disease, but it is, in fact, a disease in which a virus attacks and devastates the immune system. Hence it is an immunodeficiency virus or deficiency syndrome. In contrast, in an autoimmune disease, the attack is led by the immune system itself. To think of it in political terms: in HIV/AIDS, the immunity-security system that protects the body is under attack from a foreign enemy, but, in autoimmune diseases, the body's security-immunity system attacks the body.

44 Martin Hägglund, *Radical Atheism: Derrida and the Time of Life* (Stanford, CA: Stanford University Press, 2008), 9. For the stretching and weak correspondence see W. J. T. Mitchell, *Cloning Terror: The War of Images, 9/11 to the Present* (Chicago: University of Chicago Press, 2011), 48, and Zahi Zalloua, *Theory's Autoimmunity: Skepticism, Literature, and Philosophy* (Evanston, IL: Northwestern University Press, 2018), 13.

'autoallergic' to describe pathogenic immune reactions.[45] Yet although these suggestions make sense for those philosophers and social scientists who wish to write solely of 'autoimmunity', they are up against the entrenched language of medical science and the biological humanities. For these reasons, I will retain the concept of autoimmune disease or disorder, and when referring to 'autoimmunity' as a problem it will be in citations from the work of others. A further reason to do so lies in the powerful political connotations of 'disease' and 'disorder'. When the body operates as one of the most potent concepts through which politics is imagined, disease becomes a lens through which disorder, crisis, decline, and decay are understood; it becomes a reminder that what once seemed 'orderly' can quickly become 'disorderly' – *chaotic*, perhaps, as we shall discuss in chapter 4 – and that seemingly self-regulating systems turn out to be not so self-regulating after all. The autoimmune *disease* or *disorder* therefore reminds us that there is a political issue at stake.

Despite the comments in the previous few paragraphs, Derrida's work is important for stressing that the logic of autoimmunity (that is, autoimmune *disease*) is one of internal threat. Everything is threatened from within. It is this that sets Derrida's work apart from Esposito's. (The latter's claim that he is in 'complete agreement' with Derrida is not entirely true: his continual references to the 'positive' dimensions of immunization are evidence of his general tendency to hold on to the affirmative powers of *immunitas*.) This threat from within is one reason why the question of the death drive is highly relevant, politically as well as psychoanalytically. What I am interested in with the category of the autoimmune disease, then, is a situation in which the body (and thus *the Self*) is threatened, damaged, and ultimately destroyed from within by its own immunity-security, and a parallel in which the body politic (and thus *the System*) is threatened, damaged, and ultimately destroyed from within by its own security-immunity. If, for the individual body, the autoimmune disease involves a fragmentation of Self to the point of uncontrolled self-destruction, a manifestation of the death drive and form of suicide (a discussion we will have in chapter 2), then, imagined

45 Tauber, *Immunity*, 116; Tauber, 'Reconceiving Immunity', 54–57; Michal Schwartz, *Neuroimmunity* (New Haven: Yale University Press, 2015), 69–71, 120–27, 197–98; A. B. Poletaev, L. P. Churilov, Yu. I. Stroev, and M. M. Agapov, 'Immunophysiology versus Immunopathology: Natural Autoimmunity in Human Health and Disease', *Pathophysiology*, vol. 19, no. 3 (2012), 221–31.

politically, the autoimmune disease is the self-destruction of the body politic by that body's own protection. Security systems run amok, overreacting to imagined threats to the extent that they can no longer tell friend from enemy, turning self-defence into self-destruction and undoing themselves with their own hyper-intensified security measures.

If we accept that there is indeed a political analogue of immunity, then there must also be a political analogue for autoimmune disease. The politics of immunity is thus a politics of the autoimmune disease, because such disease points us to the idea that even if all external threats are evaded, the body (politic) can still destroy itself through its uncontrollable search for security. To confront the autoimmune disease in a book on the politics of immunity is to confront what is at stake in security's destructive power.

This destructive power reaches its crescendo towards the end of chapter 5, and then in the final chapter, where the focus shifts back from biological to legal immunity. As we have already noted, immunity stands with one foot in biology and one in law, and its politics emerges from tacking back and forth between those fields. Yet many of the scholars who have written about immunity tend to focus on the former rather than the latter. In a long footnote outlining the roots of some of immunity's lexical resources in legal history, Derrida still claims that it is 'especially in the domain of biology that the lexical resources have developed their authority'.[46] Likewise Esposito, equally aware of immunity's rich legal history, nonetheless builds almost entirely on its biomedical history; after all, Esposito's key term 'immunization' might make sense when thinking biologically, but one rarely finds 'immunization' in the law books. One does, however, find that certain non-biological notions of immunity do still resonate through the law books, and one finds it most often concerning questions of violence carried out in the name of security.

On the one hand, we find international lawyers, moral philosophers, and politicians insisting that something called 'noncombatant' or 'civilian' immunity exists and that good liberal states respect it. 'Noncombatant immunity has served as a fundamental limit on political violence', we are told, and this immunity is 'a basic human right codified in international

46 Derrida, 'Faith', 80.

law'.[47] The second part of this claim is true. The first part is a fiction. If noncombatants have such immunity, then why are there so many damaged and dead civilian bodies produced by political violence? To give the most commonly cited statistic: during the century in which civilian immunity is said to have become properly established and codified, the number of civilian deaths in war went from between 5 and 10 percent of deaths in World War I to over half the deaths in World War II and now to around 90 percent of the deaths in wars fought by the end of the twentieth and into the twenty-first century. So, on the one hand, we must deal with the fiction of noncombatant immunity.

On the other hand, as well as the law of noncombatant immunity, we have the immunity granted to certain state officials. Servants of the state who carry out acts of violence, usually in the name of security, are granted immunity from prosecution for such acts. In the case of policing, for example, the idea of a police officer's immunity is used time and again to allow police to avoid prosecution for acts of torture, violence, and killing. In the UK there have been just under 1,800 deaths in police custody or through contact with police since 1990; the number of convictions of police officers remained at zero, until, in June 2021, there was the first successful prosecution of a police officer for manslaughter. The number of aboriginal deaths in police custody in Australia in the same period is 432; the number of convictions is zero. This pattern is played out across all states. Barely a day goes by without a police officer in the US killing someone; on most days, it is more than one. In the name of security, the state kills, and to protect those of its servants who do the killing it grants them immunity. The immunity may well be 'qualified', as many states like to call it, it may sometimes appear de facto rather than de jure, but the fact that it is called immunity is important, as many movements and struggles against police power have discovered. To understand how this happens, we will, in the final chapter, bring into view the kind of immunity historically granted to the figure of the Ambassador (better known under the more commonplace 'diplomatic immunity'). The servants of the state who conduct its violence do so as its *ambassadors*. As such, they have inherited the kind of immunity historically granted to the figure of the Ambassador. They are protected,

47 Sahr Conway-Lanz, *Collateral Damage: Americans, Noncombatant Immunity, and Atrocity after World War II* (Abingdon, Oxon: Routledge, 2006), xi.

in other words, by the same kind of legal fiction that surrounds the immunity of the Ambassador. Such immunity, it turns out, is claimed by the state for itself. Hence, and contra Esposito, Sloterdijk, and Luhmann, it is not community that cannot exist without immunity, but sovereignty. This is the very reason immunity and security now go hand in hand.

In his *Commentaries on the Laws of England* (1765–1769), William Blackstone observed that students of law tend to get a little startled when they discover the existence of legal fictions. And yet, he writes, such fictions are integral to law's functioning and hence to the operation of sovereign power. They can take the form, for example, of fictitious legal persons such as the corporation or fictitious sovereign principles such as 'The King Can Do No Wrong'. One such fiction is immunity. As regards the Ambassador, Hugo Grotius observed in his long treatise on the legitimacy of liberal violence, *The Rights of War and Peace* (1625), that the Ambassador is by a 'Sort of Fiction' taken for the very 'Person' that they represent. In being taken for the 'person' of the sovereign, even when sovereignty has been depersonalized, this fiction allows certain figures to be *granted the immunity claimed by sovereign power.* The Diplomat is one such figure, but so too are the figures who, in the name of security, carry out sovereign violence.

Sovereignty thrives on fictions, most notably the fiction of immunity. But it equally thrives on the fiction of security. Perhaps that is why security and immunity live and kill together, even to the extent of destroying the very body we think they are protecting.

1

War Power, Police Power, Cell Power

Why is it so difficult to imagine that I am cared about, that something takes an interest in what I do, that I am perhaps protected, maybe even kept alive not altogether by my own will and doing? Why do I prefer insurance to the invisible guarantees of existence? For it sure is easy to die. A split second of inattention and the best-laid plans of a strong ego spill out on the sidewalk. Something saves me every day from falling down the stairs, tripping at the curb, being blindsided. How is it possible to race down the highway, tape deck singing, thoughts far away, and stay alive? What is this 'immune system' that watches over my days, my food sprinkled with viruses, toxins, bacteria?

James Hillman, *The Soul's Code* (1996)

'Our country is at war.' 'The world is at war.' 'We are under attack from an invisible enemy.' 'Ours is now a wartime government.' 'Wartime President.' 'Medical personnel are frontline workers.' 'In this fight we can be in no doubt that each and every one of us is directly enlisted.' 'We are at war, and this is our draft.' 'Raising an army of the infected.' 'A war economy.' As if to prove the truth of all these claims, measures of total war were announced for the whole of society: emergency laws, new police powers, quarantine, troops mobilized, new behaviour instilled in the population. All to conduct a war against . . . a *virus*.

Such was life during the Covid pandemic that started in early 2020. A total war against a virus capable of destroying millions of lives and therefore a war that we were told was necessary to defend the state, protect the system and secure the social order from collapse. The front-line of nurses, doctors, and the medical police was to be supported by the sturdy trench system of modern warfare formed of the institutions of civil society. It was a war fought through the people's bodies and the body of the people. It was a war of immunity: the immunity of each and every single body was integral to the defence of the realm, and the immunity of the body politic was integral to the protection of each and every one of us. Body and body politic were as one in their search for mutual security. We were encouraged to imagine the sovereign nation as a body and our body as a sovereign nation.

'Imagine That Your Body Is a Sovereign Nation'

In June 1986, an article by Peter Jaret appeared in *National Geographic*, accompanied by a series of images, called 'Our Immune System: The Wars Within'. The article was circulated widely and discussed by many. It received the most commendations of any article that year, was reprinted many times, and spawned many other journal articles along the same lines in popular publications such as *Time* and *Reader's Digest*.[1] Jaret's article contained material from his book co-authored with Steven Mizel, called *In Self-Defense*, but the text was illustrated with images from a book called *The Body Victorious*, published in Swedish in 1985 and then in English translation in 1987. That book, which also received a lot of publicity and became an international bestseller, is credited largely to Lennart Nilsson, one of the world's leading medical photographers, and it is impossible to open the book and not be stunned by the array of shapes and colours in the images. For what is laid bare is the range of processes through which the body 'victoriously' defends itself against all the forces which oppose it or threaten it. Like Jaret's 1986 article, the book suggests that the body's defence system is the site of 'life-and-death struggles between attackers and defenders, waged with a ruthlessness found only in total war'. The site of an injury is 'a battlefield

1 Peter Jaret, 'The Wars Within: Our Immune System', *National Geographic*, vol. 169, no. 6 (1986), 702–34; Emily Martin, *Flexible Bodies: Tracking Immunity in American Culture – From the Days of Polio to the Age of AIDS* (Boston: Beacon Press, 1994), 51.

on which the body's armed forces, hurling themselves repeatedly at the encroaching microorganisms, crush and annihilate them'.

> No one is pardoned, no prisoners are taken – although fragments of the invading bacteria, viruses, rickettsias, parasites, and fungal micro-organisms are conveyed to the lymph nodes for the rapid training of the defence system's true bloodhounds, the 'killer cells'. These cells learn in detail, molecule by molecule, how to recognise the adversary, whereupon they launch their offensive.

They go on:

> A cell whose identification is faulty is immediately destroyed by the armed force which is constantly on patrol . . . The human body's police corps is programmed to distinguish between bona fide residents and illegal aliens – an ability fundamental to the body's powers of self-defence.

What then follows is a series of diagrams illustrating the way the body deals with 'the foreign invader'. Light-blue paths represent 'older defences', which work in tandem with 'the newer, special defence force' (in green). Elsewhere, in the lymphoid tissue, special 'B-lymphocytes are trained', which are 'the precursors of the large plasma cells . . . which produce the body's sniper ammunition, the antibodies'. The different types of T-lymphocytes include 'aggressive killer cells, helper cells, and suppressive cells', all of which 'have specialised tasks to perform when the immune system launches a counterattack'. Despite the book's title, *The Body Victorious*, the victory is never fully won. Rather, the body is the site of perpetual war, or perhaps is even the war itself. The body is a battlefield, from cradle to grave.

At the heart of this war is the immune system:

> The organisation of the human immune system is reminiscent of military defence, with regard to both weapon technology and strategy. Our internal army has at its disposal swift, highly mobile regiments, shock troops, snipers, and tanks. We have soldier cells which, on contact with the enemy, at once start producing homing missiles whose accuracy is overwhelming. Our defence system also boasts

> ammunition which pierces and bursts bacteria, reconnaissance squads, an intelligence service, and a defence staff unit which determines the location and strength of troops to be deployed.

The war goes on:

> One type of white blood cell is the *granulocyte*. Granulocytes are small, fast-moving, and dynamic feeding cells in the blood, kept permanently at the ready for a blitzkrieg against the microorganisms or foreign particles. They constitute the infantry of the immune system. Multitudes fall in battle, and together with their vanquished foes, they form the pus which collects in wounds.
>
> Another type is the considerably larger *macrophage* (large feeding cell), the armoured unit of the defence system . . . tracking down their victims.

This 'giant army of granulocytes and macrophages is, of course, a fearsome opponent for the invading microorganisms', and so some enemy-bacteria seek other methods of invasion and destruction.

> One refined method employed by many bacteria and all viruses is to hide inside the body's own cells. They disguise themselves, as it were, in a uniform which the immune system's soldiers have learned to overlook. In this situation, the defence system deploys its special commandos or frontline troops, the *B- and T-lymphocytes*.

Lymphocytes 'kill their opponents differently: by using homing projectiles (antibodies), and with some form of poisoning (killer cells). The ability to do so requires training at technical colleges'. They continue:

> The lymphocytes which attend the technical college of the thymus are *helper*, *suppressor*, and *killer cells* called T-lymphocytes (or T-cells). They are among the most indispensable armed forces of the immune system.
>
> Helper cells constitute the defence staff unit, directing troop operations . . . Killer cells are formed in the thymus to kill those of the body's own cells which contain foreign antigens.

The defence mechanisms also include the skin, 'the first line of defence', and a form of 'chemical warfare' in the form of sweat and sebum.[2]

The Body Victorious was a bestseller, the images alone being a fascinating guide to the immune system, despite (or perhaps because of) the fact that the images show acts of violence after violence. The power of the images also no doubt helps explain why Jaret's *National Geographic* article was so widely circulated. Jaret's article tells the same story of incessant battles within, of combatants too tiny to see, of legions of specialist defence forces, of the body's munitions factories, and of chemical weapons such as antibodies. The article presents the process in diagrammatic form: 'The Battle Begins', 'The Forces Multiply', 'Conquering the Infection', 'Calling a Truce'.

These war stories will come as no surprise to anyone who has ever read a magazine article or popular book on immunity, as the same ideas run through the medical and popular literature in the field. This much has been pointed out many times before and has been subject to sustained critique for its 'militarization' of medical language and excessive use of war images. I want to briefly give an indication of the extent of these ideas and images, however, partly to show just how wide and deep their use is, but also because I want to point to some of the limitations of the critical comments made about them. Most importantly, I want to use these observations to then expand the argument beyond the narrow frame of 'war metaphors' and the militarization of medical discourse. The politics of immunity requires more than just an argument against the language of war.

'Imagine that your body is a sovereign nation', suggests Michael Weiner in his popular text *Maximum Immunity*.

> What prevents your body, the host country, from becoming a Petri dish for every kind of bacterium and virus? That is the job of your immune system. It is the immune system that protects you from attack by foreign and internal agents, that plans defensive actions, and that provides the necessary personnel and equipment.[3]

2 Lennart Nilsson, with Kjell Lindqvist and Stig Nordfeldt, *The Body Victorious: The Illustrated Story of Our Immune System and Other Defences of the Human Body* (1985), trans. Clare James (London: Faber, 1987), 20–31.

3 Michael A. Weiner, *Maximum Immunity* (Bath: Gateway Books, 1986), 22.

Text after text takes up the theme. John Dwyer's book *The Body at War* is organized around immunity as 'the programming of cells to recognise the enemy' and includes chapters on 'War and the Warriors', 'Taming the Warriors', and 'Defeat of the Warriors'.[4] *The War within Us* by Cedric Mims is organized around the idea of an 'ancient war between the invader and the defender' in which 'the defences of the host have called forth answering strategies by the parasite, which have led in turn to counterstrategies on the part of the host'.[5] Even if you think you are at peace, you are in fact at war:

> Outwardly you may be a gentle, peaceful person, a loving parent, a churchgoer, even a pacifist. Inside of you, however, whether you will it or not, an awesome fighting force is on the alert. The human immune system is an efficient war machine that never negotiates a treaty. It strikes no bargains with the enemy. It takes no prisoners.[6]

These tropes from popular health and medicine books are not simplifications for a popular audience, since the language is lifted from scientific texts within immunology. Developing the clonal selection theory in the late 1950s, Macfarlane Burnet would put it that 'when foreign and hence potentially dangerous material enters the body – classically as an invading micro-organism – it requires to be recognized as foreign . . . Any defence force must know how to distinguish friend from enemy'.[7] Stewart Sell's *Immunology, Immunopathology and Immunity*, a leading

4 John Dwyer, *The Body at War* (London: Unwin Hyman, 1988), 33.

5 Cedric Mims, *The War within Us: Everyman's Guide to Infection and Immunity* (San Diego, CA: Academic Press, 2000), viii, x, 82, 175, 237.

6 Ellen Michaud, Alice Feinstein, and the editors of *Prevention Magazine*, *Fighting Disease: The Complete Guide to Natural Immune Power* (Emmaus, PA: Rodale Press, 1989), viii, 1–11. The same ideas and images are everywhere. As a sample, see Steven B. Mizel and Peter Jaret, *In Self-Defense: The Human Immune System* (San Diego: Harcourt Brace Jovanovich, 1985), 57, 105; Marion D. Kendall, *Dying to Live: How Our Bodies Fight Disease* (Cambridge: Cambridge University Press, 1998), 9, 12, 60, 64; J. H. L. Playfair and B. M. Chain, *Immunology at a Glance* (Oxford: Blackwell, 2001), 25, 57; Lewis Wolpert, *How We Live and Why We Die: The Secret Lives of Cells* (London: Faber and Faber, 2009), 1, 2, 155–76; Catherine Carver, *Immune: How Your Body Defends and Protects You* (London: Bloomsbury, 2017), 7, 13.

7 Sir Macfarlane Burnet, 'Auto-Immune Disease I: Modern Immunological Concepts', *British Medical Journal*, vol. 2, no. 5153 (1959), 645–50 (645).

textbook in medical biology that has run into several editions, opens as follows:

> Immunology is the study of the *system* through which we identify infectious agents as different from ourselves and defend or protect ourselves against their damaging effects. From the time of conception, the human organism faces attack from a wide variety of infectious agents and must have ways to identify and react to them . . . The immune system provides us with highly specialized ways to defend ourselves against invasion and colonization by foreign organisms. This defensive ability is called *immunity*.[8]

Likewise, Edward Golub and Douglas Green's *Immunology: A Synthesis*, an equally important reference point in the field, offers a 'Machiavellian view' on how to protect the body politic from internal and external invaders. For Golub and Green, the immune system is part of the body's *armamentarium*, and they add that 'the use of the term *armamentarium* shows that this process has been visualized as a battle between two opposing forces, one good and one very bad'. Other major textbooks such as David Wilson's *Science of Self: A Report of the New Immunology* (1971) and Jan Klein's *Immunology: The Science of Self-Nonself Discrimination* (1982) follow suit.[9] The two popular books by one of the authors of *Killer Lymphocytes* (2005), William R. Clark, carry the theme over: on the one hand, immunity is *In Defense of Self* (2008); on the other hand, we are *At War Within* (1995). Such imagery and language, it should be noted, functions in the more detailed discussions of both innate and acquired immunity. Innate immunity consists of general reactions such as inflammation and fever and operates through the phagocytes and macrophages, while acquired immunity involves the work of 'killer lymphocytes'.[10]

---

8 Stewart Sell, *Immunology, Immunopathology and Immunity*, 5th edn. (Stamford, CN: Appleton and Lange, 1996), 3.

9 Edward S. Golub and Douglas R. Green, *Immunology: A Synthesis*, 2nd edn. (Sunderland, MA: Sinauer Associates, 1991), 546.

10 Gregory J. Bancroft, 'The Role of Natural Killer Cells in Innate Resistance', *Current Opinion in Immunology*, vol. 5, no. 4 (1993), 503–10; Charles A. Janeway, 'How the Immune System Recognizes Invaders', *Scientific American*, vol. 269, no. 3 (Sept. 1993), 72–79.

The same theme permeates children's books on the subject, with cartoon images of military forces fighting invasions, patrolling the body, mustering specialist troops whenever the threat of invasion looms. This began with a book by renowned physicist George Gamow and microbiologist Martynas Yčas, *Mr. Tompkins Inside Himself* (1968), a biology book for children which presented white blood cells fighting off and killing invaders, and continues with more recent literature such as *Cell Wars* (1990), by Fran Balkwill and Mic Rolph, which won the Children's Science Book Prize in 1991. For slightly older children there is the more recent comic *Immunity Warriors: Invasion of the Alien Zombies*, launched in January 2017 to encourage children in Canada to undergo immunization. The comic presents an attack on the body, 'made of billions of cities called cells', under assault, with comic figures organizing the defence: 'Commander, we're under attack. Sector 12 is down . . . The enemy has somehow infiltrated the city's defenses . . . The enemy are taking over the City's command center . . . They've turned the citizens into zombies.' As always with the zombie narrative, and as I discuss at length in *The Universal Adversary* (2016), the emergency war powers are mobilised:

> Send out the NKC [Natural Killer Cells] and the Macro Squad. Now! . . . Call in the elite forces . . . B-Team, steady your aim and target the enemy for the Macro Squad . . . There must be some sacrifices. T-squad, you know what you must do.

In the end, the battle is won, the body defended, and the project continues with a nice recognition of the role the police power might play in such wars: a 'Wanted' poster appears for the influenza virus, complete with a picture of the microscopic invader-intruder.

From popular science books to immunology textbooks, from scientific research to children's literature, a common sense about the body's immunity has emerged that runs through our imagination and the way we are expected to care for the Self. Protect, defend, and preserve one's body through the correct regimen: the care of the Self involves thinking about and operating on the Self as a body under siege in need of a general defence policy to repel invaders, fortify itself, wage campaigns, block entry points, and identify the Self as a killing machine. The *care of the self* leaves one at *war with the world*. The message is clear: *You are by nature a war machine.*

Now, this image of the body has been criticized from many quarters. Unravelling the ways in which the controlling images in accounts of diseases are drawn from the language of warfare, paralleled in and reinforced by a medicalized notion of the body politic, Susan Sontag finished her book containing her two major essays on the topic (*Illness as Metaphor* and *AIDS and Its Metaphors*), by suggesting we rethink our images and metaphors concerning bodies and illnesses.

> The one that I am most eager to see retired – more than ever since the emergence of AIDS – is the military metaphor. Its converse, the medical model of the public weal, is probably more dangerous and far-reaching in its consequences, since it not only provides a persuasive justification for authoritarian rule but implicitly suggests the necessity of state-sponsored repression and violence (the equivalent of the offending or 'unhealthy' parts of the body politic). But the effect of the military imagery on thinking about sickness and health is far from inconsequential. It overmobilizes, it overdescribes, and it powerfully contributes to the excommunicating and stigmatizing of the ill.[11]

Other thinkers have followed suit in this criticism, including some of the philosophers who have nonetheless still sought a new paradigm of immunization, and the world is now somewhat overwhelmed with articles and essays on 'war metaphors in health care', 'the use of metaphor in the discourse on bodies', 'war and the allegory of medical intervention', 'war as a medicine', 'medicine as war', 'body wars', and similar titles. None of them has had much to add to Sontag's argument. The gist of virtually all the literature is to suggest that the language and imagery of war is inappropriate for thinking about bodies, health and illness. They point variously to the idea that war rhetoric is saturated with anger, violence, hatred, and antagonism; that war imagery in medicine encourages an aggressively nationalistic sentiment ('flu' or 'Spanish flu'? 'virus' or 'kung flu virus'?); that the language of war legitimizes emergency and authoritarian measures; that the endless nature of such wars tends towards permanence; that practices of care become battles and places of care become war zones, closing down alternative ways of thinking about

11 Susan Sontag, *Illness as Metaphor* and *AIDS and Its Metaphors* (Harmondsworth: Penguin, 1991), 179–80.

caring for one other; finally, that the war rhetoric makes our crumbling healthcare systems appear to be the result of an enemy virus rather than the result of public policies.

These points and the general concern have been noted within immunology. Indeed, the points were made early in immunology's history, for example by Ludwik Fleck, in *Genesis and Development of a Scientific Fact* (1935) and again by René Dubos in *Mirage of Health* (1960). 'Emphasis on the antagonistic relationship between self and foreign can be the mainspring for a language of conflict or warfare in immunology', note Eileen Crist and Alfred Tauber, though they also claim that the view of immunology as 'cloaked politics' is misguiding.[12] Other immunologists have sought an alternative language to war-making. Francisco Varela, for example (who with Humberto Maturana coined the term 'autopoiesis', about which we shall have more to say below), rejects the war model when thinking about the immune system in his later work, making it very clear at one point that this way of imagining immune processes has a long historical dark side, underpinning the civil war in his own country, Chile, and being important in the rise of fascism. In his later work, Varela and fellow researchers such as António Coutinho and John Stewart suggest a more cognitive model and a neurophenomenology or *immunoknowledge* to think through the question of body politic (about which we will say more in chapter 5). Here, we can note their suggestion that, instead of 'war', we think of a mutual 'dance between immune system and body' that allows the immune network to learn from its dancing experience and generate the immunoknowledge.[13]

The idea of dance plays on the suggestion made by others that we imagine immunity through the image of an orchestra rather than an army. This was proposed by Richard Gershon, known for his work showing the importance of T-lymphocytes to acquired immune tolerance. 'Endocrinologists speak of an endocrine orchestra, implying an

12 Eileen Crist and Alfred I. Tauber, 'Selfhood, Immunity, and the Biological Imagination: The Thought of Frank Macfarlane Burnet', *Biology and Philosophy*, vol. 15 (1999), 509–33 (529–30).

13 Francisco J. Varela and Antonio Coutinho, 'Immunoknowledge: The Immune System as a Learning Process of Somatic Individuation', in John Brockman (ed.), *Doing Science: The Reality Club* (New York: Prentice Hall, 1988), 249; John Stewart and Antonio Coutinho, 'The Affirmation of Self: A New Perspective on the Immune System', *Artificial Life*, vol. 10, no. 3 (2004), 261–76 (262).

interreaction between the various components of the system which determines overall endocrine function. This idea might well be adopted by immunologists.' In this vision, the T-cell moves from killer to leader of the band. Playing with a similar image, Stephen Hall suggests an immunological 'disarmament campaign' replacing war with music. Given the coordinated and syncopated nature of the immune response, 'perhaps the better analogy is to a beautifully scored piece of music', an 'immune symphony' in which the concert is further enhanced with each new discovery. All 'the orchestra players' circulate in the body, gather together in the spleen or in lymph nodes, and work as an ensemble, comments Rem Petrov. Michal Schwartz in *Neuroimmunity* also describes the immune system as the body's orchestrating system. Extending the imagery, some have suggested that the orchestra might well be leading a dance. In a comment about how to live in a world of systems that we shall pick up in chapter 3, Donella Meadows encourages us to 'get the beat of the system': 'We can't control systems or figure them out. But we can dance with them!' One such system is the immune one: to dance with your immune system while the orchestra plays, tallies with the idea of the 'rhythms of the body' or the 'music of life'. In this way, we can imagine, along with Donna Haraway, the range of bacteria, fungi and protists 'which play in a symphony necessary to my being alive at all'. Better still, we might imagine, with Silvia Federici, such a dance as part of the wider reappropriation of our bodies from capital and the war machine.[14]

The very fact that such suggestions and imagery might strike some of us as a little odd is, no doubt, because we are used to imagining our bodies as spaces of war and strategic fields of battle and defence, and so used to living in a culture in which the idea of war is ubiquitous, that any other image seems a little limp, even pathetic, by comparison. And so,

14 Richard Gershon, 'The Immunological Orchestra', *Lancet*, vol. 291 (27 January 1968), 185–86; Stephen S. Hall, *A Commotion in the Blood: Life, Death and the Immune System* (New York: Henry Holt and Co., 1997), 453; Rem Petrov, *Me or Not Me: Immunological Mobiles* (1983), trans. G. Yu. Degtyaryova (Moscow: Mir Publishers, 1987), 114; Michal Schwartz, *Neuroimmunity* (New Haven: Yale University Press, 2015), 3; Donella H. Meadows, *Thinking in Systems: A Primer* (White River Junction, VT: Chelsea Green Publishing, 2008), 170; Donna J. Haraway, *When Species Meet* (Minneapolis: University of Minnesota Press, 2008), 3–4; Silvia Federici, *Beyond the Periphery of the Skin: Rethinking, Remaking, and Reclaiming the Body in Contemporary Capitalism* (Oakland, CA: PM Press, 2020), 123.

despite critique after critique and plea after plea that we leave war to the war-makers, the images of war never stop, as thinker after thinker succumbs to using the imagery, often despite themselves and often telling us that such tropes and images cannot be abandoned however much we dislike them. Even a book called *The Tao of Immunology*, which begins by pointing out the limitations of the language of war and defence and seeks instead to imagine immunity through the art of Zen and ideas about 'balance' and 'harmony', is replete with friends and enemies, friendly fire, overkill, vigilant bastions, missiles, smart bombs, laser-marked targets, modern-day gunners, and anti-terrorist squads.[15] The idea of the smart bomb as a route to harmonious living is replicated by others, such as in an essay on meditation and immunity which suggests imagining a 'return to a state of harmony' by 'destroying any foreign invaders, crushing them, repelling them'.[16]

My point here, then, is not to offer yet another critique of militarized medical discourse or the use of war to understand and describe bodies; another contribution to such critique will get us nowhere. What I want to suggest is that the problem is precisely *not* simply one of *militarized* thought and is in fact far broader than the conception of the body as a strategic system. Or rather, if the body is imagined as a strategic system then the 'strategy' is one of security rather than war. In other words, the issue is a far broader *logic of security*. This point is sometimes noted by commentators but then ignored. Andrew Goffey, for example, notes that 'it is the aggressively imagistic language of security and warfare, which runs throughout the historical development of immunology, that has proved of most interest to critical researchers',[17] but he then goes on to say plenty about war and next to nothing about security. This simply replicates a broader liberal position which is happy to critique war and its tropes but less comfortable with a *critique of security*. We need an analysis of the politics of immunity that grapples with the ways in which it is imagined through the broad lens of security rather than the narrow

15 Marc Lappé, *The Tao of Immunology: A Revolutionary New Understanding of Our Body's Defenses* (New York: Plenum, 1997).

16 Margo Adair and Lynn Johnson, 'Applied Meditations for Healing', in Jason Serinus (ed.), *Psychoimmunity and the Healing Process* (Berkeley, CA: Celestial Arts, 1986), 186–88.

17 Andrew Goffey, 'Homo Immunologicus: On the Limits of Critique', *Medical Humanities*, vol. 41, no. 1 (2015), 8–13 (9).

lens of 'militarized' life. Such an analysis allows us to unravel the conjoined articulation of *immunity-security*. This turn to 'security' helps us better understand why the political and cultural imagery that I have been outlining thus far relies on the idea that the body needs *policing* as well as *defending* from invaders. As one immunological textbook puts it, in a rather twee comment: those of us who want to imagine immunity but 'who don't like violence' might 'picture something less disturbing, like little police cars escorting the diseased cells out of the body'.[18] This move from the violence of militarized destruction to the supposed comfort and 'security' of policing is common. 'The analogy between the immune system and a police force seems helpful', notes Wilson, having already said plenty about war.[19] The politics of immunity, then, as a politics of security, concerns the police power as well as the war power.

'A Natural Policing Thing That Happens in the Body'

Early in the history of biopolitical thinking and well before the recent post-Foucauldian 'immunitarian turn', Morley Roberts sought to incorporate an argument about immunity into a wider frame of 'bio-politics'. In a series of publications through the 1920s and 1930s Roberts suggests that life itself depends on an immunization as 'active warfare', with each cell a 'fortress' or 'armed city' in possession of 'cell tools and weapons'. Yet Roberts also stresses the immune process as an organized police power necessary to avoid disorder and anarchy. When considering the immune process, he writes, 'It is impossible not to see deep social analogues in the police', especially when 'the social police [are] so greatly occupied with one disturbance that they are unable to deal with another': just as 'the constable with a prisoner in custody will not attend to his ordinary duties' so 'the cells of the reticulo-endothelial system may be a like case'.[20]

18 Kenneth Bock and Nelli Sabin, *The Road to Immunity: How to Survive and Thrive in a Toxic World* (New York: Simon and Schuster, 1997), 387.

19 David Wilson, *The Science of Self: A Report of the New Immunology* (London: Longman, 1971), 267.

20 Morley Roberts, *Warfare in the Human Body: Essays on Method, Malgnity, Repair and Allied Subjects* (London: Eveleigh Nash, 1920), 137–38; *Bio-Politics: An Essay in the Physiology, Pathology and Politics of the Social and Somatic Organism* (London: Dent, 1938), 68–69, 157, 164.

Roberts's suggestion has resonated through immunological discourse, though this has gone largely unacknowledged, and certainly without anything like the kind of commentary that exists on the trope of war. Yet once we do start acknowledging it, we find time and again claims for 'the police function of immunity', as Petrov puts it. In *The War within Us*, Mims imagines 'a dozen devious devices' an invader might use to evade the immune system. First, an invading virus would need to be 'tucked away from police surveillance', and the virus would want to keep changing appearance to 'make things harder for the police'. Second, because 'communication and collaboration between immune cells (policemen) is the foundation of effective response', the invading virus would need to 'interfere with police communications'. Third, the virus might even 'invade police headquarters' by attacking the immune system as a whole. Fourth, the virus might decide to 'set the alarm bells ringing, call out the police, create a diversion' in order to 'get the police to make the wrong sort of response'. Finally, any good invading virus should 'have something ready up [its] sleeve in case of local encounters with police'. Irun Cohen suggests that 'the immune system is not only a department of defense, it also functions as a department of internal welfare', in the form of 'cells that patrol the body systematically . . . similar to the strategy of the police, sanitation and fire departments and the board of health'. Marc Lappé in *The Tao of Immunology* imagines a 'cellular police system' that is 'directed to stop and identify adolescents who wore certain age-related clothing (e.g., gang colors), while being told to ignore "more appropriately" attired adults'. Edward Bullmore speaks of macrophages as the 'border guards' and 'robocops' of the immune system. The immune system 'patrols the body' (Niels Jerne) and 'polices the toxic' (Tauber).[21]

Such ideas are commonplace, as anthropologists who have sought to get people to talk about their immune systems have found:

---

21 Petrov, *Me or Not Me*, 24, 173; Mims, *War within Us*, 109–32; Irun R. Cohen, *Tending Adam's Garden: Evolving the Cognitive Immune Self* (Dan Diego: Academic Press, 2000), 5, 118; Lappé, *Tao of Immunology*, 88; Edward Bullmore, *The Inflamed Mind: A Radical New Approach to Depression* (London: Short Books, 2018), 28, 148; Niels Kaj Jerne, 'The Immune System', *Scientific American*, vol. 229, no. 1 (July 1973), 52–60 (52); Alfred I. Tauber, *Immunity: The Evolution of an Idea* (Oxford: Oxford University Press, 2017), 20.

> I mean, there's a natural policing thing that happens in the body right? And, I mean it's an incredible policing thing, when you think about it, or at least it's incredible in my mind, because it's a system-wide authority that works . . . It's easy to talk about policing systems and white blood cells being the good guys and this kind of stuff.[22]

The idea of immune process as policing and immunity as security is what underpinned the idea of 'immunosurveillance' that came to the fore in the 1970s. At an international conference on 'Immune Surveillance' held in May 1970, Burnet, by that time probably the world's leading immunologist, commented that 'immunological surveillance is . . . well established'.[23] Burnet had earlier suggested the *foreignness* of disease, as we have seen, but as well as treating phagocytic cells as the final defenders of the body against foreign invaders he also describes antibodies as the body's 'reminders'.

> The 'reminders' are like plain-clothes detectives with perfect memories for criminal faces; when they come into contact with their 'opposite number', this is recognized as a dangerous individual which has on a former occasion penetrated the defences either of the body or of society. It or he must be apprehended . . . Just as in human communities we have a policeman at the gate of the police barracks, so in the body those cells which produce antibody are themselves provided with the 'reminder' they produce. Contact with the 'remembered' antigen stimulates a rapid liberation of further antibody (police reinforcements) to deal with the emergency.[24]

It is this kind of police trope that helped Burnet pursue the idea of immunosurveillance.

---

22 Cited in Bjorn Claeson et al., 'Scientific Literacy, What It Is, Why It's Important, and Why Scientists Think We Don't Have It', in Laura Nader (ed.), *Naked Science: Anthropological Inquiry into Boundaries, Power, and Knowledge* (New York; Routledge, 1996), 109–10.

23 F. M. Burnet, 'Impressions and Comments', in Richard T. Smith and Maurice Landy (eds.), *Immune Surveillance: Proceedings of an International Conference Held at Brook Lodge, Augusta, Michigan, May 11–13, 1970* (New York: Academic Press, 1970), 512.

24 F. M. Burnet, *Biological Aspects of Infectious Diseases* (Cambridge: Cambridge University Press, 1940), 125.

Burnet had commented in his 1960 Nobel Prize lecture that 'it would profit the organism to maintain a surveillance over the orthodoxy of its chemical structure and to stamp out heresy before it could spread', and a few years later in his 1968 autobiography he suggested that developments within the field of applied immunology are 'just beginning to be spoken of under the name of "immunological surveillance"'. By 1970, the same year as the conference on 'Immune Surveillance', Burnet was confident enough in this idea to publish a book called *Immunological Surveillance*.[25] The 1970 conference opened matter-of-factly: 'What is surveillance? The word itself raises images, most frequently of police actions.'[26] Later in the conference proceedings, 'immune surveillance' is described in the following way:

> One can view the Mafia and the police officers as opposing social adaptations. The effects of a long-standing interrelationship between the two forces are apparent. On the one side of the police are gun-wielding, club-carrying authorities, undercover agents and detectives, specialized forces for infiltrating the ranks of the opposition, IRS [Internal Revenue Service] officers skilled in detecting, quantifying and reporting illicit income and many other adaptations to the need for despoiling the Mafia. On the opposite side, equally ingenious and sometimes even more skillful adaptive mechanisms for getting around the defenses developed by the law enforcement agencies and designed to avoid the effectiveness of the suppressive machinery can be observed.[27]

From thereon, it became common to imagine the immune system performing 'a surveillance function perpetually patrolling the body, as it were, for evildoers'.[28]

25 Frank M. Burnet, 'Immunological Recognition of Self', Nobel lecture, 12 Dec. 1960; *Changing Patterns: An Atypical Autobiography* (London: Heinemann, 1968), 187; *Immunological Surveillance* (Oxford: Pergamon Press, 1970), 161.

26 Richard T. Smith, comments as session chairman in 'Organization and Modulation of Cell Membrane Receptors', in Richard T. Smith and Maurice Landy (eds.), *Immune Surveillance: Proceedings of an International Conference Held at Brook Lodge, Augusta, Michigan, May 11–13, 1970* (New York: Academic Press, 1970), 3.

27 Robert A. Good, comments as session chair and discussant of 'Evaluation of the Evidence for Immune Surveillance', in Smith and Landy (eds.), *Immune Surveillance*, 439–511.

28 Macfarlane Burnet, *Genes, Dreams and Realities* (Harmondsworth: Penguin, 1973), 169.

In effect, 'immune surveillance theory' became and has remained a major immunological paradigm: essays appeared on the surveillance capabilities of specific parts of the immune system such as the thymus, on 'molecular surveillance' and on cancer as a failure of surveillance; international conferences contained sections on 'immune surveillance'; dictionaries of philosophical biology such as Peter Medawar and J. S. Medawar's *Aristotle to Zoos* (1984) came to include entries on 'immunological surveillance'; and popular texts now reiterate time and again the idea of the body's 'surveillance policemen' and immunosurveillance agencies. 'Inside each of us is a surveillance network that would make the NSA green with envy', observes one popular text.[29]

In his overview of the history of the idea of immunity, Tauber suggests that 'surveillance may well be the original function of the immune system'.[30] Or, as Mark Taylor puts it, the immune system involves a sophisticated 'intelligence community' engaged in 'perpetual surveillance' as part of a 'complex game of espionage'.[31] What was taking place with this naturalization or biologization of surveillance was a strengthening of the idea of the immune system as police power, but also as a power comprehensible through a general security logic, with the immune system a game of surveillance and counter-surveillance, a kind of paranoid system of corporeality. How far could this idea be pushed? In one of the leading immunological textbooks in the discipline, Jan Klein insists that we must use our imagination when thinking about immunity:

> Imagine a totalitarian city that George Orwell might have created for *1984*. The city has a strict code governing the appearance of its buildings and a squad of building inspectors who enforce this code. Suppose that one home owner has become dissatisfied with the appearance of his house and begins to remodel it to his liking . . . As soon as he touches the façade and alters it so that his house is different from all the other houses on the street, the inspectors become alarmed. They run to their headquarters with the news, mobilize a demolition squad, and return to the offending house, which they proceed to tear

29 Carver, *Immune*, 25.
30 Tauber, *Immunity*, 111, 117.
31 Mark C. Taylor, *Nots* (Chicago: University of Chicago Press, 1993), 239–40.

> down. Some of the inspectors will take part in the demolition, while others will coordinate the squad's work to make it the most efficient with the least damage to the adjacent 'conforming' houses.

One might observe in passing that enforcing building regulations is hardly the main form of power for which the police system of *1984* is known, but let us stick with the point:

> According to the immune surveillance hypothesis, the cells of the immune system are like the building inspectors: they patrol the tissues of the body, searching for nonconforming alterations to the cell surfaces . . . When they spot a cell with an unfamiliar plasma membrane, they become activated and organize an all-out attack on the strange cell, destroying it before it can spread through the body.[32]

The violence is never far away, and so the police power is always at war just as the war power turns out to be a police power. In this case, natural killer cells engage in killing but also conduct 'constant surveillance, looking for cells that have gone bad'. Hence 'the immune system's cellular troops have your entire body under constant surveillance'.[33]

As powers of war and police coincide, so the language of the 'invader' slips into the language of the 'intruder', and both together often fold into the figure of the 'illegal alien'. When 'an immune cell bumps into a bacterial cell and says "Hey, this guy isn't speaking our language, he's an intruder", the immune system acts accordingly'.[34] This is a natural and instinctive police response on the part of the body: 'That we are not overwhelmed is due to nature-divine providence having endowed us with a nonspecific, first-line defense system that "instinctively" recognizes the hostile foreigners', and this defence system is composed of 'specialized "policeman" cells'.[35] In this light, we can return to *The Body Victorious* that we discussed above and note the fact that the immune process to distinguish between bona fide residents and illegal aliens is

32 Jan Klein, *Immunology: The Science of Self-Nonself Discrimination* (New York: John Wiley, 1982), 647.

33 Michaud et al., *Fighting*, 5, 10.

34 Jaret, 'Wars', 733.

35 Robert S. Desowitz, *The Thorn in the Starfish: The Immune System and How It Works* (New York: W. W. Norton and Co., 1987), 105.

described as a *police programme*. The task of this specialist police function is not easy, because 'recognising them [invaders] as foreign can be very difficult', not least because 'many invaders look very much like us'. 'It can be as difficult for our immune system to detect foreignness as it would be for a Caucasian to pick out a particular Chinese interloper at a crowded ceremony in Peking's main square.'[36]

The whole system of immunity-security operates with police intelligence units gathering information. If we imagine the immune system like the security system of an important and secret industrial complex, for example, we find that

> all the authorised personnel in the plant and all the goods that have been legitimately brought into the plant display, in an appropriately prominent situation, a vivid identification tag. Anything or anybody that does not display this vivid ID will be regarded as a danger that must be rapidly removed.

'The cells of the immune system are constantly exchanging information with other cells', and this information is central to the surveillance process. The policing of immunity involves an 'identification system' wherein every cell has an 'identification badge' so that 'patrolling cells' know whether the cell is friend or enemy. This identification process is reinforced by the fact that 'every one of the body's many billions of cells is equipped with "proof of identity" . . . On the cells of all living creatures, the molecules form cell-specific structures. These constitute the cell's identity papers, protecting it against the body's own police force, the immune system', and reinforced by that system's 'vast criminal records'. Hence even a 'demilitarised' body still needs 'police intel'.[37]

The police role does not end with surveillance and intelligence, as there are even more specialized police functions. First, there are those dealing with serious crimes. The major criminal-disease will try to hide, which means that, as Robert Desowitz puts it,

36 Dwyer, *Body*, 29.

37 Respectively: Dwyer, *Body*, 29; Jaret, 'Wars', 733; Michaud et al., *Fighting*, 5, 10; Nilsson, *Body*, 21, 28; Ed Yong, *I Contain Multitudes: The Microbes within Us and a Grander View of Life* (London: Vintage, 2016), 90–92.

> the process of discovery reads more like the plot of a mystery novel. First there is the killing, after which the killer and his modus operandi are described. However, in the current edition of our mystery the detective-immunologists are still not satisfied as to the nature of the actual weapon and whether or not the killer has any accomplices.

Desowitz admits that the reader of his book on immunity might mistake it for a book on criminal psychopathology.[38] Second, there is the policing of the workplace, as we have seen with the industrial complex described in Dwyer's *The Body at War*. 'Many weapons systems are available to the security forces of the complex', including 'vital "first line of defence" security men' which police the body like a workplace:

> Should a would-be saboteur enter the establishment in the early hours of the morning, our spotters would recognise the likelihood that the saboteur is an intruder and move in closer for a better look using a closed-circuit TV camera that locks on to the suspect and freezes a close-up image of his face on one half of a monitor screen. A computer then runs through all the physical appearances of the plant's legitimate employees on the other half of the screen, so that with incredible accuracy the physical features of the foreigner and members of the legitimate family are compared.[39]

This idea of saboteurs at work folds into 'subversive agents' and microorganisms that try to 'subvert' or escape the immune defences.[40]

This police logic is where we can situate recent developments within immunology which have sought to move beyond the military model of foreign invaders. Polly Matzinger, for example, has developed the idea that immunity is less a question of Self versus non-Self (an issue to which I turn in chapter 2) but of danger. In this model, it is *danger* rather

38 Desowitz, *Thorn*, 102, 112.

39 Dwyer, *Body*, 30.

40 Gerald B. Pier, Gloria J. Small, and Henry B. Warren, 'Protection against Mucoid *Pseudomonas aeruginosa* in Rodent Models of Endobronchial Infections', *Science*, vol. 249, no. 4968 (1990), 537–40 (537); Luis P. Villarreal, 'From Bacteria to Belief: Immunity and Security', in Raphael D. Sagarin and Terence Taylor (eds.) *Natural Security: A Darwinian Approach to a Dangerous World* (Berkeley: University of California Press, 2008), 50.

than *foreignness* that initiates an immune response. This would explain why some foreign entities do not trigger an immune response and why some non-foreign entities do trigger such a response. Matzinger makes the point that this allows us to dispense with the problematic language of war. 'For half a century we have studied immunity from the point of view of . . . an army of lymphocytes patrolling the body for any kind of foreign invader', but we need to 'stop running a cold war with our environment and focus instead on a new view of global inter-relationships'. Immunity in this view becomes a conversation based on the idea of achieving some kind of 'harmony' rather than a war.[41] With such arguments, Matzinger has been hailed as 'blazing an unconventional trail' to a new 'immunological paradigm', as 'immunology's dangerous thinker', and as 'tearing up immunity's rulebook' through a 'Copernican revolution' within immunology; she was the subject of a BBC documentary called *Turned On by Danger* in 1997, as well as other films.[42] And yet Matzinger rejects the war model only to replace it with a 'police model'. Asked in an interview about how her 'danger model' differs from the standard immunological approaches, Matzinger answers,

> It's just a different way of looking at things. Let me use an analogy to explain it. Imagine a community in which the police accept anyone they met during elementary school and kill any new migrant. That's the Self/Nonself Model.
>
> In the Danger Model, tourists and immigrants are accepted, until they start breaking windows. Only then, do the police move to eliminate them. In fact, it doesn't matter if the window breaker is a foreigner

41 Polly Matzinger, 'The Danger Model in Its Historical Context', *Scandanavian Journal of Immunology*, vol. 54 (2001), 4–9 (8); Polly Matzinger, 'An Innate Sense of Danger', *Immunology*, vol. 10 (1998), 399–415; Polly Matzinger and Ephraim J. Fuchs, 'Beyond "Self" and "Non-Self": Immunity Is a Conversation, Not a War', *Journal of NIH Research*, vol. 8, no. 7 (1996), 35–39; Polly Matzinger, 'Friendly and Dangerous Signals: Is the Tissue in Control?', *Nature Immunology*, vol. 8, no. 1 (2007), 11–13.

42 Claudia Dreifus, 'A Conversation with Polly Matzinger: Blazing an Unconventional Trail to a New Theory of Immunity', *New York Times*, 16 June 1998; Marilynn Larkin, 'Polly Matzinger: Immunology's Dangerous Thinker', *Lancet*, vol. 350, no. 9070 (5 July 1997), 38; Philip Cohen, 'Tearing Up Immunity's Rulebook', *New Scientist*, vol. 149, no. 2023 (30 March 1996), 14. For a more judicious discussion highlighting the gender politics surrounding some of the discussions of Matzinger's work, see Lisa Weasel, 'Dismantling the Self/Other Dichotomy in Science: Towards a Feminist Model of the Immune System', *Hypatia*, vol. 16, no. 1 (2001), 27–44.

> or a member of the community. That kind of behavior is considered unacceptable, and the destructive individual is removed.
>
> The community police are the white blood cells of the immune system. The Self/Nonself Model says that they kill anything that enters the body after an early training period in which 'self' is learned. In the Danger Model, the police wander around, waiting for an alarm signaling that something is doing damage. If an immigrant enters without doing damage, the white cells simply continue to wander, and after a while, the harmless immigrant becomes part of the community.[43]

Thus, even those seeking to imagine immunity without recourse to the trope of militarized violence fall back on other tropes of violent powers of elimination, in this case, the 'danger model' and its far from 'unconventional' assumptions about the police power.

Other writers who have distanced themselves from the military model also turn to police and security. 'The war metaphor is misleading, incomplete', writes Matt Richtel, and then continues to use it over and over again while also combining it with a police model of border security.[44] For Wade Sikorski, the idea of the body at war 'is not entirely accurate because the body is not repulsing an invader so much as it is cultivating self'. And yet the immune system can be seen as 'the ecopolity's police force'. Just as the Self 'needs a memory to police it, to remember what it was, what it is, so that it can keep it the same', so too the body 'maintains itself by policing its inner ecology'. The violence of the war model thus gets transposed onto the police model: the mobilization of a vast forces of violence such as 'natural killer cells' now appear as forms of police engaged in the war of life.[45]

We have then, in the immunological literature, an overwhelming police power working alongside and as part of the war power. Like society, the body must be defended. Immunity is a police power as well as war power. Immunity is security. From where did this idea of the body as a *security system* emerge? The answer to this is complex. To begin to answer it, we need to first understand the emergence of the idea in the

43 In Dreifus, 'Conversation'.

44 Matt Richtel, *An Elegant Defense: The Extraordinary New Science of the Immune System* (New York: Harper Collins, 2019), 7–8, 69, 135.

45 Wade Sikorski, *Infected with Difference: Healing Dis/Ease in the Body Politic* (no publisher, copyright Wade Sikorski, 2011), 17, 22–24, 30–42, 144.

late nineteenth century of the body possessing something called 'immunity'. That, in turn, will push us back into the nineteenth century and the idea of the 'cell' as a political as well as biological idea.

'It Is to This Resistance . . . That We Give the Name of Immunity'

As noted in the introduction, immunology did not exist as a distinct field of study until the very end of the nineteenth century. It is generally held that the history of immunology consists of a first phase, known as a 'physiological' period, running from roughly 1880 to 1910, followed by a 'chemical' period from 1910 to 1950, and then a 'biological' period from 1950 onwards. The last of these periods is often seen as the 'golden age' of immunology, the dawn of a 'new immunology' which saw a proliferation of journals and conferences and the creation of international societies, along with major theoretical developments including, as we shall see in chapters 2 and 3, the invention of the immune *Self* and the immune *System*. The middle 'chemical' period is often described as immunology's 'dark ages', a period of intellectual indolence in which the research was rather pedestrian.[46] The first phase is crucial, however, not least because this was the period when the concept of immunity moved from law and politics, which in turn had a profound impact on the way 'immunity' was imagined thereafter. A central figure in this period is Elie Metchnikoff, widely regarded as the 'father of natural immunity'.[47]

Metchnikoff received the Nobel Prize in 1908 for his work on the role of phagocytes in the inflammatory process, sharing that prize with Paul Ehrlich. It is conventional to treat the early phase of immunology as a struggle between cellular theory and the humoral school, represented by Metchnikoff on the one side and Ehrlich on the other. The joint award of the Nobel Prize masked the differing ways in which the two thinkers

46 F. M. Burnet, 'The Impact of Ideas of Immunology', *Cold Spring Harbor Symposium on Quantitative Biology*, vol. 36 (1967), 1–8; Arthur M. Silverstein, *A History of Immunology* (San Diego, CA: Academic Press, 1989), 172, 305–06, 329–30; Anne Marie Moulin, *Le dernier langage de la médecine: Histoire de l'immunologie de Pasteur au sida* (Paris: Presses Universitaires de France, 1991), 390–91; Pauline M. H. Mazumdar, *Species and Specificity: An Interpretation of the History of Immunology* (Cambridge: Cambridge University Press, 1995).

47 Siamon Gordon, 'Elie Metchnikoff: Father of Natural Immunity', *European Journal of Immunology*, vol. 38, no. 12 (2008), 3257–64.

thought about immunity. In contrast to the cellular approach, Ehrlich in the 1890s coined the term *Antikörper* ('antibody', though he originally used this interchangeably with 'antitoxin') to describe an aspect of the immune process that appeared distinct from the cellular process. Ehrlich's approach to immunology was to treat the immune process as fundamentally 'chemical' in nature, rather like life itself; 'life', he once commented, is 'a chemical incident'.[48] As such, he sought to place immunology on a chemical plane, with toxin and antitoxin influencing one another through direct chemical interaction. Ehrlich pictured antibodies as groups of atoms found in the protoplasm of cells, and so, adopting the nomenclature of organic chemistry, he described them as 'side-chains' and, in the process, considered the organism to be 'naturally' immune. But this natural immunity was essentially a *passive* condition, in the sense that it was merely the consequence of the absence of any chemical interaction between cellular and toxic components. This continued the line of thought in early texts such as André-Thérèse Chrestien's *De l'immunité et de la susceptibilité morbides au point de vue de la clinique médicale* (1852), which also lacks anything like an active conception of immunity.[49] As Ed Cohen points out, rather than an active engagement or defence of the organism, Chrestien's conception of immunity merely denotes a failure to contract an illness suffered by others.[50]

In contrast to these accounts, Metchnikoff saw the immune process as the result of organismal *activity*. This conception meant treating the phagocyte as an *active agent* and imbuing it with an immanent purpose; the phagocyte became more or less an *organism in itself*.[51] This conception arose from an understanding of phagocytic cells as active markers of selfhood able to define the self-constituents of the developing organism and thus actively defend it from attack; the roots of the concept of an 'immunological self', as a crucial addition to the liberal vision of selfhood and about which we shall have more to say in chapter 2, can be

48 Cited in Arthur M. Silverstein, *Paul Ehrlich's Receptor Immunology: The Magnificent Obsession* (San Diego: Academic Press, 2002), 2.

49 Moulin, *Dernier Langage*, 23–24.

50 Ed Cohen, *A Body Worth Defending: Immunity, Biopolitics, and the Apotheosis of the Modern Body* (Durham, NC; Duke University Press, 2009), 214.

51 Élie Metchnikoff, *Immunity in Infective Diseases* (1901), trans. Francis G. Binnie (Cambridge: Cambridge University Press, 1905), 521, 539, 545.

found in this conception. In the context of nineteenth-century thought, we can describe this conceptualization of the active phagocyte as essentially 'vitalist', but the point here is that it helped form what became one of the most familiar idioms in immunology, namely intentionality. It also placed a logic of security at the core of this intentionality: the phagocyte was conceived of not simply a cellular agent but as a cellular agent actively engaged in acts of security to defend the body.

During the early 1880s, Metchnikoff developed a new hypothesis concerning the role of leukocytes in immunological responses, proposing that leukocytes were able to recognize some microorganisms as 'foreign' and able to destroy them. This is the process he called 'phagocytosis'. Metchnikoff tells the story of observing the response of a starfish larva to a thorn: 'It struck me that similar cells might serve in the defence of the organism against intruders.' In lower organisms, the phagocyte might serve a simple nutritive function (*phagocyte* coming from the terms *phagein*, to eat, *kytos*, cell), but in higher organisms with a more complex digestive process the phagocyte's function to 'eat or be eaten' takes on a more defensive role vis-à-vis foreign invaders. The phagocyte not only 'eats' but also continues to eat in order to *protect*. With this, Metchnikoff felt that he had the explanation of all immunity. Without much in the way of research or evidence, Metchnikoff connected the way a starfish deals with a thorn to the way humans deal with invading bodies and parasites, and he did so by imagining 'immunity' in ways which remain familiar to us now.[52]

Although the idea of 'phagocytosis' was rooted originally in the idea of intracellular digestion, the idea of defending the self against intruders meant that 'digestion' could be understood as a form of devouring and devouring understood as a militant defence. In his essay on a yeastlike infection among some daphnia (a water flea), he describes the infection as 'a battle between two living beings – the fungus and the phagocytes'. Playing on the idea of the power of the phagocyte as a mechanism akin to eating, Metchnikoff observes that the phagocyte 'can act because of this as destroyers of parasites'. The trope of digestion is a war trope.

52 Alfred I. Tauber and Leon Chernyak, *Metchnikoff and the Origins of Immunology: From Metaphor to Theory* (Oxford: Oxford University Press, 1991), xiv–xv; Scott H. Podolsky and Alfred I. Tauber, *The Generation of Diversity: Clonal Selection Theory and the Rise of Molecular Immunology* (Cambridge, MA: Harvard University Press, 1997), 14–16.

'Hardly has a piece of the spore penetrated into the body cavity, than one or more blood corpuscles attach to it, in order to begin the battle against the intruder.' From this, he says, it is clear 'that spores which reach the body cavity are attacked by blood cells, and – probably through some sort of secretion – are killed and destroyed'.[53] Metchnikoff expands on this idea in one of his later lectures:

> If we examine the organisation of an animal or a plant, we find that their most characteristic features are their organs of attack and defence. The carapace of the crayfish, the shell of molluscs and the teeth of the vertebrates, as well as many other organs, are so many means of protection to these animals in their perpetual warfare.

The phagocyte and, by implication, the immune system as a whole, conduct 'war against the aggressor by devouring, englobing and digesting it', and hence they become 'a more or less highly organised army' engaged in perpetual warfare against 'invading objects' and 'foreign bodies'. 'Immunity against infective diseases should be understood as the group of phenomena in virtue of which an organism is able to resist the attack of the micro-organisms that produce these diseases'.[54] Thus Metchnikoff sought to show that the host actively defended itself not simply by *eating* the foreign invader but by engaging in a 'veritable battle that rages in the innermost recesses of our beings'. This is a *permanent conflict situation*, and as he insisted in a later book, *The Nature of Man*, 'the word conflict is not used metaphorically'.[55]

The power of the conception of phagocytes being developed by Metchnikoff lay in the inscription of a set of ideas about the phagocyte's strategy, volition, and will, which allow it to act as the agent of the organism as a whole, securing and maintaining the integrity of the organism

53 Elias Metchnikoff, 'Ueber eine Sprosspilzkrankheit der Daphnien. Beitrag zur Lehre über den Kampf der Phagocyten gegen Krankheitserreger' ['A Disease of Daphnia Caused by a Yeast. A Contribution to the Theory of Phagocytes as Agents of Struggle Against Disease-Causing Organisms'], *Archiv für Pathologische Anatomie und Physiologie und für Klinische Medicin*, vol. 96, no. 2 (1884), 177–95.

54 Elias Metchnikoff, *Lectures on the Comparative Pathology of Inflammation, Delivered at the Pasteur Institute in 1891*, trans. F. A. Starling and E. H. Starling (London: Kegan Paul, 1893), 2; Metchnikoff, *Immunity*, 10, 566–67.

55 Élie Metchnikoff, *The Nature of Man: Studies in Optimistic Philosophy* (1903), trans. P. Chalmers Mitchell (New York: G. P. Putnams, 1903), 239.

by protecting it from foreign invaders. It is easy to put this down to what René Dubos calls the 'gory phase of Darwinism' when 'one had to be friend or foe, with no quarter given'; immunology has 'always and necessarily had a Darwinian basis', Burnet once observed. But Metchnikoff suggests that the struggle also takes place *within the organism itself*. The whole logic of the struggle for existence gets written into the conceptual foundations of cellular immunology, which turns out to be resolutely social in its outlook. Some of the opposition to Metchnikoff's arguments centred on this very point. 'Metchnikoff's view of inflammation as a combat between phagocytes and the causes of disease . . . is not even "permissible as a poetical conception"', as one writer put it in the *British Medical Journal* in 1896, adding that 'the view that phagocytosis is a phenomenon of warfare between cells and foreign bodies must, therefore, be rejected'. Dubos picks up the point that the idea of 'a kind of aggressive warfare against the microbes, aimed at their elimination from the sick individual and from the community', meant that there was no place for the biological concepts that had prevailed in other fields, 'according to which different species can reach a modus vivendi compatible with their co-existence' that might be based on, and contribute to, imagining 'ecological equilibrium'.[56]

Dubos's point is important, for several overlapping reasons. As we have seen in this chapter, it can often seem that military language and imagery is a natural way of thinking about the body and its ills. Yet this way of thinking rarely appears in the older Hippocratic and Galenic traditions. As L. J. Rather has shown, militarized language concerning the health of the body makes an occasional appearance in the sixteenth and seventeenth centuries, such as in the work of Jacques Pierre Brissot, Tommaso Campanella, and Thomas Sydenham, but for the most part other ideas were far more prevalent, such as a 'balance' between the 'four

56 Élie Metchnikoff, 'The Struggle for Existence between Parts of the Animal Organism' (1892), in *The Evolutionary Biology Papers of Elie Metchnikoff*, eds. Helena Gourko, Donald I. Williamson, and Alfred I. Tauber (Dordrecht: Kluwer, 2000), 207–16; René Dubos, *Mirage of Health: Utopias, Progress and Biological Change* (London: George Allen and Unwin, 1960), 64; anonymous review of E. Ziegler, 'Entziyndung [Inflammation]', entry in *Real-encyclopädie der gesammten Heilkunde*, 3rd edn., in *British Medical Journal*, vol. 1, no. 1846 (1896), 1209–10; Sir F. M. Burnet, 'The Darwinian Approach to Immunity', in J. Sterzl et al., *Molecular and Cellular Basis of Antibody Formation: Proceedings of a Symposium Held in Prague on June 1–5, 1964* (New York: Academic Press, 1965), 18.

humors' of blood, phlegm, black, bile and yellow bile, derived in turn from the four elements of earth, air, fire, and water. Disease was a sign of a lack of physiological balance (or 'contamination', 'excess of irritation', or 'loss of vital force'), rather than physiological warfare, and this rebalancing was understood as part of the healing powers of nature.[57] This has its parallel in the political discourse concerning the body politic, which also focused heavily on the idea of a 'balance', either between the various political versions of the humours or between the structural and functional differentiation of the parts of the body, as in the 'balance of powers', a point which we shall take up in chapter 4. The body corporeal and body politic might suffer from dropsy, consumption, palsy, pestilence, weakness, and frenzy, but such things are indications of imbalance, discord, or disproportion. The key point made by Rather and others is that the idea of war against disease does not really become central until the second half of the nineteenth century, at which point balance and healing were usurped by war and conflict. Disease moved from being part of nature to being something 'unnatural'. Hence the logic of war came into its own in medicine from the mid-nineteenth century, once bacteria had been identified as the agents of disease, 'invading' and 'infiltrating' the body like an invisible enemy and requiring the mobilization of the body's forces as in a war.

In this context, the 'germ war theory of disease' emerged out of the research of Louis Pasteur in France and Robert Koch in Germany. This constituted a watershed, paving the way for a new medical order by demonizing germs and bacteria as enemy agents engaged in war against the body. 'War on Bacteria!' sang the participants of a medical congress in 1886: 'Let's wage a joyful war / On these bacteria / Victory is beckoning us' and 'Once the enemy's face is unveiled / Means will certainly be found'.[58]

57 L. J. Rather, 'On the Source and Development of Metaphorical Language in the History of Medicine', in Lloyd G. Stevenson (ed.), *A Celebration of Medical History* (Baltimore: Johns Hopkins University Press, 1982), 141–43; L. J. Rather and J. B. Frerichs, 'On the Use of Military Metaphor in Western Medical Literature: The *bellum contra morbum* of Thomas Campanella (1568–1639)', *Clio Medica*, vol. 7, no. 3 (1972), 201–08.

58 L. J. Rather, *Addison and the White Corpuscles: An Aspect of Nineteenth-Century Biology* (London: Wellcome Institute, 1972), 18; Nancy Tomes, *The Gospel of Germs: Men, Women, and the Microbe in American Life* (Cambridge, MA: Harvard University Press, 1998), 5–7, 25, 44; Christoph Gradman, 'Invisible Enemies: Bacteriology and the Language of Politics in Imperial Germany', *Science in Context*, vol. 13, no. 1 (2000), 9–30.

'Life' was to be fought and secured against countless enemies, known and unknown. Dubos writes,

> The germ theory was formulated at a time when many biologists and social philosophers believed that one of the fundamental laws of life is competition, a belief symbolized by phrases such as 'nature red in tooth and claw' and 'survival of the fittest'. The ability of an organism to destroy or at least to master its enemies or competitors was then deemed an essential condition of biological success. In the light of this theory . . . aggressive warfare against microbes was particularly the battle cry of medical microbiology, and is still reflected in the language of this science.[59]

The germ theory of disease thus became a theory of universal 'war against the smallest but most dangerous enemies of mankind', as Koch put it in a lecture to the International Medical Congress (IMC) in 1890.[60] Viruses and bacteria came to be regarded as the universal adversary: invisible and ubiquitous, nonhuman and inhuman, hostile and of untold number, able to nestle everywhere, and possessing the power to destroy humanity. The germ was a Demon against which a perpetual war must be fought (and in chapter 4 we shall connect the Demon with the Parasite).[61] The germ theory demanded a total war against the enemy of all mankind, a war involving Cell, Self, System, and Sovereign.

When Koch finished his lecture to the IMC with a comment on the war on the universal enemy, it was to encourage the medically advanced nations to work together for 'the benefit of all mankind'. Koch was building on what was by then a history of such international work. In response to the spread of cholera earlier in the century, especially the devastation wrought by the epidemic of 1865, an International Sanitary Conference took place in 1866 in Constantinople. The International Sanitary

59 René Dubos, *Louis Pasteur, Free Lance of Science* (New York: Scribners, 1960), xxxvi.

60 Robert Koch, *Über bakteriologische Forschung* (Berlin: August Hirschwald, 1890), 15.

61 Ludwik Fleck, *Genesis and Development of a Scientific Fact* (1935), trans. Fred Bradley and Thaddeus J. Trenn (Chicago: University of Chicago Press, 1979), 59–60; Elias Canetti, *Crowds and Power* (1960), trans. Carol Stewart (London: Victor Gollanz, 1962), 47.

Conference was a series of fourteen conferences held between 1851 and 1938, generally regarded as the precursor to the World Health Organization. The key question dealt with by the organization's 1866 conference was quarantine. How far should we go in controlling the movement of goods and people to stop the spread of a violent and devastating disease 'invading' the territory? One section of the report produced by the commission to the conference is called 'Immunity with Regard to Cholera':

> The Commission would not consider that it had faithfully performed its task, if, after having proved the transmissibility of cholera, and indicated, as far as possible, the conditions which favour its propagation, it took no notice of the resistance which certain countries, certain localities, and the majority of people oppose to its development. It is to this resistance . . . that we give the name of immunity.[62]

The commission was giving 'resistance' to disease a name: 'immunity'. The report discusses 'relative immunity' and the 'temporary immunity' that exposure to an epidemic can create. But it notes that the resistance of any individual to cholera is heavily dependent on the hygienic conditions and relative health of the population that exists in various localities. It is immunity of the people (as a locality or country) as a whole that is important. As Cohen points out in his analysis of the committee's work, the significance of individual immunity appears both subsequent and secondary to immunity's geopolitical importance.[63] What was at stake was the resistance to cholera of certain countries, localities, towns, and cities. There were places whose location, size, and conditions would appear to make them susceptible to the disease, yet which appear to have escaped. Such places do not prove that cholera is not transmittable. Instead 'they prove only that there are localities, as well as individuals, which enjoy a certain immunity against its transmission; an immunity, which for these localities, may be complete, or partial, permanent, or temporary'. In this early account of immunity – so early that the

62 International Sanitary Conference, *Report to the International Sanitary Conference of a Commission from that Body* (1866), trans. Samuel L. Abbot (Boston: Alfred Mudge, 1867), 87.

63 Cohen, *Body*, 233.

commission felt the need to spell out that it was giving resistance to the disease a name – the question of quarantine and thus, in effect, the question of immunity, was a question of police: How do we better police a territory in order to improve security and hence immunity? From plague to Covid, the quarantine has always been a police tool. Quarantine was historically used as a mechanism for regulating the movement of the paupers and beggars, who would eventually become the working class, and in the colonies as a means of managing indigenous peoples. In the context of the nineteenth century, this police mechanism of containment, assigning bodies to spaces in the name of security, coincided with germ warfare. Cholera is a question of hygiene as well as biology, but hygiene is a social question. Hygiene is political, legal, territorial; it is a question of medical police.

It is significant that, despite articulating a notion of resistance through the lens of immunity in a way that makes us think in biological terms, the commission is actually concerned with quarantine and territory; the notion of immunity invoked by the commission plays heavily on a much older logic of immunity: diplomatic immunity. Given cholera's invasive character and the fact that it is therefore no respecter of territorial borders, the question of how states deal with quarantine is a question of diplomacy, and so states turned to the history of the diplomatic mission as a way of grappling with how spaces and places might be 'protected': the 'Ambassador' and 'Embassy' had for centuries possessed a form of 'immunity' by virtue of representing a certain kind of sovereign power. We shall have more to say about the immunity of ambassadors and embassies in chapter 6. Here, we need only note that the appearance of immunity as a biological concept in the 1860s played heavily on the long history of sovereign power to exclude, ban, quarantine, and control.[64] Within this nexus of biology, law, and politics, immunity could not help but be attached to notions of police.

Although 'immunity' was for the International Sanitary Conference a juridico-political term as much as a medical one, for Metchnikoff it was a term used to describe a biological capacity of the organism alone. That said, the juridico-political is never far away in Metchnikoff's conception. Biological life could increasingly be characterized as war and defence, as we have seen, but Metchnikoff also considers the possibility

64 Cohen, *Body*, 228.

of 'harmony' despite life's cellular components being in conflict. But, because Metchnikoff's *starting point* was disharmony – 'I wish only to point out the frequency of the natural occurrence of disharmony' – so disharmony needs to be managed, and Metchnikoff comes to offer what Tauber calls a 'profoundly novel concept of health', in which 'health' is not a given but must be actively sought and the disharmonious constituents must somehow be brought into harmony.[65] As a consequence, physiological mechanisms articulated through the 'disharmony' of war, such as inflammation, can also be articulated as a means of policing the conflict in question in order to achieve harmony once more. Something must be in control of the body's inflammation just as something must manage inflamed situations within the body politic. This is why we find in Metchnikoff's work a desire to think of the phagocyte's agency in terms of its police function, and why we frequently encounter references to 'Metchnikoff's policemen' in the literature.[66] This was the basis of criticism of Metchnikoff at the time, and Tauber has commented that although Metchnikoff was able to deal with many criticisms, the one he found difficult to deflect was 'the charge that he assigned phagocytes an unwarranted autonomy to police the organism'.[67] For Metchnikoff, the phagocyte patrols the body and clears it of unwanted cellular debris in the form of dead or injured cells like a corporeal police power, maintaining a certain kind of 'harmony' and 'cleanliness'. The harmonizing role played by the phagocyte involves patrolling the body and *managing order* in a state of *permanent conflict*. One of the phagocyte's main activities is, in effect, the police function of order and harmony between conflicting forces. In policing order as well as fighting wars, the

65 Metchnikoff, *Nature*, 37; Alfred I. Tauber, 'The Immunological Self: A Centenary Perspective', *Perspectives in Biology and Medicine*, vol. 35, no. 1 (1991), 74–86 (78).

66 Charles J. Brandreth, 'The Man Who Prolongs Life', *London Magazine*, Jan. 1910, 578–84 (579); Eileen Crist and Alfred I. Tauber, 'The Phagocyte, the Antibody, and Agency in Immunity: Contending Turn-of-the-Century Approaches', in Anne-Marie Moulin and Alberto Cambrosio (eds.), *Singular Selves: Historical Issues and Contemporary Debates in Immunology* (Paris: Elsevier, 2001), 130, 131; Luba Vikhanski, *Immunity: How Elie Metchnikoff Changed the Course of Modern Medicine* (Chicago: Chicago Review Press, 2016), 248; James A. Stefater III, Shuyu Ren, Richard A. Lang, and Jeremy S. Duffield, 'Metchnikoff's Policemen: Macrophages in Development, Homeostasis and Regeneration', *Trends in Molecular Medicine*, vol. 17, no. 12 (2011), 743–52.

67 Alfred I. Tauber, 'Metchnikoff and the Phagocytosis Theory', *Nature Reviews: Molecular Cell Biology*, vol. 4, no. 11 (2003), 897–910 (900).

phagocyte maintains the integrity of the body. The whole problem of order thus appears to reside in the immune cells.

In *The History of Biological Theories* (1909), Emanuel Rádl observed that 'it is as if the clue to all living problems were hidden in the cell, as if the microscope could disclose to us all the unknown springs of "being" '.[68] This would certainly appear to have been true of the birth of immunology and Metchnikoff's cellular approach. Paraphrasing Rádl, we might say that, as it developed in the late nineteenth century, immunology believed that all the problems of organismic order – that is, all the problems of how bodies are policed – were solved by the work of cells. Maybe the key to why this happened lies in the very concept of the cell. Donna Haraway once commented on the method of taking an intensely physical entity of biology and then teasing out from it a larger narrative.[69] Let us do this for the cell; perhaps the larger narrative will turn out to be one of security.

Life and Death in a Cell

It is hard for us now, accustomed as we are to thinking in 'cellular terms', to realize just how novel the idea of the cell once was. It is also hard for us to appreciate that to think in cellular terms was to *think politically*.[70] Marx's comment, in the preface to the first edition of *Capital* (1867), that 'for bourgeois society, the commodity-form of the product of labour . . . is the economic cell-form', is now usually read with little or no discussion, as though we all know exactly what he means. But Marx was employing the language of a new science that had emerged in the previous thirty years. 'Everything consists of cells', Engels had pointed out to Marx in a letter dated 14 July 1858, and Marx wanted to imagine the commodity as the economic *cell-form* as a way of imagining the basic *structure of the body of capital*.[71] In so doing, he was reflecting the way

68 Emanuel Radl, *The History of Biological Theories* (1909), trans. E. J. Hatfield (Oxford: Oxford University Press, 1930), 231.

69 Donna Haraway, *How Like a Leaf: An Interview with Thyrza Nichols Goodeve* (New York: Routledge, 2000), 24.

70 Hannah Landecker, *Culturing Life: How Cells Became Technologies* (Cambridge, MA: Harvard University Press, 2007).

71 Karl Marx, *Capital: A Critique of Political Economy*, vol. I (1867), trans. Ben Fowkes (Harmondsworth: Penguin, 1976), 90.

that cell theory was itself inherently political. Perhaps, then, there is some mileage in considering the politics of the 'cell'.

The emergence of the cell as a biological entity can be traced back to the microscopic work of Robert Hooke in the mid-seventeenth century. In his *Micrographia* (1665), Hooke explains how he made a razor-sharp slice into a piece of cork, observed that it possessed a partitioned structure like pores, and adds that they 'were indeed the first microscopical pores I ever saw, and perhaps, that were ever seen'.[72] Hooke compares the plant object to a honeycomb, and hence the work of the bee, and then the honeycomb to the work of a human in the form of the small room that went by the name 'cell'. Nonetheless, Hooke's discovery did not initiate very much interest and not a lot happened with biological cell theory for over a century and a half. Biological cell theory only really comes into its own from around 1830: 'cell membrane' dates from 1837, 'cell division' from 1846, 'cell body' from 1851, and the 'cell' gradually displaced 'fibre' as the primary structural unit of living organisms. Matthias Schleiden announced in 1838 that the cell was the basic living unit of all plant structures, Theodor Schwann made the same claim about the animal kingdom a year later, and scientists started to argue that diseases and resistance to diseases could be understood in terms of how they affect different cells. Schwann was able to show that all organic tissues have one common principle of development, namely the formation of cells, and from this, a whole branch of biology developed around 'cellular therapeutics' (as William Addison called it in a book of that title published in 1856) or 'cellular pathology' (the title of Rudolf Virchow's key text of 1858). Regardless of the fact that 'cell' was very quickly seen to be a misnomer, on the grounds that, whatever it was, the cell was far from being a hollow chamber surrounded by walls and thus nothing like the small room discussed by Hooke, the idea of the cell nonetheless quickly came to be used to explain everything, from inheritance to variation, reproductive biology to the nature of sex, neurology and brain anatomy, and even immaterial phenomena such as the 'soul'.

In the view of Schwann, who coined the term 'cell theory', the cell is 'the elementary part of organised bodies'. 'All organised bodies are composed of essentially similar parts, namely, of cells', and these cells 'grow in accordance with essentially similar laws'. Moreover, each cell

72 Robert Hooke, *Micrographia* (London: Royal Society, 1665), 112–13.

> is, within certain limits, an Individual, an independent Whole. The vital phenomena of one are repeated, entirely or in part, in all the rest. These Individuals, however, are not ranged side by side as a mere Aggregate, but so operate together, in a manner unknown to us, as to produce an harmonious Whole.

Each cell possesses 'a power to take up fresh molecules and grow', which means that the cell 'possesses a power of its own, an independent life, by means of which it would be enabled to develop itself independently'. Such claims were reinforced by research in the 1840s and 1850s showing sperm cells to be mobile single cells, the amoeboid movement of white blood cells and the development of the animal embryo from repeated binary cell divisions. For Schwann, as for many others, 'the question . . . as to the fundamental power of organized bodies resolves itself into that of the fundamental powers of the individual cells'.[73]

In this context, the 'cell' also became the grounds for rethinking the social, psychological and political world. We have noted Marx's use of the concept of the cell, for example, and commentators have noted the presence of some of the key themes of Romantic thought in the work of Schwann and other early cell theorists, identifying a key point of unity and diversity for all life. Philip Ritterbush, for example, observes that cell theory's representation of the organism as an assembly of essentially similar structural units which always arose from pre-existing cells was not only scientific, in that it referred to structure and function, but also 'fulfilled the esthetic requirements of the idea of organic form'. Hence 'in the development of the cell theory we witness the transformation of esthetic presuppositions into scientific knowledge in a manner that parallels Kant's statement that the sense of beauty is an aid to the discovery of truth'.[74] The cell enabled a rethinking of what it means to have a body and to be a body in a living world. To the extent that, as Michel Foucault points out, medical science was formed from the beginning of the nineteenth century as a political as well as biological field, with the body inscribed politically, so this inscription was increasingly centred

73 Theodor Schwann, *Microscopical Researches into the Accordance in the Structure and Growth of Animals and Plants* (1839), trans. Henry Smith (London: Sydenham Society, 1842), 2, 91–92, 166, 186, 193.

74 Philip C. Ritterbush, *The Art of Organic Forms* (Washington, DC: Smithsonian Institute, 1968), 33–34.

on the thing called the cell, with the political field imagined as residing in the body's cellular structure.[75] Living subjects could now be conceived of through the lens of power *right down to their cellular core*.

Cell theory revolutionized conceptions of the individual, the body and therefore the body politic. There now existed the assumption that each cell is itself a perfectly individualized independent whole, to the extent that body and Self become systems of cells, each of which begins to appear to be an organism in its own right: eating, living, willing, dying. But, on this view, one should speak not of a life made up of cells but of *the lives of cells*, and hence, possibly, of the non-existence of our Self or perhaps the retreat of the Self into body's cells, the latter conceived of in turn as possessing their own independent agency.[76] We take up the question of the (immune) Self in the chapter to follow, but, here, we can note that 'independence' is a political rather than a scientific term, and as much as cells were imagined as independent, so they were also imagined in relation to others and constituting a greater whole. As such, the vision of the body as an association of cells could be applied to the long-standing tradition of thought about the body politic. If, as Laura Otis puts it, 'cell theory relies on the ability to perceive borders',[77] it might equally be said that cell theory relies on the ability to perceive *orders*. Borders of territory, orders of independent but connected cells: *cell theory was political theory*. Most notably, the cell came close to representing bourgeois ideals of self-contained and self-regulating units: the cell and/as a Self. The organism could be understood as a socio-political arrangement of cellular parts, rather like the state, which could in turn be understood as a socio-political arrangement of cellular parts, rather like the organism. Biological cell theory came to form the basis of a political theory of the cell-state.

Just how far such thinking could go can be seen in the work of Rudolf Virchow and Ernst Haeckel. In an essay from 1855 called

75 Michel Foucault, *The Birth of the Clinic: An Archaeology of Medical Perception* (1963), trans. Alan Sheridan (London: Tavistock, 1973), 35; 'On the Archaeology of the Sciences: Response to the Epistemology Circle' (1968), in Michel Foucault, *Essential Works*, vol. II, *Aesthetics, Method, and Epistemology*, trans. Robert Hurley et al. (London: Penguin, 1998).

76 Lewis Thomas, *The Lives of a Cell* (London: Futura, 1974), 2.

77 Laura Otis, *Membranes: Metaphors of Invasion in Nineteenth-Century Literature, Science, and Politics* (Baltimore: Johns Hopkins University Press, 1999), 5.

'Cellular Pathology', developed into a book with that title three years later, Virchow sought to show that diseases might originate in single cells. In the process, Virchow reiterated the basic idea that all life must be understood through the cell and cell formations: 'The cell is the locus to which the action of mechanical matter is bound, and only within its limits can that power of action justifying the name of life be maintained.' Such a claim places Virchow at the heart of the development of ideas about 'the human motor': within the locus of the cell '*it is mechanical matter that is active – active according to physical and chemical laws*'. We will take up the question of the human motor in chapter 5. For the moment, we should focus on Virchow's stress on the fact that all life is bound up with the existence and development of cells, for it is '*the cellular elements that are the true vital units*'. Virchow's claim reiterates Schwann's argument about the cell's centrality to the *composition* of organisms, but Virchow also sought to use cell theory to explain their *genesis*. For Virchow, *omnis cellula a cellula*: all cells come from cells. Cellular development is the process through which organization is generated and the 'life-force' transmitted; the idea of an 'organism' begins to really take hold at this point. The idea that each cell is an autonomous unit manifesting all the characteristics of life itself is then developed at greater length in the book *Cellular Pathology* (1858). In effect, the cell is to be the fundamental starting point of all 'biology' (itself a new science in the nineteenth century), with complex organisms made up of a larger or smaller number of similar or dissimilar cells.[78]

'What the individual is on a large scale, the cell is on a small scale', Virchow states. This conception and its political implications come to the fore in the book, where the *structural composition of a body* is a 'so-called individual' forming a *social arrangement of parts*. A body is

> an arrangement of a social kind, in which a number of individual existences are mutually dependent, but in such a way, that every

78 Rudolf Virchow, 'Cellular Pathology' (1855), in Rudolf Virchow, *Disease, Life, and Man: Selected Essays*, trans. Lelland J. Rather (Stanford, CA: Stanford University Press, 1958), 84, 88, 99; Rudolf Virchow, *Cellular Pathology as Based on Physiological and Pathological Histology: Twenty Lectures* (1858), trans. Frank Chance (London: John Churchill, 1860), 3, 13.

> element has its own special action, and, even though it derives its stimulus to activity from other parts, yet alone effects the actual performance of its duties.

What Virchow is grappling for here is a 'theory of *free* cell formation', with the 'freedom' intended to be understood politically; the same can be said of his use of 'equal' when describing the body as composed of parts of 'relatively equal value'. The ultimate composition of the body lies in the fact that 'an infinite quantity of cellular elements manifest themselves side by side, more or less autonomous, and in a great measure independent of one another'.[79]

Autonomous, free, equal: the cellular organism begins to look like the liberal state. This becomes clear in an essay called 'Atoms and Individuals', written the year after the publication of *Cellular Pathology*, in which the 'cell territories' discussed in the book come to be treated as forms of commonwealth: '*The individual is . . . a unified commonwealth* in which all parts work together for a common end.' The organism is 'a society of cells, a tiny well-ordered state, with all of the accessories – high officials and underlings, servants and masters, the great and the small'. On this basis, the state 'can be termed an organism, since it consists of living citizens; conversely the organism can be termed a state, or a family, since it consists of living members of like origin'.[80] Atomism and organicism come together: 'Are the cells or the human beings individuals? Can a single answer be given to this question? I say No!' Why? Because 'the "I" of the philosopher is a consequence of the "We" of the biologists', and so there is no good reason 'to distinguish between collective individuals and single individuals'.[81] This is a landmark moment in the development of atomistic and organismic views of life, observes Bernard Cohen.[82] On the one hand, the individual cell is endowed with personhood and thus appears as a microcosm of Self; on the other hand, the cell is to be seen as part of an organic whole. Virchow was treating cells as essentially sovereign entities, as

79 Virchow, 'Cellular', 84; *Cellular*, 10, 14, 229.

80 Virchow, *Cellular*, 14–15, 94, 243–46; 'Atoms and Individuals' (1859), in *Disease, Life*, 123–24, 130, 138–40.

81 Virchow, 'Atoms and Individuals', 139–40.

82 I. Bernard Cohen, *Interactions: Some Contacts between the Natural Sciences and the Social Sciences* (Cambridge, MA: MIT Press, 1994), 50–54.

both a kind of Leibnizian 'monad' and also as the grounds of the Cell-State. From Cells to Selves to Sovereigns.

Such an approach to the cell is part of Virchow's belief that medicine is a form of social science and politics is a form of medicine. The doctor's role is to improve the body politic, and the politician's role is to keep the social body in good health. Virchow's political biography makes this clear: he fought at the barricades in 1848; was dismissed by the Prussian authorities from a post in Berlin early in his career; advocated for doctors to recommend measures such as the German state taking responsibility for improving the sewage disposal system in Berlin; and helped found the liberal German Progress Party in 1861, which he then represented in the Prussian House of Representatives and then the Reichstag, from where he regularly opposed Bismarck's policies. His combined work in biology and politics reflects his belief in their fundamental connection and is what underpins his conception of the cell. His biographer Erwin H. Ackerknecht puts it as follows:

> Cellular pathology was to Virchow far more than a biological theory. His political and biological opinions reinforced each other at this point. Cellular pathology showed the body to be a free state of equal individuals, a federation of cells, a democratic cell-state. It showed it as a social unit composed of equals, while an undemocratic oligarchy of tissues was assumed in humoral, or solidistic (neuro) pathology. Just as Virchow fought in politics for the rights of the 'third estate', thus in cellular pathology he fights for a 'third estate' of cells.[83]

With that in mind, comparison is often made between Virchow's generally liberal concept of the cell-state and Haeckel's more reactionary and authoritarian twist. Set against a background of an increasingly strained application of Darwin's concept of the evolution of species to the social world – Haeckel was one of the first to try to properly bring together Darwin's arguments in *On the Origin of Species* within cellular biology – Haeckel's synthesis of evolutionary theory, biologism, and romantically inclined folkish nationalism is often contrasted to Virchow's liberal

83 Erwin H. Ackerknecht, *Rudolf Virchow: Doctor, Anthropologist, Statesman* (Madison: University of Wisconsin Press, 1953), 45.

republicanism; it is also sometimes posited as providing part of the ideological roots of National Socialism.[84]

For Haeckel, the fact that 'tissue is made up of the same microscopic particles, the *cells*', is the basis for understanding political as well as biological life. 'We can only arrive at a correct knowledge of the structure and life of the social body, the State, through a scientific knowledge of the structure and life of the individuals who compose it, and the cells of which they are in turn composed.' The biology of cells is a metaphysics of souls: 'Living cells possess a certain sum of physiological properties to which we give the title of the "cell-soul".' For Haeckel, 'Every cell has a "soul"', but the psychic activity of the cell-soul is at the same time connected to the psychic activity of the 'communal soul of the entire colony'. The 'colony' (a term from earlier cell theory that had by then become commonplace) refers to the 'cell-state' as 'the true "cell-commonwealth"'. The colony 'controls all the separate "cell-souls" of the social cells – the mutually dependent "citizens" which constitute the community'. Cells are thus not 'elementary organisms' but are 'real, self-active citizens which, in combinations of millions, constitute the "cellular-state", our body'.[85]

In his 1874 book *Anthropogenie*, Haeckel describes the cell as 'an individual of the first order', created in accordance with the same laws as a state, so that the body, like the state, consists of 'many different citizens engaged in different occupations [but] united towards a common purpose'. He goes on: 'The further the division of labour advances, the more perfect or "civilised" the multicellular organism will be, and the more differentiated the Cell-State.'[86] In his *General Morphology* (1866), in the popular text *The History of Creation* (1868), and in the later text *The Wonders of Life* (1904), he likewise imagines organisms as a 'state of cells', 'cell-society', 'cell-state' and 'republican state of cells'.[87]

84 Daniel Gasman, *The Scientific Origins of National Socialism: Social Darwinism in Ernst Haeckel and the German Monist League* (London: Macdonald, 1971), xiv; Paul Weindling, *Health, Race and German Politics between National Unification and Nazism, 1870–1945* (Cambridge: Cambridge University Press, 1989), 27–32, 41–43; Andrew Reynolds, 'Ernst Haeckel and the Theory of the Cell State: Remarks on the History of a Bio-Political Metaphor', *History of Science*, vol. 46 (2008), 123–52 (132–33).

85 Ernst Haeckel, *The Riddle of the Universe* (1895–1899), trans. Joseph McCabe (London: Watts and Co., 1900), 8, 26–27, 139, 155, 160.

86 Ernst Haeckel, *Anthropogenie* (Leipzig, 1874), 120, 129.

87 Ernst Haeckel, *Generelle Morphologie der Organismen* (Berlin, 1866), 270; *The History of Creation* (1868), trans. E. Ray Lankester (London: Henry King, 1876), 301;

Now, in a lecture to the Medical-Scientific Society of Jena in 1877, Haeckel observed that although the comparison between cells and citizens might lend itself to the idea of a 'cell-republic', along the lines outlined by Virchow, where the organism is only loosely centralized, one might equally envisage a far more centralized animal body as a 'cell-monarchy'. This centralized and monarchical conception of the cell-state was compounded by his description of the cells of the brain and central nervous system as 'aristocratic', on the grounds that they are found only in the most evolved organisms. Hence, instead of Virchow's vision of autonomous cells, one gets cells as obedient law-abiding citizens. The background to this was the social and political crisis of the German state, the rise of the Social Democratic Workers' Party, and what Haeckel considered to be 'Utopian doctrines of social democracy'. Haeckel regarded the attack on the German Empire made by Social Democracy as 'insane' and the working-class takeover of Paris in 1871 as a 'horror'. If the theory of natural selection corresponds to any political tendency 'that tendency can only be aristocratic, certainly not democratic, and least of all socialist'. Socialism, according to this cell theory, is nothing less than unnatural. 'Darwinism, I say, is anything rather than socialist!' The theory of natural selection teaches us about a 'cruel and merciless struggle for existence which rages throughout all living nature, and in the course of nature *must* rage', and not a political struggle for socialism. Darwinism 'is aristocratic in the strictest sense of the word', meaning that only a 'picked minority' – picked by nature, but also therefore picked to survive by the political system – can and should survive.[88] Haeckel was, of course, far from alone in making such links at the time. Nietzsche, for example, incorporates ideas from cell theory into his notes and writings to reinforce his concept of the aristocracy of the body and to render the will to power as biological as well as metaphysical and political, identifying 'a kind of *aristocracy* of "cells" in which mastery resides'.[89]

It is because he believed that the theory of natural selection points to the aristocratic anti-democratic aspects of nature that Haeckel liked to

---

The Wonders of Life: A Popular Study of Biological Philosophy (1904), trans. Joseph McCabe (London: Watts and Co., 1905), 135, 142.

88 Ernst Haeckel, *Freedom in Science and Teaching* (1878), trans. T. H. Huxley (London: C. Kegan Paul, 1879), xxvii, xxv, 92–94.

89 Friedrich Nietzsche, 'Notebook 40' (1885), in *Writings from the Late Notebooks*, trans. Kate Sturge (Cambridge: Cambridge University Press, 2003), 42.

claim that there is an evolutionary tendency towards more centralized, hierarchical organizations. Thus, in response to Virchow's republican cell-state, Haeckel suggests that a consideration of the 'cell-soul' shows a fundamental differentiation between types of organism:

> The only difference is this, that in the organism of the higher animals and plants the numerous collected cells, to a great extent, give up their individual independence, and are subject, like good citizens, to the soul-polity which represents the unity of the will and sensations in the cell community. We here also must distinguish clearly between the central soul of the whole many-celled organism or the personal psyche (the person-soul), and the particular individual soul or elementary soul of the individual cells constituting that organism (the cell-soul).[90]

The point of this discussion is not to adjudicate on a debate within German cellular biology of the late nineteenth century. Neither is it to suggest that there is something peculiar to cell theory that contributed to German nationalism. In England, Herbert Spencer's account of the division of labour as 'the principle of all organisation' is illustrated by him through the division between cells, and one can find British accounts of the 'cell-soul' as the root of the 'cell society'.[91] The point, rather, is to unravel a little the fact that the cell has never just been a biological 'thing-in-itself'. It is not clear if anything ever could be. Rather than a name for a thing in and of itself, the 'cell' points to a certain kind of social relation of power, a name for a kind of interaction that is immediately political.

'Who could tell whether one is a republican because one is a partisan of cell theory, or rather a partisan of cell theory because one is a republican?' Georges Canguilhem asks. Equally, who could tell whether one is a monarchist because one is a partisan of cell theory, or rather a partisan of cell theory because one is a monarchist? As we have seen and as Canguilhem spells out, the history of the concept of the cell is

90 Haeckel, *Freedom*, 58.

91 Herbert Spencer, 'The Social Organism' (1860), in Herbert Spencer, *The Man versus the State* (Harmondsworth: Penguin, 1969), 195–233; Herbert Spencer, 'Professor Weismann's Theories', *Contemporary Review*, vol. 63 (1893), 743–60; George J. Allman, 'Protoplasm and Life', *Popular Science Monthly*, vol. 15 (Oct. 1879), 721–49.

inseparable from the history of the concept of the individual. In the chapter to follow, we will pick this up through the concept of the Self. Here, however, I want to take the argument in a different direction and highlight the fact that by the time of its consolidation as a biological concept, the politics of the cell was also already a security concept. After all, what do we imagine when we imagine a cell? 'The word "cell" makes us think not of a monk or a prisoner but of a bee', claims Canguilhem.[92] But this surely concedes too much to Hooke's original work. If the word 'cell' makes us think of anything, it is more likely to be a prisoner than a bee. In other words, when we imagine a 'cell', we might well imagine something intrinsic to immunity and biological selfhood, but we cannot help but also imagine one of security's key technologies.

Stretching back to the early twelfth century, 'cell' referred originally to a small or subordinate monastery, and then later, circa 1300, to a small room for a monk or a nun; it also referred to a hermit's dwelling. The word stems from the Latin *cella*, meaning 'small room', 'storeroom' or 'hut', but is also related to the Latin *celare*, meaning 'to hide' and 'to conceal'. The longer history hints at the Latin *cella* as the name of the small room which held the image of God, picking up on the earlier Greek word which described the unknown chamber within the inner sanctum of the temple. The word was extended to describe the segregation, classification and management of those within the small and discrete spaces of confinement developed for the inmates of prisons and asylums. On this reading, it is perhaps no small coincidence that not long after describing the pore-like 'cells' in cork, Hooke took responsibility for redesigning some of London's buildings following the Great Fire, including Bridewell Prison and Bethlem Hospital (the latter being the asylum known as 'Bedlam'). Hooke's biographer Stephen Inwood points out that Hooke's building played a significant part in the development of the asylum and influenced asylum design for at least 150 years thereafter. And at the heart of this design were long galleries, probably the longest in England at that time, lined with small rooms: 'cells'.[93] The cell was being consolidated as a political site of enclosure and

92 Georges Canguilhem, 'Cell Theory', in *Knowledge of Life*, trans. Stefanos Geroulanos and Aniela Ginsberg (New York: Fordham University Press, 2008), 30, 48.

93 Stephen Inwood, *The Man Who Knew Too Much: The Strange and Inventive Life of Robert Hooke 1635–1703* (London: Pan Books, 2003), 217.

confinement, training and discipline, at the very moment of its discovery and rise in the realm of physiology.

According to Darlene Brooks Hedstrom, in monastic confinement the cell was firstly a space for teaching, training, disciplining one's own ascetic development, and reaffirming the values of the community. It was also a sacred space, rendering it simultaneously a form of social space. And it was likewise a space of performative ritual, facilitating an internalization of the cell within the soul of the monk. The monastic cell was thus not only a space for the performative acts of devotion and the teaching of discipline but was also somehow 'an extension and reflection of the interior life of the monk'.[94] In other words, the bordered space of the monastic cellular enclosure was intended to have as its correlate the bordered enclosure of the soul, such that the reciprocal movement between the order of the soul and the order of secure confinement could help enclose the movements of the soul within the borders of the body and administer both body and soul within a secure space. At the same time, as a 'secure space' the monastic cell was always already a space of battle: first, a space of battle over its order and border as a secure space; second, a space of battle in which the soul fights against the flesh; and hence, third, a space of battle against the universal adversary (in the form of Demons and the Devil, not Germs) that threaten the path to the security of God. As Hedstrom suggests, individuals enclosed by the cell undertake a training in both exterior life and interior will. The cellular form becomes a model of social reality.

When that security architecture par excellence known as the prison began to be structured around the individual cell, it was with a clear understanding of the cell's monastic history as a space of enclosure, confinement and disciplinary technique: individuals isolated in their existence but brought together in a strict hierarchical framework to reinforce the moral teaching and sustain the well-ordered society of which the cell is but one part. The cell is containment. The idea in Roman law that the prison was intended for 'the safe-keeping of men, and not for their punishment'[95] gradually waned with the coming of modernity,

94 Darlene L. Brooks Hedstrom, 'The Geography of the Monastic Cell in Early Egyptian Monastic Literature', *Church History*, vol. 78, no. 4 (2009), 756–91; Thomas Nail, *Theory of the Border* (Oxford: Oxford University Press, 2016), 99–101.

95 *Digest of Justinian*, bk. 48, title 19, 8.9.

not least as the number of imprisonable offences on the statute books increased. The cell became a form of punishment and then a technology for the reform of the inmate. The cell became a key disciplinary space for the containment of the individual: the expression of Sovereign power and the reform of the Self.[96]

The development of the cell as a system of enclosure, containment, punishment, and reform gets fully realized in the 1830s and 1840s, by which point it had come to combine its religious roots with the power of law. 'The isolation of the man and his soul from the outer world, the combination of legal punishment with theological torture', Marx noted in the 1840s, 'finds its ultimate expression in the *cellular system*.'[97] A number of writers have noted the importance of this period for the consolidation of cellular power: in 1840 Tocqueville observed the flurry of prison growth that had taken place in America over the previous decade, with the two main systems – the Pennsylvania system of solitary confinement and solitary work, and the Auburn system in the state of New York which provided for solitary confinement by night and collective work in the day – both structured around the cellular system; Georg Rusche and Otto Kirchheimer point to the rise of solitary confinement combined with new techniques of prison labour as consolidating the cellular structure of prisons in the nineteenth century; and Foucault suggests that the laws and principles of cellular internment get developed and passed in the early 1840s.[98] And all agree with Marx that what was at stake was the isolation of the soul in the cell. This is one reason why Foucault presents *Discipline and Punish* as a history of the modern soul. Paraphrasing Foucault, we might say that the historical reality of the soul in relation to the cell is that in contrast to its monastic form

96 This politics of the cell gets reinforced when, much later, in the 1920s, the meaning of 'cell' as a small group of people organized as a political group then comes into play, connoting the smallest structural unit of a larger organism but capable of independent functioning, and remains with us now in the references to 'terror cells'.

97 Karl Marx and Frederick Engels, *The Holy Family; or Critique of Critical Criticism* (1845), ch. 5, sect. 3.a. The English translations of this tend to offer 'solitary confinement' for Marx's *Zellularsystem*.

98 Alexis de Tocqueville, *Democracy in America*, vol. I (1840), trans. George Lawrence (London: Fontana, 1968), 308–09; Georg Rusche and Otto Kirchheimer, *Punishment and Social Structure* (New York: Columbia University Press 1939), 127–34; Michel Foucault, *Discipline and Punish: The Birth of the Prison* (1975), trans. Alan Sheridan (London: Penguin, 1977), 23, 28–29, 143, 235, 250, 293.

where the soul is subject to the methods of theology, in its modern form it is subject to the methods of security and police, giving new meaning to the idea of the 'cell-soul'.

All of which is to note that this cell-soul, born out of the security of cellular confinement, gets fully realized in the heyday of biological cell theory and the creation of the biological cell-soul. As the treatment of the body narrowed more and more from the body as a whole to the organ, then to tissue, then to cell, so the diseases of the social body were to be treated within the cells of prisons and asylums. A security architecture par excellence, the cell is imagined as the container of life (and hence death, as we shall discuss in chapter 2). If there is 'truth', it was thought to lie in the cell-soul.

In *Chance and Necessity* (1970), his major work on the philosophy of biology, written in the golden age of immunological research and systems theory (about which we will have much more to say in chapter 3), Jacques Monod suggested that we should imagine the cell as a *biological machine*.[99] One might equally suggest that one must imagine the cell as a *political machine*. The cell is 'law-code and executive power', as Erwin Schrödinger put it in *What Is Life?* (1944), a book which loomed large over the work of Monod and many other thinkers in the twentieth century. This defines the biological cell using a phrase that is pretty much as good a definition of police power as one could hope to find. But Schrödinger turns, instead, to the idea of war: he compares the cell's law-code and executive power with General Montgomery's strategy of having the cell of his army, the individual soldier, meticulously informed of all his designs. Nonetheless, for Schrödinger, the cell as law-code and executive power means that the cell resembles a *political power dispersed through the body*.[100] The implication is clear: the powers of war and police get written into the very texture of cellular life. The cell acts, and its acts are not solely of life but of life organized politically for the security of an organism imagined as a body politic. Cellular discourse could not help but serve explicitly political functions. It is precisely this concept of the cell that permeated thinking about immunity. As a

99 Jacques Monod, *Chance and Necessity: An Essay on the Natural Philosophy of Modern Biology* (1970), trans., Austryn Wainhouse (London: Penguin, 1997), 111.

100 Erwin Schrödinger, *What Is Life?* (1944), in *What Is Life? with 'Mind and Matter' and 'Autobiographical Sketches'* (Cambridge: Cambridge University Press, 1992), 23, 79.

political machine, the cell functions as police technology, security mechanism and a means of containment.

We hear little about the cell-soul these days, not least because we do not really hear much about the soul in general. What we do hear far more about, however, is the Self. Canguilhem points out that the science of immunological life appeared as an extension of the science of cellular life and even the life of the cell-soul.[101] (Virchow would eventually endorse Metchnikoff's work on phagocytosis, in an essay entitled 'The Battle of the Cells and Bacteria' [1885], and Metchnikoff would in turn endorse the idea of the cell-soul.) The history of modernity is a history whereby 'the Soul' became 'the Self'. The Cell-Soul was a major biologistic step in this transition: from Soul to Self via the Cell. This is one reason why the Self became one of immunity's key ideas.

101 Georges Canguilhem, *The Normal and Pathological* (1943/1950), trans. Carolyn R. Fawcett (New York: Zone Books, 1991), 207–25.

2

Imagining an Immune Self

> *In the secret places of her thymus gland Louise is making too much of herself. Her faithful biology depends on regulation but the white T-cells have turned bandit. They don't obey the rules. They are swarming into the bloodstream, overturning the quiet order of spleen and intestine. In the lymph nodes they are swelling with pride. It used to be their job to keep her body safe from enemies on the outside. They were her immunity, her certainty against infection. Now they are the enemies on the inside. The security forces have rebelled. Louise is the victim of a coup . . .*
>
> *The faithful body has made a mistake. This is no time to stamp the passports and look at the sky. Coming up behind are hundreds of them. Hundreds too many, armed to the teeth for a job that doesn't need doing. Not needed? With all that weaponry?*
>
> *Here they come, hurtling through the bloodstream trying to pick a fight. There's no-one to fight but you Louise.* You're the foreign body now.
>
> Jeanette Winterson, *Written on the Body* (1992), italics added

In a chapter on structures of feeling in his book *Marxism and Literature* (1977), Raymond Williams distinguishes between practical and official consciousness and suggests that they are always different. This difference, he suggests, is perhaps an indication of relative freedom or control. Two of the most dominant concepts that persist in both practical and official consciousness are 'Self' and 'System'. We imagine ourselves as

Selves and interpellate the world in that way, and we imagine ourselves as a series of systems and as part of a series of systems. We are admonished time and again to perform 'self-control' but also told that we cannot control 'the System'. The System, we are told, controls itself (its 'Self'?) and runs the way it does because that is how systems run and hence how the System runs. We are told that our own systems are self-regulating, and yet also told that we should be doing all we can to maintain their self-regulation. It is no wonder, then, that we sometimes feel that we have lost our Self in Systems.

In this chapter and the three which follow, I want to think about immunity through two of its core concepts: Self and System. I want to also use immunity as a springboard to a consideration of what is at stake politically when we think about and through notions of Self and System. 'Here's one of the central questions in immunology: how does the immune *system* know what is *self* and what is foreign?'[1] But such a question, from a book titled *In Defense of Self: How the Immune System Really Works*, really requires asking a number of rather important prior questions: What is this thing called *Self* that the immune system is said to know? What is this *System* that is said to know the Self? If the immune system *knows* things, then does it have the powers of cognition? What then is the relationship between this system and other physiological systems? Indeed, what *is* a system? I will save most of the discussion of system for the next three chapters. Here, I want to develop the previous chapter's discussion of the political technology of the Cell into an account of the political technology of the Self, by situating the account within the wider assumptions about selfhood in Western culture and, more specifically, into the wider assumptions about the security of the Self. Doing so opens the space for a consideration of the autoimmune disease: if the immune system is a defence of Self, then the autoimmune disease must be a form of Self-destruction. Such a claim demands that we think both psychoanalytically and politically about the autoimmune disease.

---

1 William R. Clark, *In Defense of Self: How the Immune System Really Works* (Oxford: Oxford University Press, 2008), 177, emphases added.

The Mystique of the Self

Immunology stands and falls on a few basic assumptions. One such assumption is the distinction between Self and non-Self. A 1971 report on the 'new immunology', described as 'the most exciting of the scientific disciplines at the moment', was titled *The Science of Self*. 'This science of self is a creation of the last twenty years', the report observed, and constitutes nothing less than 'the start of modern immunology'.[2] The report's reference to the organization of immunology around the 'self' in the previous twenty years is due to the fact that in immunology's earlier periods the focus was on 'organism' rather than 'self'. The problem that later became understood as the immunological recognition of the self arose occasionally but gained little attention, and immunology remained largely based within the 'chemical approach' of 'humoral immunology' and focused on the study of antibodies. But, as Alfred Tauber has shown at length through a series of works, the emergence of 'Self' as the operative concept for immunology in the second half of the twentieth century was a crucial moment in the development of research in the field, helping to unify the differing strands of immunological thinking around infection, protection, information, tolerance, and surveillance.[3] In the process, the concept of Self became central to immunological thought. Hence by the 1970s, textbooks and other publications frequently reiterated the point made in *The Science of Self*, often in their very titles and often repeating the language of security, protection, and defence we discussed in the previous chapter, for example the books *In Defense of Self*, cited above, and *In Self-Defense: The Human Immune System*. The major textbook by Jan Klein, *Immunology* (1982) has as its subtitle *The Science of Self-Nonself Discrimination* and begins by stating that 'recognition allows organisms to discriminate between *self* – that is, everything constituting an integral part of a given individual – and *nonself*, or all the rest'. Melvin Cohn's foreword to Rodney Langman's *Immune System* (1989) starts by stating categorically that 'two components characterize an immune system: the self-nonself discrimination and the

2 David Wilson, *The Science of Self: A Report of the New Immunology* (London: Longman, 1971), 3, 13.

3 Alfred I. Tauber, *The Immune Self: Theory or Metaphor* (Cambridge: Cambridge University Press, 1997).

determination of effector class'. The terminology of Self/non-Self discrimination has been so entrenched that in the year 2000 Cohn and Langman would write in an editorial introduction to a major journal that 'after several spirited attempts to discard this terminology no new consensus has emerged'.[4] To put it bluntly, wherever one looks in immunology one gets the same message: 'The story of immunity is the story of the self'.[5] Such ideas concerning the Self work their way into wider discourses about identity, playing on a certain 'common sense' about Self and Other.

Central to this focus on Self is the work of Frank Macfarlane Burnet, pioneer of the clonal selection theory that became the paradigm or framework for virtually all immunological research in the second half of the twentieth century. That theory emerged out of three overlapping and interlocking strands of research. First, Burnet and his colleagues in Australia had been working on the idea that a body has a means for distinguishing Self from 'not-self', such that antigens that were present before birth were accepted as 'self', meaning that the body made no antibody to them; the body possessed 'self-tolerance'. Second, a group in the UK led by Peter Medawar were working on the idea that the reason skin from a donor grafted onto the wounded or burned body of a person was rejected by the recipient's body was in some ways an immunological phenomenon. Third, in 1955 the Danish immunologist Niels Jerne put forward a 'natural selection theory' which proposed that 'natural antibodies', the specificity of which had for many years been doubted, were constantly being formed before any antigen was presented, and the antigen acts as a 'selective carrier' of this natural antibody.[6] In other words, an antigen would 'match' one of the range of antibodies to form an antigen-antibody complex, which would then be taken up by a phagocyte, triggering further production of the antibody. This was in accord with Burnet's concept of self-tolerance. The three strands eventually came together in Burnet's reworking of some of Jerne's ideas. Instead of the

4 R. E. Langman and M. Cohn, 'Editorial Introduction', *Immunology*, vol. 12 (2000), 159–62 (159).

5 Irun R. Cohen, *Tending Adam's Garden: Evolving the Cognitive Immune Self* (Dan Diego: Academic Press, 2000), 3.

6 Niels K. Jerne, 'The Natural-Selection Theory of Antibody Formation', *Proceedings of the National Academy of Sciences of the United States of America*, vol. 41, no. 11 (1955), 849–57 (849, 856).

idea that phagocytising an antigen-antibody complex causes cells to produce antibody, Burnet proposed that the natural antibody was in fact attached to the cell surface as a 'receptor'. This was then selected by the cell that carried the suitable antibody, which was the stimulus for the cell to start proliferating. This became the clonal selection theory, developed through the late 1950s and then the 1960s, and given its ultimate vindication as the central topic of immunology at the 1967 *Cold Spring Harbor Symposia on Quantitative Biology*. By the time his autobiography was published in 1968, Burnet was willing to declare the clonal selection theory as 'my most important scientific achievement', going so far as to suggest that although his 1960 Nobel Prize was for the work on immunological tolerance, that work was itself 'essentially a way-station' on the road to the clonal selection theory. Speaking of the general vindication of the theory in the closing paper of the 1967 symposium, Jerne would state that 'each of these cells . . . is committed to the production of antibody molecules of one particular specification, and each cell will transmit this single commitment to its progeny of antibody secreting cells and memory cells'. In effect: *one cell, one antibody*. And he adds: 'This is Burnet's Dogma in its uncompromising form.'[7]

We will return to some of the politics of an immunological 'dogma' in chapter 3. Here we can focus on the fact that the clonal selection theory dominated the intellectual agenda of the 'new immunology' and that at the core of the theory is the Self. But this centrality of Self became clear only as the theory developed. In his early career Burnet kept returning to the problem of Self. In *Biological Aspects of Infectious Diseases* (1940), written for a popular audience, he starts to refer to 'self and 'not self', but his use of inverted commas around the terms suggests a critical ambivalence about them. 'Any organism which lives by digesting the substance of other organisms must in some way be able to distinguish between "self" and "not-self"'. He adds that this 'primitive differentiation' is the basis of the mechanism of immunity. His 1941 book *The Production of Antibodies* makes no mention of Self, although it does suggest that immunology is founded on an intolerance of 'foreign matter'. In a lecture

7 F. M. Burnet, *The Clonal Selection Theory of Acquired Immunity: The Abraham Flexner Lectures of Vanderbilt University, 1958* (Cambridge: Cambridge University Press, 1959); Sir Macfarlane Burnet, *Changing Patterns: An Atypical Autobiography* (London: Heinemann, 1968), 190; Niels Kaj Jerne, 'Summary: Waiting for the End', *Cold Spring Harbor Symposia on Quantitative Biology*, vol. 36 (1967), 591–603 (593, 601).

published as a paper in 1948 he claims that 'recognition of "self" from "not self" . . . is probably the basis of immunology'. The terms 'self' and 'not-self' are still in inverted commas, and the 'probably' hints at an apprehension about the claim; in an article co-authored with Frank Fenner the same year, the quotation marks are still there, as they also are in the second edition of *The Production of Antibodies* (1949) co-authored with Fenner.[8]

Tauber observes that, however apprehensive Burnet was, he clearly needed to capture an idea for which he could find no better concept and was in effect 'forced to adopt this philosophical term and use it as a placeholder for a scientific language he did not possess'.[9] That said, Warwick Anderson and Ian R. Mackay have pointed to the importance to Burnet of Alfred North Whitehead's writings on an embodied Self in the process of becoming through its engagement with the objective world. Burnet met Whitehead personally and regarded him highly. 'If nothing else', write Anderson and Mackay, 'Whitehead's influence seems to have confirmed in the 1940s the importance of the idea of biological individuality and given Burnet a convenient term to express it: "self"'.[10] This perhaps explains why Burnet's discussions of Self and non-Self become far more confident and direct as his work progresses. In his Nobel Prize lecture, delivered in December 1960 with the title 'Immunological Recognition of Self' and the first of his publications to have 'Self' in its title, Burnet describes his work as 'concerned with immunological theory primarily only in so far as it deals with the problem of self-recognition'. Two years later, in *The Integrity of the Body* (1962), 'Self' and related terms such as 'Self-recognition' appear time and again without inverted commas around them: 'it is one of the concise statements of modern immunology that the body will accept as itself only what is genetically indistinguishable from the part replaced'; 'somehow we must find an explanation of how during embryonic life

8 F. M. Burnet, *Biological Aspects of Infectious Diseases* (Cambridge: Cambridge University Press, 1940), 29; 'The Edward Stirling Lectures. I. The Basis of Allergic Disease', *Medical Journal of Australia*, vol. 1, no. 2 (1948), 29–35; Macfarlane Burnet and Frank Fenner, 'Genetics and Immunology', *Heredity*, vol. 2, no. 3 (1948), 289–324 (318).

9 Tauber, *Immune Self*, 140.

10 Warwick Anderson and Ian R. Mackay, 'Fashioning the Immunological Self: The Biological Individuality of F. Macfarlane Burnet', *Journal of the History of Biology*, vol. 47, no. 1 (2014), 147–75 (154).

the body acquires or generates the information which allows it to differentiate immunologically between what is self and what is not self.' Seeking in the book to define the clonal selection theory in a 'collection of English words of four letters or less', Burnet offers the following: 'To know self from not self is a main need for life. In the womb it is laid down what is to be let live in the body. Self is what no cell dare act upon to harm; any cell that may harm self must die.' In his 1963 text *Autoimmune Diseases*, written with Ian Mackay, he writes that 'the essence of autoimmune disease is probably the failure at some point of differentiating between the body's own material (self) and foreign material (not self)'. A few pages later, they revert back to the use of inverted commas, but in a further article the following year, called 'The Darwinian Approach to Immunity', Burnet refers to his work as 'concentrated on the immunological significance of self and not self . . . It is fair to say that, at the present time, the most important unsolved problems of immunology are in just this field of self-not self discrimination'. So entrenched had this idea of Self and non-Self become that a few years later Burnet was willing to publish a book with those very terms as the title: *Self and Not-Self* (1969). This was the first volume of a two-volume textbook called *Cellular Immunology*. Burnet makes clear that the key aim of the single volume *Self and Not-Self* was to present the clonal selection theory to a general audience, but he suggests in the preface a

> hope that the central theme that gives the title *Self and Not-Self* to the book is everywhere to the fore. The need and the capacity to distinguish between what is acceptable as self and what must be rejected as alien is the evolutionary basis of immunology.

In the opening chapter on the history of immunological ideas he claims that by the end of the 1950s the advances in immunology had been so great that the importance of 'self' and 'not-self' to immunology had become apparent to all. So confident was he in such claims that in his autobiography of 1968 he retrospectively claims that he had introduced the self and not-self distinction in 1937.[11]

11 Sir Macfarlane Burnet, *The Integrity of the Body: A Discussion of Modern Immunological Ideas* (Cambridge, MA: Harvard University Press, 1962), 13, 19, 96; Ian R. Mackay and F. Macfarlane Burnet, *Autoimmune Diseases: Pathogenesis, Chemistry and Therapy* (Springfield, IL: Charles C. Thomas, 1963), 14, 29; Sir F. M. Burnet, 'The

Now, some have suggested that Burnet was never fully comfortable with 'self' as a scientific term. Tauber and Scott Podolsky point out that in *Self and Not-Self* the term 'self' barely makes it past the title page of the book and gets nothing by way of explanation in the book. The title *Self and Not-Self* may have implied that immunology deals with distinguishing the organism from its world, but the focus of the book is on the immune mechanism, not the immune Self as an entity. And Thomas Pradeu suggests that the 'self' is not really the object of Burnet's demonstration but rather serves to designate a physiological given.[12] Further evidence for Burnet's lack of comfort with 'self' as a scientific term comes from Burnet's own reflections on the concept: in 1965, for example, he commented that his own interest 'has been concentrated on the immunological significance of self and not self at a highly academic level'. Anderson, Myles Jackson, and Barbara Gutmann Rosenkrantz have therefore suggested that the terms 'Self' and 'non-Self' were for Burnet simply a metaphorical resource.[13] That said, immunology is fairly replete with metaphor, as we have already seen, and while some immunologists might regard metaphors such as 'Self' as an obstacle to 'scientific' thought – well before Burnet embarked on his long route towards the concept, Ludwik Fleck had suggested, in an argument that the concept of 'immunity' is probably best abandoned, that an organism should not be construed as self-contained – the metaphors themselves are always revealing. Burnet could have used a term such as 'organism', which might have been more consistent with biological thought of the previous decades. The fact that 'Self' was used tentatively in quotation marks and then liberated from the quotation marks and made its way into titles of books and articles indicates that the notion of Self that was

Darwinian Approach to Immunity', in J. Sterzl et al., *Molecular and Cellular Basis of Antibody Formation: Proceedings of a Symposium Held in Prague on June 1–5, 1964* (New York: Academic Press, 1965), 17; Sir Macfarlane Burnet, *Self and Not-Self: Cellular Immunology*, bk. I (Victoria: Melbourne University Press, 1969), vii, 7; Burnet, *Changing Patterns*, 190.

12 Alfred I. Tauber and Scott H. Podolsky, 'Frank Macfarlane Burnet and the Immune Self', *Journal of the History of Biology*, vol. 27, no. 3 (1994), 531–73 (567); Thomas Pradeu, *The Limits of the Self: Immunology and Biological Identity* (Oxford: Oxford University Press, 2012), 55.

13 Burnet, 'Darwinian Approach', 17; Warwick Anderson, Myles Jackson, and Barbara Gutmann Rosenkrantz, 'Toward an Unnatural History of Immunology', *Journal of the History of Biology*, vol. 27, no. 3 (1994), 575–94 (588).

implicit within immunological thought in the 1940s and 1950s became, through the 1960s, far more *explicit* as the defining idea of immunological thinking, moving gradually from metaphor to theory to paradigm.[14] This much is evident from the appearance of the term 'Self' in the titles and subtitles of vast swathes of immunological literature, as we have noted.

Despite the fact that 'the Self' is seen by many immunologists as a scientifically weak concept, leading many to also argue that, in Tauber's words, 'there is no such thing as *the* immune self', at least not as a 'definable scientific construct', and despite the fact that, as some commentators have noted, 'Self' is at best a heuristic tool with decidedly imprecise scientific definitions, immunology has nonetheless for over half a century sought to build an account of the immune system with that very tool. On this basis, writers such as Tauber suggest that the very concept of 'immune selfhood' is not really of much operative use.[15] And yet it is abundantly clear that 'Self' and, of course, the idea of one's immune system protecting one's Self, is the common sense of our cultural, political and biological modes of thought.

Why has such an imprecise, nebulous, and, as we shall see, atheoretical concept such as 'Self' taken on such an unassailable status within immunology? How has a concept that is at best simply a metaphor or, at worst, an impediment to immunological thought, been reified into perhaps the most significant concept of immunological theory? The question is especially pressing given that the language of Self and non-Self became central to immunology at the very moment that autoimmune diseases were being identified. As explained in the introduction, autoimmune disease involves immune cells attacking the body they are meant to protect. In an autoimmune disease, the immune system is unable to distinguish between healthy body tissue and antigens. The result is an immune response that attacks and destroys normal

14 Ludwik Fleck, *Genesis and Development of a Scientific Fact* (1935), trans. Fred Bradley and Thaddeus J. Trenn (Chicago: University of Chicago Press, 1979), 60–62; Alfred I. Tauber, 'Reconceiving Immunity: An Overview', *Journal of Theoretical Biology*, vol. 375 (2015), 52–60 (53); Eileen Crist and Alfred I. Tauber, 'Selfhood, Immunity, and the Biological Imagination: The Thought of Frank Macfarlane Burnet', *Biology and Philosophy*, vol. 15 (1999), 509–33; Pradeu, *Limits of the Self*, 42, 50–60, 74, 80–81.

15 Tauber, *Immune Self*, 7–8, 139–41; Alfred I. Tauber, *Immunity: The Evolution of an Idea* (Oxford: Oxford University Press, 2017), 3, 13.

body tissues. In the language of this chapter, then, an autoimmune disease appears to be 'self-inflicted', an act of 'self-destruction', and Burnet's early suggestion that 'Self is what no cell dare act upon to harm' is completely overturned. Burnet initiated the Self/non-Self research agenda in order to explain the apparent paradox of why some of us do and some of us do not succumb to autoimmune disease, and 'autoimmunity' along with 'autoimmune disease' gradually rose to prominence as immunological terms through the 1950s. This makes the use of such a nebulous concept even more interesting. One reason might be that, as Tauber notes, despite being rather imprecise as a scientific term, the idiom of Self resonates with our understanding of core identity as we think about this within Western culture: the 'Self' that runs through immunology is more or less the same as the concept of identity derived from the Latin *idem*, meaning same. 'Self' was imported into immunology with a rich conceptual history connected to the questions of identity in the sense of the 'sameness' of an entity. Burnet and other immunologists were therefore using a concept with rich philosophical, psychological and political implications. The 'strength of loose concepts', to use Ilana Löwy's phrase, lies perhaps in the extent to which the concepts resonate through the wider culture; imprecise scientific concepts often perform important social roles.[16] Yet perhaps there is more to say.

We know that there is no such thing as a 'Self' that is unproblematic or a permanent given. A large number of diverse philosophers and philosophical traditions have challenged the very concept of the Self; literary traditions and the biological humanities have explored the way that 'I contain multitudes'; more than enough critically minded psychotherapists have suggested that the Self is always already divided; more than enough microbiologists have pointed out that the sum total of bacteria, fungi, protozoa, archaea, viruses and other organisms within each of us constitutes genetic material that far surpasses our 'own', to the extent that the body which we take to be integral to Self is itself 'but a fiction of a

16 Tauber, *Immune Self*, 7; Leon Chernyak and Alfred I. Tauber, 'The Dialectical Self: Immunology's Contribution', in Alfred I. Tauber (ed.), *Organism and the Origins of Self* (Dordrecht: Kluwer Academic, 1991), 125–26; Ilana Löwy, 'The Strength of Loose Concepts – Boundary Concepts, Federative Experimental Strategies and Disciplinary Growth: The Case of Immunology', *History of Science*, vol. 30, 1992, 371–96 (372–73); A. David Napier, 'Nonself Help: How Immunology Might Reframe the Enlightenment', *Cultural Anthropology*, vol. 27, no. 1 (2012), 122–37.

self'.[17] This is symbiosis, which includes living with parasites, amid parasites, and as parasites to each other, as we shall see in chapter 4. In other words, we all know that the life (and death) of any given Self is always already implicated in the lives of other Selves; that 'Selves', like bodies, do not pre-exist as such but materialize in social interaction; that in a social system constituted through perpetual transformation, permanent revolution and human alienation, in which 'all that is solid melts into air' (to use Marx's phrase), there can be no stable Self; that Selves are formed through a variety of 'political technologies of the self' (as Foucault calls them), from schools to hospitals, prisons to asylums, bureaucracies to armies, as well as therapeutic and medical interventions. Immunologists have recognized this more than enough times: 'the whole dear notion of one's own Self – marvellous old free-willed, free-enterprising, autonomous, independent, isolated island of a Self – is a myth'; 'all of the reflective traditions in human history . . . have challenged the naïve sense of self'.[18] The suggestion that the thing called 'Self' is a thing that appears and then disappears feels familiar and yet elusive, seems constant and yet shatters at the slightest examination, applies to the immunological Self as much as it does to the neurological Self, endocrinological Self, emotional Self or psychological Self. Indeed, like many of these Selves, the immunological Self appears to be the master of its own downfall, as we shall see with the autoimmune disease. The science of immunology and thus, in a roundabout way, the politics of immunity, appears to be dealing with a *mystique of the immunological self*.[19]

In the light of this, some immunologists have pursued a different line. Tauber's observations about Burnet's use of a philosophical term as a kind of immunological placeholder has been the springboard for his own argument, developed over a series of publications, that the borders of Self and identity are dynamic, inconstant, and often elusive. As soon as one realizes the greater ecology of the immune system and that much of what passes as 'foreign' to the Self is either ignored or incorporated, it

17 Dorion Sagan, 'Metametazoa: Biology and Multiplicity', in J. Crary and S. Kwinter (eds.), *Incorporations* (New York: Zone Books, 1992), 368–70.

18 Respectively: Lewis Thomas, *The Lives of a Cell* (London: Futura, 1974), 167; Francisco J. Varela, Evan Thompson, and Eleanor Rosch, *The Embodied Mind: Cognitive Science and Human Experience* (Cambridge, MA: MIT Press, 1991), 59.

19 Arthur M. Silverstein and Noel R. Rose, 'On the Mystique of the Immunological Self', *Immunological Reviews*, 159 (1997), 197–206.

becomes clear that the borders of Self and other are open in such a way as allows movement between the immune system and its environments. Moreover, the autoimmune disease makes very clear just how ambiguous the idea of immune selfhood really is, a point which will become a major focus of this chapter. 'Briefly put, the issue for me has been neatly divided between those who adopt a notion of the Self, namely that there indeed is a "Self", and the contesting argument that the self is an artifice, a conceit, a model at best.'[20] For Tauber, although the idea of an immune Self plays naturally and conveniently into one of our central cultural understandings about selves, this has been one of immunology's biggest errors. The central issue for immunology, Tauber suggests, is not to discern the basis of Self/non-Self discrimination but, rather, to establish the nature of organismal identity; we might well be 'selves', but the immune process is a biological activity. Tauber therefore argues for an expanded 'ecological perspective' which allows us to move beyond the immune Self, based on an idea of symbiosis which more easily points to the importance of a community of others that contribute to the welfare of the organism and in which immunity involves mediating the body's participation in this community, in a way that makes it impossible to identify a circumscribed entity that is the Self. In so doing, the immune system learns through evolution which organisms to exclude and resist and which to allow entry and support. 'From this vantage, there is no circumscribed, autonomous entity that is a priori designated "the self". What counts as "self" is dynamic and context-dependent.' This would then reflect the wider 'ecologic sensibility' of contemporary biology. Tauber himself identifies Jerne's idiotypic network theory and the latter's desire to understand the immune system as a cognitive enterprise as the most significant shift. The idiotypic network theory holds that antibodies form a highly complex system. Under the general rubric of 'cognition', Jerne posits the immune system as capable of recognizing not only foreign antigens but also self constituents as antigens (the 'idiotopes'), meaning that there is no essential difference between 'recognized' and 'recognizer'. Any given antibody might serve either or both functions. In such an account there is no 'Self' and hence no foreign 'Other' but simply an immune response to the introduction of a substance that is

20 Alfred I. Tauber, 'The Elusive Immune Self: A Case of Category Errors', *Perspectives in Biology and Medicine*, vol. 42, no. 4 (1999), 459–74 (461).

'recognized'. In the network conception, the only thing close to a 'self' is *the immune system itself*.[21] Other criticisms from within immunology include C. A. Janeway's work in the 1980s. For Janeway, the immune system responds to relevant 'non-Self' bodies such as infectious agents but not to innocuous bodies, and so the issue is a question of whether the agent is 'non-self and threat'. This argument became better known as the 'danger model' developed by Polly Matzinger and which we encountered in chapter 1.[22]

In similar fashion, some have argued for an immunological sense of a 'liquid self', turning arguments from within social theory about 'liquid modernity', associated with the work of Zygmunt Bauman, into an argument about immunological process.[23] Pradeu, in *Limits of the Self*, likewise argues for a 'continuity theory' to supplant the conceptual vagueness of the Self/non-Self approach, although his stress on 'individuality' instead of 'Self' might appear to some to be trying to smuggle the Self in through the back door. For Varela and his colleagues, the immune system is a closed network of interactions which self-determines its ongoing pattern of stability and its capacities of interaction with its environment. In this approach, that also puts the emphasis on the 'cognitive' nature of immune events (and about which we will have more to say in later chapters), such immune events are understood as a form of self-recognition, and whatever falls outside this domain is nothing less than *nonsensical*. Varela and his colleagues have sought to change our focus from an antigen-centred immunology to an organism-centred immunology using integrative, adaptive concepts which emphasize the stability of autonomous autopoietic networks.[24] Varela's arguments are rooted in what he

21 Alfred I. Tauber, 'Expanding Immunology: Defensive versus Ecological Perspectives', *Perspectives in Biology and Medicine*, vol. 51, no. 2 (2008), 270–84 (280); Alfred I. Tauber, 'Moving beyond the Immune Self?', *Immunology*, vol. 12 (2000), 241–48 (243); Scott F. Gilbert, Jan Sapp, and Alfred I. Tauber, 'A Symbiotic View of Life: We Have Never Been Individuals', *Quarterly Review of Biology*, vol. 87, no. 4 (2012), 325–41 (333).

22 C. A. Janeway, 'Approaching the Asymptote? Evolution and Revolution in Immunology', *Cold Spring Harbor Symposia on Quantitative Biology*, vol. 54 (1989), 1–13.

23 Andrea Grignolio et al., 'Towards a Liquid Self: How Time, Geography and Life Experiences Reshape the Biological Identity', *Frontiers in Immunology*, vol. 5 (2014), 1–17.

24 N. M. Vaz and F. J. Varela, 'Self and Non-Sense: An Organism-Centered Approach to Immunology', *Medical Hypotheses*, vol. 4, no. 3 (1978), 231–67; Francisco J. Varela and Mark R. Anspach, 'The Body Thinks: The Immune System in the Process of Somatic

regards as a need to consider some of the *imaginary dimensions* surrounding the organism and autopoiesis; his reference point is likewise to the world of social theory, this time to the work of Cornelius Castoriadis on the imaginary institution of society.[25] Varela's idea on this score is surely a provocative one; indeed, one might say that without its imaginary dimension, 'immunity' would fail to have developed as a political trope with such a profound cultural reach.

Yet despite these criticisms, challenges, and attempts to move beyond the Self, the concept retains its place at the heart of the immunological imagination. Perhaps its mystique, rooted in the wider sense of 'Our Bodies, Ourselves', to borrow the title of the feminist classic, is precisely what allows it to retain such power. Certainly, one reason it retains its power as a way of thinking about immunity lies in the fact that immunology's concept of Self and non-Self reinforces a fantasy of agency and will, being the basis of dominant cultural assumptions about Self-making, Self-management and Self-control. As such, it reinforces our core concept for thinking about identity and consciousness, rationality and choice, and moral autonomy and political liberty. What sense might we make of this, politically?

Donna Haraway observes that it is in part the immunological language of Self that allows the immune system to appear as an icon for systems of symbolic and material 'difference' in late capitalism: the immune system appears as 'a map drawn to guide recognition and misrecognition of self and other in the dialectics of Western biopolitics'. Ed Cohen has taken that argument further and connected the immunological Self – and thus, in effect, the whole logic of immunity – to the long history of the natural law notion of Self-defence.[26] The notion of Self-defence draws us back once more into the imagery of war and

Individuation', in Hans Ulrich Gumbrecht and K. Ludwig Pfeiffer (eds.), *Materialities of Communication* (Stanford, CA: Stanford University Press, 1994), 283; Francisco J. Varela and Antonio Coutinho, 'Immunoknowledge: The Immune System as a Learning Process of Somatic Individuation', in John Brockman (ed.), *Doing Science: The Reality Club* (New York: Prentice Hall, 1988).

25 Francisco J. Varela, 'Organism: A Meshwork of Selfless Selves', in Tauber (ed.), *Organism*, 79–107.

26 Donna J. Haraway, *Simians, Cyborgs, and Women: The Reinvention of Nature* (London: Free Association, 1991), 208, 204, 224; Ed Cohen, *A Body Worth Defending: Immunity, Biopolitics, and the Apotheosis of the Modern Body* (Durham, NC; Duke University Press, 2009).

police discussed in the previous chapter. Yet the fact that the Self/non-Self distinction emerged at the very moment immunology was struggling with the idea of autoimmune disease sends us in what appears to be the very opposite direction: to self-destruction rather than self-defence. Self-destruction, however, undermines the whole notion of a secure and defended Self. It also poses some rather profound philosophical and political issues. We thus need to consider in the rest of this chapter *both* self-defence *and* self-destruction as part of the mystique of the immunological Self; in the final section of chapter 5 and then into chapter 6 this will generate a wider political discussion of the self-destruction of the body politic.

From Self-Defence . . .

Much is made of the roots of 'immunity' in Roman law, as we have seen in the introduction and will further examine in chapter 6. Yet the Romans had no concept of 'Self'. They had *persona*, a fundamentally legal category. 'Self' is a modern invention, emerging initially through early modern philosophical reflection about whether the individual soul is a substance or something else that is supported by a substance, then morphing into the entity identified as having self-knowledge and psychological consciousness, and thus becoming a key political notion. This becomes the Self that occupies our moral universe, a figure that becomes most acutely obvious in the seventeenth century. To offer from the *Oxford English Dictionary* just a few examples of first uses of 'self-' as a prefix, we get: self-protection (1635), self-deification (1637), self-dislike (1640), self-preparation (1642), self-advancement (1645), self-blame (1645), self-vindication (1647), self-sure (1652), self-promotion (1653), self-enrichment (1659), and self-expansion (1668). It is with that backdrop in mind that John Locke's account of consciousness as the criterion of personal identity is often treated as the first full articulation of selfhood.

In chapter 27 of book II of *An Essay Concerning Human Understanding* (1689) Locke suggests that 'to find wherein *personal identity* consists, we must consider what *person* stands for'. A person, he writes, is

> a thinking intelligent being that has reason and reflection and can consider itself as itself, the same thinking thing in different times and

> places; which it does only by that consciousness which is inseparable from thinking and, as it seems to me, essential to it: it being impossible for anyone to perceive without perceiving that he does perceive. When we see, hear, smell, taste, feel, meditate, or will anything, we know that we do so. Thus it is always as to our present sensations and perceptions, and by this everyone is to himself that which he calls *self* . . . For since consciousness always accompanies thinking, and it is that that makes everyone to be what he calls *self*, and thereby distinguishes himself from all other thinking things: in this alone consists *personal identity*.

Ellen Meiksins Wood, Neil Wood, and Étienne Balibar have, in different ways, shown that the theoretical revolution brought about by Locke's rethinking of the Self is nothing less than 'the decisive moment of *the invention of consciousness* as a philosophical concept', as Balibar puts it, and thus a decisive moment in the development of the Self. Their analysis builds from Marx's insight that the ideal of the Self that emerges in the seventeenth century has its material roots in the development of bourgeois power, and that this Self comes to possess a claim to some kind of 'sovereignty' within the liberal order. One reason this ideal of the Self is understood in relation to an increasingly commercial order is because Locke also fused a very modern problematic of (Self-)identity with a much older problematic of appropriation. The Self is not only embodied, but the body is also the Self's prime possession. The foundation for legal property is the natural property of the Self inherent in the body. This is given an explicit political grounding in the *Two Treatises of Government*. 'Every Man has a *Property* in his own *Person*. This no Body has a Right to but himself.' Man, for Locke, is 'Master of himself, and *Proprietor of his own Person*'. The Self is imagined as a property-owning body even in the state of nature, and the idea of ownership is conceived of as a social relation of one's Self (as subject) to one's own (as property) and therefore to other selves as owners of property. The Self resides in a body owned by the Self; the body is constituted as the property of the Self, and property becomes property through the labour of the body and the work of hands.[27] Locke had been medically trained, worked as a

27 John Locke, *Two Treatises*, ed. Peter Laslett (Cambridge: Cambridge University Press, 1988), 287, 298. Etienne Balibar, *Identity and Difference: John Locke and the*

doctor and was close friends with leading physicians such as Thomas Sydenham, but the importance of the body to his work is also evident from the fact that one's body becomes a metonym for one's Self: the body performs the important role of *being* property, *producing* property, and *embodying the property-owning Self in law*, the name for which is 'Person'. For Locke, my right to my body is a way of implying that if others claim power over my body then they have infringed my right to my own Self as my property. This ownership of Self *and* property and Self *as* property implies a fundamental protection; it implies a need for a system of what Locke calls 'mutual security'.

If it is the case that the state of nature has been a foundational political fiction of bourgeois modernity, then the Self that was imagined as existing within it has been even more so, becoming ingrained in the cultural norms and ideological assumptions of Western modernity. 'Even those who reject many of Locke's doctrines feel the power of his model', observes Charles Taylor, adding that Locke's model of a Self that is able to make and remake itself permeates contemporary discussions of identity, modern psychology, education, and culture and helps explain 'why we still think of ourselves as "selves" today'.[28] A simple list of terms prevalent in our culture captures this generalized and almost compulsory worship of Self: self-government (as the foundation of political liberty), self-formation (as the foundation of social liberty), self-ownership (as the foundation of absolute property), self-reliance (as the foundation of autonomy), self-control (as the foundation of good order and discipline), self-esteem (as the foundation of personal value). These components combine into a dream of mastery, of an autonomous and rational self-contained, self-sufficient, and self-conscious subject: the individual as master of themselves, as a Sovereign Self. This figure which comes to dominate the social imaginary will also become Self-centred, an adjective that captures most completely the egoistic possessive individualism of bourgeois society and which appears for the first time in English in the very period Locke was writing. But if it is indeed a model

Invention of Consciousness (1998), trans. Warren Montag (London: Verso, 2013), 1; Ellen Meiksins Wood, *Mind and Politics* (Berkeley: University of California Press, 1972), 47–73; Neil Wood, *The Politics of Locke's Philosophy: A Social Study of an Essay Concerning Human Understanding* (Berkeley: University of California Press, 1983), 155–60.

28 Charles Taylor, *Sources of the Self: The Making of Modern Identity* (Cambridge: Cambridge University Press, 1989), 171–74.

of Self for the contemporary world, it is so because such a model was from the outset 'an essential requirement in a capitalist socio-economic system in which self-ownership is assumed to be the fundamental social relation', as Silvia Federici puts it.[29] This Self will eventually become core to the idea of the immune process, in the form of the Immune Self, albeit heavily mediated by developments in the intervening centuries, as discussed in the previous chapter, from the vision of the independent cell in nineteenth-century cell theory, through the concept of the organism in the same period, and on to Metchnikoff's presentation of the phagocyte as more or less an organism in itself. This Self is a figure whose art or fashioning of selfhood is expected to take account of their own immune power (or lack of it) and to fashion themselves as a good Self by working towards their own good health. In its construction of and focus on an immunological Self imagined as protecting its autonomy, immunology contributes to the indispensable myth of the bourgeois subject. The logical and philosophical core of Western liberalism became the biological and philosophical core of immunology.

One can pursue this further by adding 'defence' to the list of words that can be prefixed with 'Self-'. In so doing, it becomes clearer that what was and remains at stake in the prevalent notion of Self is the idea of security that we identified in chapter 1 as the dominant trope in immunological discourse. The idea of threat and danger intrinsically connected to the Self/non-Self distinction is a version of the more directly political conceptual dichotomy of friend and foe. Hence, we find in immunology formulations such as the idea that 'the main question is how the body distinguishes friend (self) and foe (nonself)'; or that 'the immune system needs to be able to correctly determine friend from foe, or "self" from "not-self"'; or that Self and non-Self are 'ideas that are core to how the immune system polices our bodies but also how we define and police our societies'.[30] It would appear that the security of the healthy Self, and

29 Silivia Federici, *Caliban and the Witch: Women, the Body and Primitive Accumulation* (New York: Autonomedia, 2004), 149.

30 Respectively: Scott H. Podolsky and Alfred I. Tauber, *The Generation of Diversity: Clonal Selection Theory and the Rise of Molecular Immunology* (Cambridge, MA: Harvard University Press, 1997), 4; Kenneth Bock and Nelli Sabin, *The Road to Immunity: How to Survive and Thrive in a Toxic World* (New York: Simon and Schuster, 1997), 28; Matt Richtel, *An Elegant Defense: The Extraordinary New Science of the Immune System* (New York: Harper Collins, 2019), 69.

perhaps 'health security' in general, is impossible to imagine without one of the founding political binaries of friend and foe, which might also then explain why the healthy Self is also always imagined as a sovereign entity requiring permanent policing. For this reason, Cohen in *A Body Worth Defending* rightly points to the ways in which the Self/Friend and non-Self/Foe dichotomy plays on a concept at the heart of Western political thought and jurisprudence: Self-defence.

By the time Locke was articulating the conception of the Self that would become so powerful, other writers such as Hobbes had already articulated an entrenched notion of defence that could be brought to bear on this very figure. The term 'self-defence' emerged in the first half of the seventeenth century, and Hobbes's *Leviathan* (1651) is cited by the *Oxford English Dictionary* (*OED*) as one of the earliest texts using the term. In *Leviathan*, Hobbes grounds the commonwealth upon a natural desire and liberty for self-preservation. 'The Right of Nature, which Writers commonly call *Jus Naturale*, is the Liberty each man hath, to use his own power, as he will himselfe, for the preservation of his own Nature; that is to say, of his own Life.' From this there arises a right '*by all means we can, to defend our selves*'.[31] Hobbes imagines people 'naturally' possessing their own bodies, but the fact that these bodies are in perpetual motion and in constant desire means that they are at war with one another. Left at this, life would be 'nasty, brutish and short', so the contract is formed. The social contract is produced out of the desire of each Self to *live* and to *stay alive*. Protection from death and thus the security of one's life – 'self-preservation' – is the reason we give up the state of nature and create the Leviathan. This preservation of a person's life amounts to the preservation of the person as a Self and the body as a body. Even after the social contract is formed and the sovereign offering them security has been created, each person's excessive opinion of themselves, their envy and rage, keeps each Self at war with others, just as bodies always collide.[32] But each body, each Self, always retains a right of self-defence.

Seventeenth-century natural law in all its guises mandates defence of the Self as a foundational right. The right of self-preservation is the

31 Thomas Hobbes, *Leviathan* (1651), ed. Richard Tuck (Cambridge: Cambridge University Press, 1991), 91–92.

32 Hobbes, *Leviathan*, 54.

ground of all other rights. 'No man is supposed at the making of a Common-wealth, to have abandoned the defence of his life', Hobbes writes.[33] Hence the right of self-preservation persists even against the command of a Sovereign possessing absolute power:

> If the Soveraign command a man (though justly condemned,) to kill, wound, or mayme himselfe; or not to resist those that assault him; or to abstain from the use of food, ayre, medicine, or any other thing, without which he cannot live; yet hath that man the liberty to disobey.
>
> If a man be interrogated by the Soveraign, or his Authority, concerning a crime done by himselfe, he is not bound (without assurance of Pardon) to confess it; because no man . . . can be obliged by covenant to accuse himself.
>
> Again, the Consent of a Subject to Soveraign Power, is contained in these words, *I Authorise, or take upon me, all his actions*; in which there is no restriction at all of his own former naturall Liberty: For by allowing him to *kill me*, I am not bound to kill my selfe when he commands me. 'Tis one thing to say, *Kill me, or my fellow, if you please*; another thing to say, *I will kill my selfe, or my fellow*. It followeth therefore, that . . . no man is bound by the words themselves, either to kill himself, or any other man.[34]

Or, as Locke was to later put it, 'the law, which was made for my preservation . . . permits me my own defence'.[35]

In effect, the striving for self-preservation which grounds the very social contract and the creation of sovereign power is retained regardless of the creation of that power in the name of protection. Despite the deference to Law in the name of protection, the Law is expected to recognize the natural desire for preservation that lies within each individual Self, which is itself grounded on a physiological desire for bodily security.

What this means, as Cohen points out, is that 'Self-defence' appears as a foundational principle of natural law at least two centuries before

33 Hobbes, *Leviathan*, 206.
34 Hobbes, *Leviathan*, 151.
35 Locke, *Two Treatises*, 280.

biological science adopts the politico-juridical term 'immunity' to describe how the human organism defends itself.[36] The *Self*-defence that emerges as fundamental to immunology in the late nineteenth century, and which gets increasingly consolidated through the twentieth century to the point where it becomes part of both our cultural and our biological 'common sense', has its roots in the early political philosophy of bourgeois modernity. The idea of the protective activity of the organism that develops in the nineteenth century reinforces the political claims made for the autonomous political subject, and the idea of the autonomous political subject reinforces the assumption that our self-defence has deep roots in biological protection. Bourgeois natural law philosophy constituted self-defence as a primary political imperative and natural act. It is for this reason that a sovereign power then comes into existence for the security of the Self, for such security is impossible in the state of nature. But what also comes into existence with the logic of the self-defence of individual bodies is the 'self-defence' of the collective body. The security of the Self gets transposed onto the security of the state that is said to exist solely for the security of the Selves out of which it has been constituted, as witnessed by the countless acts of war and police carried out for the security of the system. This will, in turn, generate a fundamental problem when it comes to political immunity that we will discuss in chapter 6: When the state engages in acts of violence in its 'self-defence', who or what is being rendered immune from what or whom? In the seventeenth century, immunity meant liberty from the service of the Commonwealth, as Hobbes puts it, appropriating the Roman conception. By the twenty-first century, however, immunity had become entrenched as an idea used by the state to defend it from accusations of excessive violence; immunity used in defence of the state's own security.

Elsa Dorlin suggests that the history of 'self-defence' is marked by the continuous opposition of two antagonistic expressions of this defence. On the one hand, there is the whole tradition of legitimate defence grounded in law and articulated in relation to practices of power that operate with various modes of violence. On the other hand, there is the submerged history of a martial ethics of the Self that traverses political

36 Cohen, *Body*, 54–55.

movements and counter-conduct.[37] These two sides together constitute an overarching security logic of the Self that runs parallel with, is dependent on, and reinforces the security logic of the Cell discussed in chapter 1. The point here is that it is this notion of self-defence as a fundamental and natural function of how living organisms protect and preserve themselves as Selves that eventually gets built into the immunological imagination. When Metchnikoff came to focus on the ways that the phagocyte carries out a basic task of organismic self-preservation, a new science was being created which would give rise to the idea of an immune Self and its defence, but it was a science which spoke to a much older and decidedly political idea of Self and security.

In his book *The Integrity of the Body* (1962), Burnet suggests that 'it is axiomatic that a virus getting into the body and damaging cells will automatically be recognized as something foreign and inimical', adding that the antibody is the most important element for 'not only killing but also disintegrating the invader'. This killing and disintegration involves the body directing its 'destructive activities . . . only against foreign material or cells and not against the body's own components'. He then makes the following wry comment: 'Perhaps one could draw a political moral here.'[38] Now, one political moral to draw will be the one identified by Burnet, that 'the defensive response is more dangerous than the virus it is directed against'. In other words, the body's defence system might actually be of greater danger to the body than 'foreign viruses'. This will be the autoimmune disease, about which we will shortly have something to say. Yet that is not quite the point of Burnet's comment. What then was the point?

Burnet could take for granted that his readers would know exactly what he meant in referring to the need to deal appropriately with foreign invaders. The passage appears in a book published in 1962. With a sensitive ear to broader thought styles – recall his comments that we discussed in the introduction, to the effect that immunity is a problem that is far more philosophical than biological – Burnet adopted and adapted the dominant tropes and concepts used by Western states during the Cold War. It was left to others to state the point more categorically. One textbook on immunology published in 1971, Wilson's *Science of Self*,

37 Elsa Dorlin, 'What a Body Can Do', *Radical Philosophy*, no. 205 (Autumn 2019), 3–9 (8).
38 Burnet, *Integrity*, 39, 82–83, 96–97.

observes that the 'self' that is being defended in the immune process has a political grounding:

> From the Renaissance and the Reformation to this side of the French Revolution, the main drive of Western history can be seen as a drive for individual freedom. But in the twentieth century, revolutions are not made for individual freedom – neither the Russian, the Chinese, nor the Cuban revolutions has aimed at individual freedom. And in the non-Communist countries it is no longer reactionary or illiberal to think of limiting individual freedoms for the sake of the good of the community. If the thinking of our culture is nowadays dominated by the 'crisis of the individual', there is a parallel trend in science. Immunology, the science of self, may help provide a cool and reasonably firm foundation upon which to erect the mental structures of the next stage of our cultural evolution.[39]

Wilson's reference to earlier historical periods tallies with the discussion we have been having about seventeenth-century political thought, but he goes much farther and suggests that developments within immunology during the 1950s and 1960s parallel developments in the cultural and political struggles vis-à-vis the communist bloc. That is, this aspect of immunity needs to be understood in terms of the security politics of the Cold War.

In an earlier book, *Critique of Security*, I discuss at length the ways in which security needs the fabrication of a non-Self ('foreign' or 'Other') as distinct from Self and the ways in which this need intensified during the rise of the national security state from 1947 onwards. In this period, the logic of security came to delineate and assert modes of selfhood which served in turn to delineate and assert the logic of security. It was for this reason that the hegemonic rise of the national security state in the 1950s and 1960s occurred in tandem with, depended on, but also underpinned the proliferation of professional accounts of the Self. In psychoanalysis, for example, some practitioners sought to replace 'ego' with 'self', and some writers in the psychotherapeutic field, such as Donald Winnicott and Carl Rogers, focused on the idea of a 'true Self' even while acknowledging the term's philosophical limitations and

39 Wilson, *Science*, 30.

psychoanalytical ambiguities. The term also became prevalent in popular discourse.[40] Security intellectuals worked hand in hand with social scientists in focusing on individual character as a means of understanding the kind of 'mass society' that was emerging. The increasingly common assumption was that the roots of social problems and the threats to social order had psychological rather than structural causes. This coincided with a concern expressed by psychologists, psychiatrists, and sociologists that citizens in the West no longer knew who they were and were losing their sense of 'Self'. Hence the proliferation of widely read and influential books on the theme of the 'real self': Arthur Jersild, *In Search of Self* (1952), Rollo May, *Man's Search for Himself* (1953), Theodor Reik, *The Secret Self* (1953), Harry Fosdick, *On Being a Real Person* (1954), Allen Wheelis, *The Quest for Identity* (1958). Even texts that were not obviously about 'Self' turned out to be so: Erik Erikson's influential book *Childhood and Society* (1950), for example, placed utmost emphasis on the psychic achievement of a 'Self' and encounters with 'intruders' into this Self. The need to *find* oneself, to *become* oneself and to *be* oneself became a dominant cultural imperative. This was not simply because such activities were considered worthwhile but also because the *security of the System was thought to depend on it*. If 'psychological weapons' were integral to the security wars being fought, then one of these weapons was the idea of a nation of psychologically well-developed and well-rounded Selves, concrete entities that could also be measured and tested and about which social scientists and psychologists (and, of course, state and capital) could claim knowledge, and thus entities that could be developed into the kind of Selves needed for the capitalist world order. The literature of the Cold War was saturated with these ideas, from Walt Rostow's *The Stages of Economic Growth* (1960), which ends by arguing that the fundamental difference between communism and capitalism lies in human psychology; to David McClelland's *The Achieving Society* (1961), on the idea that the shortest way to achieve economic objectives could be to change people first; and on to Gabriel Almond and Sidney Verba's *The*

40 Eli Zaretsky, *Secrets of the Soul: A Social and Cultural History of Psychoanalysis* (New York: Vintage, 2005), 311–12; Elizabeth Lunbeck, 'Identity and the Real Self in Postwar American Psychiatry', *Harvard Review of Psychiatry*, vol. 8, no. 6 (2000), 318–22 (318).

Civic Culture (1963) on the psychological orientation of the kind of Selves that make up such culture.

The good, true, genuine Self was thus the very entity around which the body politic was to be secured, which helps explain why the security state was at various points spending almost all of its social science research budget on 'psychological' work.[41] Building *national identity* meant building *self-identity*, because both were central to *national security*. Technologies of the Self were understood as technologies of security. Just as Western security states could build on ideas of 'self-esteem', 'self-confidence', 'self-assurance', 'self-representation', 'self-respect' and 'self-improvement', so security threats were said to come from societies and movements lacking in those very qualities. The politics and psychology of the Self, and hence, by definition, a logic of Self versus non-Self, Other and foreign threat, was central to the political culture of the developing security state. The 'true' Self is a patriotic, loyal, anti-communist Self as well as stable and vigilant, but also a Self that is able to recognize itself as a good, stable, and vigilant Self committed to a System that values those very Selves.[42]

Looked at from the other direction, the communist system was described as a mass society which lacked strong and true Selves, evidence for which is the fact that they were unable to withstand communist ideology. Their weakness as Selves made them liable to domination, in complete contrast to the stable and strong Selves being nurtured in the capitalist West. Those citizens in the West tempted by communism were thus thought to have defective Selves, unable and unwilling to defend Sovereign or System. The security threat par excellence was a mass of undeveloped or poorly developed Selves.

On this view, an exemplary, well-developed and vigilant Self would be *immune to political viruses and diseases* such as communism. So long as the idea that communism was a problem over there, the other side of

41 Ellen Herman, 'The Career of Cold War Psychology', *Radical History Review*, 63 (1995), 53–85 (57).

42 Catherine Lutz, 'Epistemology of the Bunker: The Brainwashed and Other New Subjects of Permanent War', in Joel Pfister and Nancy Schnog (eds.), *Inventing the Psychological: Toward a Cultural History of Emotional Life in America* (New Haven, CT: Yale University Press, 1997), 245–46; Philip Cushman, *Constructing the Self, Constructing America: A Cultural History of Psychotherapy* (Boston: Da Capo Press, 1995), 74–90; Jamie Cohen-Cole, *The Open Mind: Cold War Politics and the Sciences of Human Nature* (Chicago: University of Chicago Press, 2014), 1–4.

the border, in the form of an Other that could be contained by being kept at a distance, the theory of immunity could be understood as coinciding with the philosophy of the nation-state. Yet, here, we encounter once again the problem of thinking of immunity simply through the question of borders. For the virus needs cells within the body (politic) in order to survive and thrive, and so as well as dealing with threats from external viruses and the need to maintain the body's protective shield, the image of immunity paralleled the image of security in highlighting threats that exist within the body (politic) itself.[43] The security state that developed during the Cold War reiterated time and again the idea of communism as, among many other things, 'bearing within itself germs of creeping disease' (George Kennan), 'the germ of death for society' (Attorney General J. Howard McGrath), 'a plague – an epidemic' (Hubert Humphrey), and a 'bloody virus' (Herbert Hoover).[44] The key issue that emerged was clear and has remained central to security politics ever since: to ensure that, as Hoover put it, 'the rank and file of our people are immune from this infection'. This is why many of the films that reinforced the ideology of capitalist states during the period show, in Fred Inglis's words, 'what can go wrong . . . when the inner life is made unsound or is invaded by political virus'.[45] Hence, films such as *The Manchurian Candidate* (1962), one of the most popular and sophisticated films of the period, pick up and reinforce the ways in which the security state adopted a viral or germ theory of political ideas and behaviour, in a merger of security and virology. The virus, like the communist and the terrorist, is absolute threat: disorder, dissolution,

43 Daryl Ogden, 'Cold War Science and the Body Politic: An Immuno/Virological Approach to *Angels in America*', *Literature and Medicine*, vol. 19, no. 2 (2000), 241–61 (248); Kenneth Dean and Brian Massumi, *First and Last Emperors: The Absolute State and the Body of the Despot* (New York: Autonomedia, 1992), 100; John Protevi, *Political Physics: Deleuze, Derrida and the Body Politic* (London: Athlone, 2001), 101.

44 George Kennan, 'The Long Telegram' (1946), in Thomas H. Etzold and John Lewis Gaddis (eds.), *Containment: Documents on American Policy and Strategy, 1945–1950* (New York: Columbia University Press, 1978), 50–63; Herbert Hoover, 'The Protection of Freedom', 10 August 1954, in Hoover, *Addresses upon the American Road* (Stanford, CA: Stanford University Press, 1955), 79–80. The others cited Robert Justin Goldstein, *Political Repression in Modern America: From 1870 to the Present* (Boston: G. K. Hall and Co., 1978); Joel Kovel, *Red Hunting in the Promised Land: Anticommunism and the Making of America* (London: Cassell, 1997).

45 Fred Inglis, *The Cruel Peace: Living through the Cold War* (London: Aurum Press, 1992), 99.

decline, chaos, and ultimately the end (ideas to which we shall turn in chapter 4). The virus thus became one of the most dominant tropes through which the body and the body politic could be imagined. Protect the System, Defend the Self. Defend the Self, Protect the System. Immunity: security. Security: immunity. On and on, round and round, leading to the key question of the time: How do we immunize ourselves against such political diseases and viruses?

In 1955, a new journal was launched called *Virology*, and the decade saw a growing number of textbooks and research studies on viruses. The mainstream press also highlighted how the virus harnesses a cell's apparatus in order to survive and duplicate, making the virus a 'sinister enemy', an 'assassin' with a 'killer instinct', as an article by microbiologist Paul de Kruif put it in *Today's Health* in 1953. Research by Priscilla Wald and others shows a cumulative amplification in tropes of aggression within virology through the 1950s.[46] Security threats came to be regarded as a virus and the security system imagined as a large-scale immune system. The idea of *immunological surveillance* discussed in chapter 1 also now makes much more sense, as does the idea of the 'political cell', as security documents of the time would often identify the 'cell system' as integral to communist organization and infiltration.[47] In this light, the idea of placing the body under permanent surveillance in order to discover the Other within was a means of fending off the virus of communism, reiterating the powerful conjunction of immunity and security.

In an essay on Cold War science and the body politic, Daryl Ogden points to the way that 'immunologists claimed that for the body to succeed against disease it had to . . . extirpate "effete" self-marked cells that weakened the body's own civil and border defenses'. The reference to 'effete' picks up on Burnet and Fenner's use of the term in the second edition of *The Production of Antibodies*, where they suggest that one of the immune system's key functions is 'the disposal of effete body cells . . . and of foreign organic material'. 'Effete' does not appear in the first edition, but by the time of the second edition, during the rise of the security state, such terms had become a key means of identifying weak

46 Priscilla Wald, *Contagious: Cultures, Carriers, and the Outbreak Narrative* (Durham, NC: Duke University Press, 2008), 162–63, 170–73, citing De Kruif at 171.

47 For example, CIA, *The Communist Party Underground* (Feb. 1957), 15.

'Selves' who could be a security threat. Ogden suggests that immunological discourse of the period employs a rhetoric 'that nearly jumps off the page if read within the terms of anti-communist, anti-homosexual [and, we should add, anti-feminist] political discourses of the post-World War II era'.[48] In *Critique of Security*, I examine how security documents of the time expressed a generalised cultural fear of a nation gone soft, impotent or effeminate through the degeneration and perversion of sexual norms. Executive orders and legislative measures highlighted the dangers of 'sexual perversion' and the idea that 'perverts in Government constitute security risks'. 'Can you think of a person who could be more dangerous to the United States of America than a pervert?' asked one senator. The stated assumption was that sexual perversion made someone susceptible to blackmail, but there was also a widespread belief that 'perverts' shared with communists a tendency towards subversion. Populist journalism reinforced the idea, as did the liberal intelligentsia. To put it bluntly: security was assumed to require forms of sexual behaviour regarded as 'normal', the outcome of which was an extensive policing of sexual difference in the name of security. This had its corollary in immunological ideas of the period, as we can see if we sample the research from the period on just one autoimmune disease, rheumatoid arthritis.

I have chosen rheumatoid arthritis as an example because it was one of the first and most prominent diseases to be defined as autoimmune (having originally been considered an infection). A number of developments occurred to make this the case: a 'rheumatoid factor' was identified in the 1940s; the International Congress on Rheumatic Diseases began a series of meetings running in the late 1940s and into the 1950s; an Arthritis and Rheumatism Foundation took shape; a National Arthritis and Rheumatism Act was passed in 1949 in the US, followed the next year by a new National Institute of Arthritis and Metabolic Diseases; in 1948 a patient known as 'Mrs G' was the first person to be administered something then known as 'Compound E' at the Mayo Clinic in Rochester, Minnesota, as an experiment to help with her rheumatoid arthritis, and the startling results led to the compound being developed as cortisol; a specific protein in the serum of patients with

48 Ogden, 'Cold War', 245, 247. Frank Macfarlane Burnet and Frank Fenner, *The Production of Antibodies*, 2nd edn. (London: Macmillan, 1949), 85, 94, 126.

rheumatoid arthritis was discovered in 1957, allowing researchers to hypothesize that the blood of rheumatoid arthritis patients contained antibodies that reacted with other antibodies as though they were antigens, resulting in the formation of complexes that caused joint inflammation. As a result of this general interest, rheumatoid arthritis was a key focus of immunological research.

One article from early in the rise of the national security state held that there was a clear relation between rheumatoid arthritis and psychological character concerning gender and sexual difference. Those diagnosed have often exhibited a 'tendency toward bodily activity'. The women are often 'tomboyish', 'hysterical' and exhibit a tendency to reject the feminine role, take on masculine attitudes, and compete with men. Women diagnosed with rheumatoid arthritis seem to have had 'an *unconscious* rebellion and resentment against men'. This is manifested in sexual proclivities: 'the defence mechanism utilized by all of them is hostile masculine identification' producing a 'pronounced bisexual attitude'. At the same time, 'some of the patients show their masculine identification in a predilection for the masculine posture in the sexual relationship'. 'The rejection of the feminine role, the wish to be a man, is in many cases naively and openly expressed', while 'in other cases it shows itself more in derivatives: being head of the house, controlling the environment and making the decisions'. At the same time, however, they also often 'express the masochistic need to serve'. The obvious contradictions in these claims are never dealt with.[49] Citing and building on this research, another academic article a few years later held that those diagnosed with rheumatoid arthritis exhibit 'inadequate sexual adjustment'. It was said that in their homes and upbringing 'the girls were tomboys, the boys sissies', and 'both preferred mother to father'.

> Two thirds of the men showed an interest in athletics as a compensation for feelings of masculine inadequacy; half of the women were interested in sports to such an extent as often to overexert themselves . . . The boys were unable to be masculine because they were

49 Adelaide Johnson, Louis B. Shapiro, and Frank Alexander, 'Preliminary Report on a Psychosomatic Study of Rheumatoid Arthritis', *Psychosomatic Medicine*, vol. 9, no. 5 (1947), 295–300.

> afraid to compete with father. The women tend to avoid pregnancies, had numerous miscarriages, and were resentful of their children.

And: 'The men showed a common conflict in that all were defending themselves against profound passive-feminine wishes . . . the women showed phallic trends. They openly rejected the feminine role in a masculine protest reaction.' 'The sexual adjustment was . . . poor.' 'The psychiatric diagnoses made on extended appraisal included schizoid personality, compulsive personality, hysterical character and conversion hysteria.'[50]

Such findings set the stage for research article after research article in scientific journal after scientific journal, establishing the idea that there is a 'rheumatoid personality', a certain kind of Self more likely to develop an autoimmune disease. This was the very same kind of Self defined as a security threat. 'Arthritic patients are sick people, psychiatrically speaking', one article found. They suffer from 'personality disintegration', as illustrated by their unconscious fantasies: 'Overtly the arthritic patient appears to be a calm, composed, and optimistic individual who rarely if ever expresses or even consciously feels hostility.' But in fact, 'they have trouble expressing their anger', they 'bottle up their anger', they suffer 'social inadequacies', they 'are constantly concerned about their looks, their clothes, and the impression they are making on others'. They have an interest in strong mother figures but question their fathers for being too strong and too weak at the same time.[51] Another article published the same year talks of the rheumatoid personality as one which used the arthritis as an 'excuse for his withdrawal from the environment and an adequate rationalization for his feelings of inadequacy'.[52] A year later, the *Journal of Chronic Diseases* ran a research article which found that rheumatoid arthritis sufferers tend to exhibit problems with 'sexual identification and adjustment',

50 James T. McLaughlin, Ralph N. Zabarenko, Pearl Butler Diana, and Beatty Quinn, 'Emotional Reactions of Rheumatoid Arthritics to ACTH', *Psychosomatic Medicine*, vol. 15, no. 3 (1953), 187–99.

51 Sidney E. Cleveland and Seymour Fisher, 'Behavior and Unconscious Fantasies of Patients with Rheumatoid Arthritis', *Psychosomatic Medicine*, vol. 16, no. 4 (1954), 327–33.

52 Edward W. Lowman et al., 'Psycho-Social Factors in Rehabilitation of the Chronic Rheumatoid Arthritic', *Annals of the Rheumatic Diseases*, vol. 13, no. 4 (1954), 312–16.

including tendencies towards exhibitionism and voyeurism. The female sufferers tended towards a 'rejection of female role' and 'masculine protest reaction', while men exhibited 'partial impotence'.[53]

To sum up, we can perhaps do no better than cite a meta-analysis of the research published in the *Journal of Chronic Diseases* in 1963 collating the 'findings' from research into rheumatoid arthritis over the previous fifteen years. The analysis suggests there are 'specific personality traits' strong enough to indicate a 'predisposition' towards this particular autoimmune disease, including poor sexual adjustment, exhibitionist, voyeuristic, and masochistic fantasies, a tendency to 'avoid closeness' and to 'outbursts of rage'. Arthritic women tend 'to reject the feminine role' by 'making the decisions' and exhibiting 'a predilection for the masculine posture in sexual relationships'.[54] A meta-analysis in the *Annals of the New York Academy of Sciences* published a few years later held that those suffering from rheumatoid arthritis 'tend to be self-sacrificing, masochistic, conforming, self-conscious, shy, inhibited, perfectionistic, and interested in sports'.[55]

It is abundantly clear that some of these supposed 'characteristics' are palpably absurd. It is also clear that, as Ann Satterfield points out, many of the traits and behaviour described as characteristic of a 'rheumatoid self' might in fact be the result of the disease and the pain that accompanies it.[56] That so much of immunology failed to see this is a telling detail about its conception of what a Self should actually be (and why other, different selves might need to be policed in certain ways). But because in immunology the immune Self becomes the physical Self which in turn becomes the psychic and personal Self, one finds suggestions such as 'by learning how to know yourself, you learn to know your enemies'.[57] Which is a nice idea that would be even nicer were it true. But what happens once it is clear just how problematic the idea of Self really is? What happens when it is discovered how difficult if not impossible it is to know

53 Stanley H. King, 'Psychosocial Factors Associated with Rheumatoid Arthritis: An Evaluation of the Literature', *Journal of Chronic Diseases*, vol. 2, no. 3 (1955), 287–302.

54 Rudolf H. Moos, 'Personality Factors Associated with Rheumatoid Arthritis: A Review', *Journal of Chronic Diseases*, vol. 17, no. 1 (1964), 41–55.

55 George F. Solomon, 'Emotions, Stress, the Central Nervous System, and Immunity', *Annals of the New York Academy of Sciences*, vol. 164 (1969), 335–43 (337).

56 Ann D. Satterfield, 'Pages from *Treatment*', *Camera Obscura*, 29 (1992), 215–24 (222).

57 Cohen, *Tending*, 218.

one's Self, liable as this Self is to shatter at the slightest examination? What happens if we discover something about our own Self, which is that despite all the efforts we put into our self-defence, from the biological to the juridico-political, we turn out to be our own worst enemy? What happens when we realize that Hobbes is simply wrong to say that self-defence is so entrenched in us that to do injury to one's Self is simply impossible? What happens if the one thing I learn about my Self is that my own 'self-defence', my own system of immunity-security, has decided that I am the enemy? For this, after all, is the autoimmune disease.

We have already noted that the language of Self became central to immunology at the very moment that autoimmune diseases were being identified as nothing less than an attack on one's own body, and that Burnet initiated the Self/non-Self research agenda in order to explain the apparent paradox of why some people succumb to autoimmune disease and some do not. If there is one thing that can be said about the autoimmune disease, it is that it is an illness which shatters the idea of immune selfhood and destroys the very distinction between Self and non-Self. Precisely *why* this happens we do not know, but we do know that the body acts as though a damaging agent is at work in the body and acts accordingly. What we do know is that we become 'foreign' to ourselves: immunity-security turns from self-defence . . .

. . . To Self-Destruction

In *Ideology and Rationality in the History of the Life Sciences* (1977), Georges Canguilhem comments on the history of the prefix 'auto':

> A remarkable and interesting fact from the epistemological standpoint is the proliferation of terms containing the prefix *auto-*, used today by biologists to describe the functions and behavior of organized systems: auto-organization, auto-reproduction, auto-regulation, auto-immunization, and so on . . . The epistemological reason for preceding these terms with the prefix auto- is to convey something about the nature of their relation to the environment.[58]

58 Georges Canguilhem, *Ideology and Rationality in the History of the Life Sciences* (1977), trans. Arthur Goldhammer (Cambridge, MA: MIT Press, 1988), 141.

As we have noted, in an autoimmune disease, the immune system is unable to distinguish between healthy body tissue and antigens, resulting in an immune response that attacks and destroys normal body tissues. It took many years for this process to become recognized, partly because of the entrenched status of the Self/non-Self distinction and partly because the language of war and police was so dominant that immunological enquiry was directed away from the destructive and disease-inducing capacity of the immune process itself. Because immunity was imagined as security, the idea that the system could actually harm the very thing it was expected to secure was essentially unimaginable. After all, is not security always a good thing? Just as people like to imagine that a security system would never turn against its own body politic, maim it and destroy it, so for many years people found it hard to believe that the system designed to protect and defend the body could also turn against that body, maim it and eventually destroy it. No one wanted to believe that *the System could turn against the Self.* If the immune system is the main thing standing between us and a plethora of microbial predators looking to wipe us out, then what do we do when that same system appears to be sending us to an ugly and early death?

In his lectures on the pathology of inflammation delivered at the Pasteur Institute in 1891, Metchnikoff had suggested the possibility of such a disease. He noted that 'cases naturally occur where the phagocytes do not fulfil their functions, a neglect followed by the most serious danger to or death of the organism', as evident by the frequency of disease and the instances of premature death.

> The phagocytic mechanism has not yet reached its highest stage of development and is still undergoing improvement. In too many cases the phagocytes *flee before the enemy* or *destroy the cells of the body to which they belong* (as in the scleroses). It is this imperfection in the curative forces of nature which has necessitated the active intervention of man.[59]

The idea of premature death is what animates Metchnikoff's work in the first years of the twentieth century, in which he presents ageing and

59 Elias Metchnikoff, *Lectures on the Comparative Pathology of Inflammation, Delivered at the Pasteur Institute in 1891*, trans. F. A. Starling and E. H. Starling (London: Kegan Paul, 1893), 194.

senility as in some sense problems of the immune process turning against its own body: the 'simpler' and more 'primitive' cells of the immune process get out of control and attack the 'higher' cells of muscles, nerves, brain and other organs. On this view, ageing becomes a form of cell war, and the various ailments associated with old age, from digestive problems and the weakening of the intellect to the reduction of oxygen in the blood are all evidence of 'a veritable battle that rages in the innermost recesses of our beings'. The atrophy of organs of the body 'is very frequently due to the action of devouring cells . . . These are the phagocytes that destroy the higher elements of the body, such as the nervous and muscular cells, and the cells of the liver and kidneys.'[60]

By the 1900s, then, there was some suggestion that 'autoimmunity' existed. The *OED* has the origins of 'autoimmunity' from the French word *auto-immunité*, which it dates from around 1896, entering English via an article in the *British Medical Journal* in 1901, where the author hesitates about using the term and refers to 'a *sort of* auto-immunity'.[61] And, yet, the idea that an organism's own tissue could be destroyed by its own immune process was still widely considered to be a fundamental contradiction. As Anne Marie Moulin points out, biologists in the late nineteenth century treated the possible existence of autoimmune disease as a major taboo, postulating instead the observation of the taboo by cellular machinery of the immune process.[62] So, the 'autoimmune disease' was essentially unimaginable. So unimaginable was it that Paul Ehrlich coined the term *horror autotoxicus* to capture the idea of immunological self-destruction.

> The organism possesses certain contrivances by means of which the immunity reaction, so easily produced by all kinds of cells, is prevented from acting against the organism's own elements and so giving rise to

60 Elias Metchnikoff, *The Nature of Man: Studies in Optimistic Philosophy* (New York: G. P. Putnam's Sons, 1903), 239; Élie Metchnikoff, *The Prolongation of Life: Optimistic Studies* (1907), trans. P. Chalmers Mitchell (New York: G. P. Putnam's Sons, 1908), 18.

61 George A. Hawkins-Ambler, 'Shock in Abdominal Operations', *British Medical Journal*, vol. 2, no. 2127 (5 Oct. 1901), 951.

62 Anne Marie Moulin, 'Multiple Splendor: The One and Many Versions of the Immune System', in Anne-Marie Moulin and Alberto Cambrosio (eds.), *Singular Selves: Historical Issues and Contemporary Debates in Immunology* (Paris: Elsevier, 2001), 228, 230.

> autotoxins . . . One might be justified in speaking of a *horror autotoxicus* of the organism. These contrivances are naturally of the highest importance for the individual.[63]

Ehrlich's failure to immunize animals against their own tissues confirmed, for him, the idea that the body was protected from reacting against itself. This fear of self-toxicity, in the sense of a presumed unwillingness or inability of an organism to damage or endanger its own existence through the formation of toxic autoantibodies, indicated not simply the paradoxical nature or abnormality of such an event but a literal *horror* entailed by the very idea. The 'horror' is interesting on both the psychoanalytical and political registers: psychoanalytically, the horror of one's own body becoming toxic to itself more or less explodes the very distinction between Self and non-Self; politically, the horror of the body's protective mechanism turning against the very thing it is meant to protect more or less explodes the whole concept of security. No wonder the autoimmune disease carries such awful significance.

Ehrlich's position held sway for a long time, with *horror autotoxicus* operating as one of immunology's founding principles and a yardstick against which new immunological thought was measured, preventing a richer understanding of the immune system's capabilities. No one wanted to believe that a system designed to offer security could in fact damage and destroy the body it is meant to secure. As a result, immunology lacked any kind of theoretical structure for thinking about a large group of diseases with implications that were philosophical and psychoanalytical as well as medical. So entrenched was Ehrlich's dictum that when, in the early 1950s, Ernst Witebsky and colleagues had compiled sufficient data about what were clearly a group of autoimmune diseases, they were unwilling to argue for the validity of their own findings. In a celebration of Ehrlich's work held in 1954, Witebsky would say that Ehrlich's *horror autotoxicus* is 'a biological concept of great importance'.

63 Paul Ehrlich and J. Morgenroth, 'On Haemolysins' (1901), in *Collected Papers of Paul Ehrlich*, vol. II, *Immunology and Cancer Research*, eds. F. Himmelweit and Martha Marquardt (London: Pergamon Press, 1957), 253.

> No living organism would be capable of producing – or would *dare* to produce, if you wish – an antibody against constituents of its own body, for this would be incompatible with life. On the basis of the side-chain theory, the concept of *horror autotoxicus* seems logical, indeed. The principle of *horror autotoxicus* is a rather burning topic in the current literature . . . The validity of the law of *horror autotoxicus* certainly should be evident to everyone interested in the field.[64]

Things gradually began to change, however, through the later 1950s, as landmark experiments by Witebsky and colleagues such as Noel Rose proved the existence of 'auto-antibodies'. Reflecting, many years later, on his own role in the period, Rose commented that 'at first, the immunologic world was suspicious of this whole business. To take one of the basic dogmas of immunology – *horror autotoxicus* – and turn it on its head, well . . . But eventually people bought into it.'[65] Eventually, indeed. The results were tested by others and attention was increasingly paid to diseases that were becoming grouped together as 'autoimmune', to the extent that, by the late 1950s, a theoretical account of autoimmune disease began to appear increasingly feasible. Macfarlane Burnet published an article on autoimmune disease in the *British Medical Journal* in 1959, had a chapter on the topic in *The Integrity of the Body* three years later (despite the same book defining 'Self' as the thing that 'no cell dare act upon to harm', as we saw above), and then, in 1963, co-authored with Ian Mackay the first monograph on autoimmune disease. In that book, Burnet and Mackay state with confidence that 'one of the greatest developments in Medicine during recent years has been the growing recognition of the importance of processes in which the immune mechanisms of the body are, as it were, turned against the body's own components'. This was followed by a large conference held on the subject in 1965 and the publication of the conference proceedings. True, as late as 1970, Burnet would say that there was still 'far from unanimity that the "autoimmune diseases" are legitimately so-called' (as an essentially evolutionary thinker, Burnet found it difficult to square

64 Ernst Witebsky, 'Ehrlich's Side-Chain Theory in the Light of Present Immunology', *Annals of the New York Academy of Sciences*, vol. 59 (1954), 168–81 (172–73).

65 Noel Rose in 'Q&A with Noel Rose: The Father of Autoimmunology', *Brigham Clinical and Research News*, 2 Oct. 2019.

autoimmune diseases with the essential defensive nature of the immune process), but it is not unreasonable to say that, even if unanimity was lacking, the idea of autoimmune disease was now widely accepted.[66]

None of which gets us closer to unravelling the real horror of the autoimmune disease. Several writers suggest that we think of auto-immune disease as a failure, an error, a mistake, an accident or an unsolved mystery.[67] But why? One reason lies in the main tropes of security, war and police that so dominate immunological thought, but with an inability or unwillingness to carry that thought through to its logical conclusion. In *The Integrity of the Body*, for example, Burnet comments that the body's defensive powers 'open the way for what may quite legitimately be referred to as a chronic civil war within the body', while Gustav Nossal, in a book called *Antibodies and Immunity*, written while he was president of the International Union of Immunological Societies, suggests that the concept of *horror autotoxicus* captures the literal horror of 'civil warfare . . . leading to anarchy'. Time and again in writing on immunity we find autoimmune disease imagined as a civil war within the body.[68] Alternatively, or perhaps equally, autoimmune disease is described as 'a mutiny in the security forces of a country'. The immune system, on this view, is 'treacherous' and acts like 'the Praetorian guard that turns its swords against the emperor'.[69] Civil war and mutiny

66 Sir Macfarlane Burnet, 'Auto-Immune Disease I: Modern Immunological Concepts', *British Medical Journal*, vol. 2, no. 5153 (1959), 645–50; Mackay and Burnet, *Autoimmune Diseases*, ix, 16; Ian R. Mackay, 'Travels and Travails of Autoimmunity: A Historical Journey from Discovery to Rediscovery', *Autoimmunity Reviews*, 9 (2010), A251–A258 (A251); Sir Macfarlane Burnet, *Immunological Surveillance* (Oxford: Pergamon Press, 1970), 186.

67 For example: Burnet, *Integrity*, 101, 165; Bock and Sabin, *Road to Immunity*, 28; John Dwyer, *The Body at War* (London: Unwin Hyman, 1988), 34, 177–78; Jan Klein, *Immunology: The Science of Self-Nonself Discrimination* (New York: John Wiley, 1982), 649, 653; Steven B. Mizel and Peter Jaret, *In Self-Defense* (San Diego: Harcourt Brace Jovanovich, 1985), 184, 188; Robert S. Root-Bernstein, 'Self, Nonself and the Paradoxes of Autoimmunity', in Tauber (ed.), *Organism*, 160, 163; Mohan Matthew and Edwin Levy, 'Teleology, Error, and the Human Immune System', *Journal of Philosophy*, vol. 81, no. 7 (1984), 351–72.

68 Burnet, *Integrity*, 174; G. J. V. Nossal, *Antibodies and Immunity* (Harmondsworth: Penguin, 1969), 210; Lennart Nilsson, with Kjell Lindqvist and Stig Nordfeldt, *The Body Victorious* (1985), trans. Clare James (London: Faber, 1987), 187; Gabor Maté, *When the Body Says No: The Cost of Hidden Stress* (London: Vermillion, 2019), 2.

69 Burnet, *Changing Patterns*, 215; Barbara Ehrenreich, *Natural Causes: Life, Death and the Illusion of Control* (London: Granta, 2018), 205–06.

are by no means the same thing, but such imagery is confused further with the idea that perhaps an autoimmune disease is the security system using 'excessive force'. For Clark, the immune system 'is capable of bringing *too much power* to bear during the course of clearing away invaders' and 'like an army lashing out blindly against an unseen and unmeasured enemy . . . is capable of using excessive deadly force in the wrong time or place'. In other words, the immune system 'is capable of overkill and the most devastating damage of all may be done to innocent bystanders',[70] perhaps killed by 'friendly fire'.[71] This is also why the idea of collateral damage is often used.

> Inflammatory wars, like military wars, inevitably cause massive collateral damage to innocent bystanders; and the weapons of the immune system, like guns and missiles, can be pointed in the wrong direction to cause casualties by friendly fire . . .
>
> Intense macrophage warfare is effectively analogous to scorched-earth or carpet-bombing tactics in human warfare. There can be massive collateral damage to non-participants in both kinds of conflict.[72]

These same tropes regarding autoimmune diseases are found in the philosophical work on immunity. Esposito, for example, points out that the literature on the immune system makes constant reference to the idea that autoimmune diseases are somehow connected to the 'imprecision' of the system's weapons. 'To stay with the military metaphor, if you drop a bomb with a hugely destructive potential from an aircraft at a high altitude, it is difficult to limit the damage to specific targets, hence the possibility of striking allies along with enemies.' This is a rather peculiar move on Esposito's part, because the debate about bomb damage has never really been a concern about 'allies'; one is not so much likely to hit 'allies', but rather, one is likely to hit 'civilians' or

70 Clark, *At War Within*, ix.

71 Bock and Sabin, *Road to Immunity*, 28; Paul Martin, *The Sickening Mind: Brain, Behaviour, Immunity and Disease* (London: Flamingo, 1998), 71.

72 Edward Bullmore, *The Inflamed Mind: A Radical New Approach to Depression* (London: Short Books, 2018), 39–40. In chapter 6 we will consider 'collateral damage' as a political category used to justify the immunity ascribed the sovereign's agents of violence.

'noncombatants'. Hence, the reason Esposito's claim is odd is that these noncombatant civilians are usually said to have 'immunity' from the violence being meted out to the combatants. Given his interest in immunity's legal history and political weight, one would have thought Esposito might make more of this possible connection with this kind of immunity (as we will do in chapter 6). Yet it is as though once Esposito has established the biological appropriation of immunity from its legal and political roots, the legal and political roots are forgotten. Instead, he has recourse to the same idea as the immunologists: the autoimmune disease evokes the 'self-dissolution of a civil war', he writes, citing in support the passage from Nossal cited above. In similar fashion, though in a more Derridean style, W. J. T. Mitchell claims that 'the image of autoimmunity would seem more strictly applicable to something like a military coup d'état, in which the armed defenders of the external borders and the internal order, the army and the police, turn against the legitimately constituted government'.[73] Yet this also makes little sense, given that autoimmune diseases attack not the governing centre of the body in order to take over that body but randomly chosen parts with no obvious rationale for those parts being chosen.

Just as the trope of 'immunity as warfare' leads to autoimmune disease imagined as civil war, mutiny, overkill, and collateral damage, so 'immunity as policing' leads to the suggestion that in an autoimmune disease the police are rioting rather than protecting: 'It's as if the body's police, rather than keeping the peace, were themselves occasionally rioting.'[74] Rioting against what? We are not told. One reason for the riot might be that police officers sometimes go a bit 'mad': 'Instead of doing their service in good faith, catching enemies or traitors (abnormal body cells), these "crazy policemen" go for "loyal citizens" (good working cells) [and] destroy them.' Elaborating on this idea, Petrov insists that the system as a whole 'works all right, faithfully fulfilling its police function', but 'one group, one "platoon" or "company", has turned traitor' and

73 Roberto Esposito, *Immunitas: The Protection and Negation of Life* (2002), trans. Zakiya Hanafi (Cambridge: Polity, 2011), 163; W. J. T. Mitchell, *Cloning Terror: The War of Images, 9/11 to the Present* (Chicago: University of Chicago Press, 2011), 46.

74 Luba Vikhanski, *Immunity: How Elie Metchnikoff Changed the Course of Modern Medicine* (Chicago: Chicago Review Press, 2016), 103; also L. J. Rather, 'Review of *Lectures on the Comparative Pathology of Inflammation* and *Immunity in Infectious Diseases*, by Elie Metchnikoff', *Medical History*, vol. 14, no. 4 (1970), 409–12 (411).

starts destroying normal body cells. 'Instead of police guarding loyal citizens, there is the "fifth column" which destroys these citizens.'[75] Maybe the police traitors are responsible for the police riots? Certainly, there must be 'subversion within the force' instigated 'at the suggestion of some foreign power'. What is to be done? 'Any attempt to get rid of the subversive policemen might be thwarted by the subversive forces themselves', and so 'unless one can find out the origin of the subversion it is impossible to reform the police force. The only alternative seems to be to dismiss every policeman and start off again with a fresh lot.'[76] And, since one cannot dismiss the whole immune-police system, one is rather stuck with 'corrupted policemen'.[77] So the system might generate dangerous police, mad police, corrupt police, and even treacherous police, but we are apparently stuck with them.

Still perplexed by the autoimmune disease? Imagine instead a 'man [who] had brutally attacked his wife after a domestic argument':

> We humans must accept the fact that millions of cells in our body would attack us mercilessly if they followed their basic inclinations. For our survival, therefore, they need to know that this is not the right thing for them to do, so they may exert a measure of self-control over their auto-aggressive natures. They need to be aware that big brother is watching them and urging passivity. It is the same situation that faces our wife basher; self-control and third-party regulatory pressures offer the best chances of avoiding dangerous escapes from acceptable behavior.[78]

Note that, despite its absolute powers and total surveillance, 'big brother' seems unable to stop the 'wife basher'. But the real point is that, in this example, the violence carried out by the autoimmune system is not the product of the system, and neither is it a product of aberrant or sociopathic elements of the system, such as the various personnel from the

75 Rem Petrov, *Me or Not Me: Immunological Mobiles* (1983), trans. G. Yu. Degtyaryova (Moscow: Mir Publishers, 1987), 262–63, 296–98.

76 Wilson, *Science*, 267.

77 Eduardo Bonavita, Maria Rosaria Galdiero, Sebastien Jaillon, and Alberto Mantovani, 'Phagocytes as Corrupted Policemen in Cancer-Related Inflammation', *Advances in Cancer Research*, vol. 128 (2015), 141–71.

78 Dwyer, *Body*, 177–78.

war power and the police power. Instead, the violence comes from a rogue 'wife basher', who simply needs to learn to exercise more self-control and be subject to better regulation.

Let us be clear: none of these arguments about civil war or traitorous police (or 'wife bashers') make any sense. More to the point, however, is the fact that they step back from imagining what is for these thinkers clearly unimaginable and hence unsayable. There is no logic in suggesting that in an autoimmune disease the immune system goes on the rampage and attacks the emperor, stages a coup or engages in a civil war. For it to be a civil war there would have to be a second fighting force, but, in the autoimmune disease, there is no such force. For the immune system to enact a mutiny or coup would suggest that it wants to take over the body's leading organ and control the body itself. But in an autoimmune disease, *the body is being destroyed*. And it is being destroyed because the body's system of immunity-security has turned against the body. Less an error and more a terror: the unremitting terror of security turning against the very thing it is meant to secure.

Now, having flagged the difficulty with the political language found in immunological discourse, we shall hold further discussion of this over until chapter 5 and then continue it in chapter 6. I want to first return to the autoimmune disease in relation to the Self, which is the task of the rest of this chapter, and then I want to take a long detour through the idea of System in chapters 3 and 4, before then picking up this thread again.

In the introduction, we cited Burnet to the effect that, although he had spent his working life in the laboratory, immunology had always struck him as more a problem in philosophy than a practical science, and we also noted his suggestion that one cannot discuss autoimmune disease without getting into deep water philosophically. But what deep water? If all disease is a relationship, as Haraway puts it,[79] then what kind of relationship is the autoimmune disease? What is it a relationship between? Certainly not Self and non-Self. It can only be Self to Self. In an autoimmune disease, the disease is no Other. To say that the immune process can spiral out of control, turning defence against the invading non-Self into an attack on the body itself, is to put one's relationship to

79 Donna Haraway, *How Like a Leaf: An Interview with Thyrza Nichols Goodeve* (New York: Routledge, 1998), 73.

one's Self in a whole new light. Indeed, it more or less shatters the core immunological concept of Self.

> Every particle of our body must bear an identification sign, or 'tag', saying 'this is me'. If something does not have such a sign, or if the tag is foreign, the particle would be saying: 'this is not me'. Everything 'other than me' is destroyed by the immune system.[80]

All well and good, but what happens when the particles who are 'me' do indeed say 'this is me' and the immune system sets about destroying them anyway? The cultural stress we place on the concept of Selfhood means that the autoimmune disease shatters the central philosophical and psychological category of Western culture as well as one of immunology's core concepts. By shattering the Self/non-Self distinction, the autoimmune disease becomes a problem not just for immunology but for modernity as a whole.[81]

Most people diagnosed with an autoimmune disease attest to the fact that finding one's protection turning to destruction is to discover something about one's Self that is truly horrifying. This is the 'horror' implied in the original idea of *horror autotoxicus*, but 'terror' might be a better word. The feeling of your own immune system turning against you is 'terrible, terrifying, terrorizing', as Derrida puts it.[82] For some reason or another, whether due to serious trauma, or stress, or for reasons that we have not even begun to understand, something happens with the Self, something new and destructive, generating an illness in which those afflicted become Others to themselves. No longer destroying the dangerous non-Self, the System appears to be destroying, disintegrating, breaking down the very thing it is meant to protect.

In a posthumously published essay called 'Fear of Breakdown', first delivered as a lecture in 1963, Donald Winnicott connects the variety of forms of disintegration with the fear of breakdown. 'Breaking down' is a

80 Petrov, *Me or Not Me*, 10.

81 A. David Napier, *The Age of Immunology: Conceiving a Future in an Alienating World* (Chicago: University of Chicago Press, 2003), 223; E. Cohen, 'My Self as an Other: On Autoimmunity and "Other" Paradoxes', *Journal of Medical Ethics*, vol. 30 (2007), 7–11.

82 Jacques Derrida, 'Autoimmunity: Real and Symbolic Suicides – A Dialogue with Jacques Derrida', in Giovanna Borradori, *Philosophy in a Time of Terror: Dialogues with Jürgen Habermas and Jacques Derrida* (Chicago: University of Chicago Press, 2003), 124.

vague term, but its deeper meaning hints at an 'unthinkable state of affairs' connected to the 'failure of a defence organisation'. To ward off the fear of breakdown and the threat of annihilation, defences arise, but the defences themselves can lead to, among other things, disintegration.[83] This goes some way to capturing the general logic of autoimmune disease and explains why those who experience such diseases often find them to be, in Derrida's words, an irreducible source of absolute terror, in that the threat comes from within and we are defenceless against it. One's vulnerability is thus suddenly *without limit*. To experience this *autoimmune self-aggression* is completely disorienting as well as disintegrating, then, as your body's security system turns against you. *You're* the foreign body now; *you're* the Other, the alien, the threat. In a literal sense, *you do not know who you are*. Or, perhaps worse, you know enough of who you are to know that you want to destroy it. *You are your own worst enemy. You make yourself sick*. In the autoimmune disease, the Self is not only divided but, to use Mark Taylor's terms, 'inherently torn, rent, sundered, and fragmented'.[84] The Self estranged from the Self, torn apart by an internal fissure and splintering its own timber to the point of annihilation. This is the autoimmune disease as a form of alienation, and hence as a deeply *political* as well as personal condition. In that sense, the autoimmune disease is symptomatic of the crisis of the Self in Western culture.

Life is what is capable of error, Canguilhem suggests, and it would certainly seem as though an autoimmune disease is a mistake of the highest magnitude, as we have already seen some thinkers suggest. Canguilhem asks whether a physiological phenomenon which ends up in the 'veritable suicide of the organism' by means of substances which are part of its own tissues could be called anything other than error.[85] Well, yes: it could be called suicide and left at that. It could be called the death drive. But neither the death drive nor suicide are definitively *errors*, are they?

In 1920, Freud published *Beyond the Pleasure Principle*, examining how, as well as being driven by the pursuit of pleasure, we also possess a

83 D. W. Winnicott, 'Fear of Breakdown' (1963), in Winnicott, *Psycho-Analytic Explorations* (London: Karnac, 1989), 88.

84 Mark C. Taylor, *Nots* (Chicago: University of Chicago Press, 1993), 253.

85 Georges Canguilhem, *The Normal and Pathological* (1943/1950), trans. Carolyn R. Fawcett (New York: Zone Books, 1991), 22, 273.

drive towards unpleasure and self-destruction. On the one side, the sexual drive, 'perpetually attempting and achieving a renewal of life'. But on the other side, a drive 'which seeks to lead what is living to death'. Eros and Thanatos, life and death.[86] Compared to the pleasure principle, which seems to connect with our internal perception of ourselves and our assumptions about what life is, the death drive is said to do its work 'unobtrusively'. Freud discusses the '*urge inherent in organic life to restore an earlier state of things* which the living entity has been obliged to abandon under the pressure of external disturbing forces'. This is 'the expression of the inertia inherent in organic life' and thus 'an expression of the *conservative* nature of living substance'. This conservative nature involves a 'compulsion to repeat'. Fish return to certain waters, birds fly along certain paths: organisms recapitulate the structures of the forms from which they have come. 'All instincts tend towards the restoration of an earlier state of things', and so the death drive can be identified as part of the general principle of stability in nature, part of the 'circuitous paths to death'. Life is a detour the inorganic takes on a route back to itself. 'If everything dies for *internal* reasons – becomes inorganic once again – then we shall be compelled to say that "*the aim of all life is death*".'[87] In this context, the pleasure principle serves the death drive. The *psyche* (or *soul*, as Freud stresses emphatically several times; *psyche* is a Greek word which may be translated as *Seele* [soul], is how he opens 'Psychical (or Mental) Treatment' in 1905, although translators often render *Seele* as 'mind'), appears to possess a drive towards its own death and 'the quiescence of the inorganic world'.[88] The 'beyond' referred to in *Beyond the Pleasure Principle* is also a synonym for the 'hereafter': what is 'beyond' the pleasure principle is a compulsion to repeat by dissolving the living substance and returning it to the inorganic state. Psychoanalysis,

86 I use the word 'drive' here despite the fact that the main English translations of *Trieb*, including the *Standard Edition* of Freud's collected works, tend to render it as 'instinct' rather than 'drive', and *Todestrieb* as 'death instinct' rather than 'death drive'. Even an essay in which Freud makes this point himself, *Triebe und Triebschicksale* (1915) is translated in the *Standard Edition* as 'Instincts and their Vicissitudes'. Bruno Bettelheim, in the most trenchant critique of the English translations of Freud, *Freud and Man's Soul* (1983), comments that the translation of *Trieb* into 'instinct' is especially regrettable. References to Freud's works are to *The Standard Edition of the Complete Psychological Works of Sigmund Freud*, listed as *SE* followed by volume number.

87 Sigmund Freud, *Beyond the Pleasure Principle* (1920), *SE18*, 20, 36, 39, 46.

88 Freud, *Beyond*, 62–63.

like disease, reminds us of the conflict between order and disorder, of the entropic nature of life, that we are bodies that die. It also reminds us that there is something fundamentally unmanageable about the Self.

There is a marked opposition between the idea that the organism possesses a drive for self-preservation and the idea that death is an internal law of life. Life is lived in the knowledge that everyone owes nature a death and that we must all expect to pay the debt, Freud comments in 'Thoughts for the Times on War and Death'.[89] The organism wishes to follow its own path to death and to ward off any way of returning to an inorganic existence other than the ways which are immanent within the organism itself. In other words, 'the organism wishes to die only in its own fashion'. The dominating tendency of mental life and perhaps even of nervous life in general is the effort to keep constant or to remove internal tension. This 'Nirvana principle', as Freud calls it, borrowing the term from Barbara Low but simultaneously acknowledging the Buddhist tradition in which 'Nirvana' involves a detachment from the passions, is 'a tendency which finds expression in the pleasure principle', but recognition of that fact is also one of the reasons for believing in the existence of the death drive.[90] Elaborating on these ideas in the years following the publication of *Beyond the Pleasure Principle* in 1920, Freud suggests in an encyclopedia entry that these two sets of drives correspond to 'contrary processes of construction and dissolution in the organism'. One set are better known as the libidinal, sexual or life drives, best comprised under 'Eros', while the other set work in silence and *lead the living creature to death*.[91] This set of drives 'manifest themselves as destructive or aggressive impulses'; 'life' consists in the manifestations of the interactions and conflicts between the sets of drives. Death is the victory of the destructive drive. And in *The Ego and the Id* (1923) he suggests that what holds sway in the superego is 'a pure culture of the death drive' which 'often enough succeeds in driving the ego into death' in what amounts to 'interminable self-torment'.[92]

Now, from around the end of World War I, the idea of aggressiveness becomes increasingly central to Freud's work, and this is, no doubt,

89 Sigmund Freud, 'Thoughts for the Times on War and Death' (1915), *SE14*, 289.
90 Freud, *Beyond*, 55–56.
91 Freud, 'Two Encyclopaedia Articles' (1923), *SE18*, 258–59.
92 Sigmund Freud, *The Ego and the Id* (1923), *SE19*, 53–54.

because he had become increasingly interested in the problem of authority, war and civilization.[93] Although, in his initial formulations, the death drive is not centrally conceived of as aggression, his account gets increasingly focused on this, albeit filtered through the lens of sadomasochism. In 'The Economic Problem of Masochism' (1924), Freud argues that psychoanalysis assumes that an extensive fusion and amalgamation of the two classes of drives takes place, meaning that we are always dealing with a mixture of the two in varying degrees; the fusion of the drives is the fragmented self. This fusion, however, is in part a 'taming' of the death drive. The libido renders the destructive drive 'innocuous' by turning it outwards into a drive for mastery or 'will to power', released in the form of a 'primal sadism' but which is in turn 'identical with masochism', which can in turn be 'once more introjected, turned inwards', producing a secondary masochism.[94] This becomes an internal principle of disunion and conflict: primal sadism has the self as its object. In *The Ego and the Id*, Freud comments that the obsessional neurotic, in contrast to the melancholic, never undertakes the 'merciless violence' that drives us to self-destruction, almost as though he is 'immune against the danger of suicide'.[95] His comparison to the melancholic in relation to self-destruction takes us back to a slightly earlier essay, 'Mourning and Melancholia' (1917), in which he comments that the ego's 'immense self-love' makes it hard to imagine 'how that ego can consent to its own destruction'.[96] The concept of melancholia allows us instead to grasp how it is that the ego can destroy itself: by treating itself as an object and directing its hostility towards it. We turn ourselves into objects that we can then hate, towards which we can direct our aggression, and which we want to destroy. Hamlet, the Shakespearean character who most interested Freud, called this 'self-slaughter'.

All these points can make Freud's concept of the death drive seem a little too broad. As Erich Fromm observes, Freud says that the death drive is originally all inside, then part of it is directed outwards in the form of aggressiveness while part of it remains internal in the form of primary masochism, but then, when the part that is directed outwards

93 Jean Laplanche, *Life and Death in Psychoanalysis* (1970), trans. Jeffrey Mehlman (Baltimore: Johns Hopkins University Press, 1976), 86.

94 Sigmund Freud, 'The Economic Problem of Masochism' (1924), *SE19*, 163–64.

95 Freud, *Ego*, 53.

96 Sigmund Freud, 'Mourning and Melancholia' (1917), *SE14*, 252.

meets obstacles too great to overcome, the drive is redirected inwards in the form of secondary masochism.[97] This breadth may well be because the death drive has no form equivalent in status to sexuality in the life drive. The difficulty, Freud notes, is that psychoanalysis 'has not enabled us to point to any drives other than the libidinal ones'.[98] Freud seeks a representative or form of the death drive, but never finds one. As he put it ten years later, in *Civilization and Its Discontents*, 'The manifestations of Eros were conspicuous and noisy enough', but it 'was not easy . . . to demonstrate the activities of this supposed death drive'.[99] 'The impossibility of giving form to the beyond of the pleasure principle is part and parcel of Freud's trouble finding a representative for the death drive that would do for it what Eros does for the life drives', Catherine Malabou notes. Without any representatives, the drive 'remains essentially *unenvisageable*'.[100] True as this is, we might note that Freud was as influenced by the *Geisteswissenschaften* ('sciences of the soul') as much as the *Naturwissenschaften* ('sciences of nature'), and so, along with *Beyond the Pleasure Principle*'s biologism, about which we shall shortly say more, and the energetics discussed in the text, about which we shall say more in chapter 4, the book is also one of his most literary, speculative, and imaginative texts. Indeed, the imaginative and speculative nature of Freud's writing on the death drive is one of the reasons why few of those in his close circle accepted the drive.[101] In the year of publication of *Beyond the Pleasure Principle*, Freud added a passage to new edition of *The Psychopathology of Everyday Life*, first published in 1901, to the effect that 'psycho-analytic observation must concede priority to imaginative writers', citing in this context Laurence Sterne's *Tristram Shandy*.[102] Might it therefore be worth exploring the speculative nature of Freud's thought to consider the idea that some portion of the death drive remains operative within the organism itself? This might then help with thinking through the problem of Self-destruction – the death drive as a

97 Erich Fromm, *The Anatomy of Human Destructiveness* (London: Penguin, 1974), 597.

98 Freud, *Beyond*, 53.

99 Sigmund Freud, *Civilization and Its Discontents* (1930), *SE21*, 119.

100 Catherine Malabou, *The New Wounded: From Neurosis to Brain Damage* (2007), trans. Steven Miller (New York: Fordham University Press, 2012), 190, 197–98.

101 For example, Wilhelm Reich, 'Interview: 19 Oct. 1952', trans. Therese Pol, in *Reich Speaks of Freud* (Harmondsworth: Penguin, 1975), 85.

102 Sigmund Freud, *The Psychopathology of Everyday Life* (1920), *SE6*, 213.

drive to eliminate the distinction between Self and non-Self – and help delineate a political understanding of the self-destructive fragmentation of the human subject in the autoimmune disease.

In *Why Do People Get Ill?*, psychoanalysts Darian Leader and David Corfield touch on the baffling nature of the autoimmune disease and pose the following question: 'What better example of the Freudian death drive than an autoimmune problem?'[103] It has been said that in the death drive it is as though the *psyche* blurts out 'Death!' when 'Life!' is the thing to say.[104] The very same point is true of the autoimmune disease: in the autoimmune disease, it is as if the body blurts out 'Death!' when 'Life!' is the thing to say. This is what is overlooked by so much of the immunological literature discussed in chapter 1 and earlier in this chapter. To focus on organismic security and defence is to assume a *logic of life*, in which life is what *is* defended and *must be* defended; life, the opposite of death, then appears a struggle against life's destruction.[105] Such a conception leaves absolutely no space for a death drive, for a living destruction, for a living self-destruction, for what Derrida provocatively calls 'life death' (in contrast to 'life *and* death'). Yet one can have an 'unconditional preference for the living body', to affirm and love life, to be 'always on the side of the yes, on the side of the affirmation of life', while also acknowledging that 'life does not go without death, and that death is not beyond, outside of life, unless one inscribes the beyond in the inside, in the essence of the living'.[106] The mutual implication of life death that is at the heart of the death drive is antithetical to the injunction 'Live!' and the trope that 'Life Must Be Defended!' which underpins the immunological thought-style, and which therefore offers little

103 Darian Leader and David Corfield, *Why Do People Get Ill?* (London: Penguin, 2007), 206.

104 Robert Rowland Smith, *Death-Drive: Freudian Hauntings in Literature and Art* (Edinburgh: Edinburgh University Press, 2010), 85.

105 François Jacob, *The Logic of Life: A History of Heredity* (1970), trans. Betty E. Spillman, in François Jacob, *The Logic of Life* and *The Possible and the Actual* (London: Penguin, 1989), 90–91.

106 Jacques Derrida, *Specters of Marx: The State of the Debt, the Work of Mourning, and the New International* (1993), trans. Peggy Kamuf (London: Routledge, 1994), 141; 'I Am at War with Myself', interviewed by Jean Birnbaum, *Le Monde*, 19 Aug. 2004, republished as *Learning to Live Finally: The Last Interview*, trans. Pascale-Anne Brault and Michael Naas (Houndmills: Palgrave, 2007), 24, 51; *Life Death* (1975–1976), trans. Pascale-Anne Brault and Michael Naas (Chicago: University of Chicago Press, 2020).

resource for grasping the autoimmune disease or for thinking politically about it and with it. The autoimmune disease, like the death drive – or even *as* the death drive – reminds us that life carries its own destruction within itself. The 'insecurity' of life is thus without limit, because even if all threats to immunity-security from outside were eliminated, the destruction can and will come from within the body being secured; destruction comes from immunity-security itself. This is the terror.

This uncanny and endless attack of Self against Self, this blurting out of 'Death!' when 'Life!' seems the thing to say, is precisely what animates those writers who have written accounts of their own autoimmune disease; is what animates my own experience and partly what pushed me into writing this book; is what animates Jacques Derrida's thinking about the death drive and autoimmunity; and is one answer to Malabou's question as to how one might render the death drive *visible*.[107] Some writers on autoimmune disease prefer to look away. Why did the disease start when it did? 'Personally I leave this one alone. I won't get near it. The answers I've heard are awful.'[108] But more than a few have seen that the autoimmune disease offers a rich conceptual frame that, as Alice Andrews puts it, 'readily lends itself to an exploration . . . of a certain pathology, an anxiety, or even a *terror of a self-destructive death-drive*'. The Self is meant to secure and protect itself and its life, and does so through an immune system, but somehow finds itself turning the immune process away from the enemy and towards one's Self, from self-defence to self-destruction. This interminable self-torment, instability and violence is experienced as a self-destructive aggression in which the body's security system slowly destroys the body, in a kind of 'suicidal assault', 'suicide mission', or 'suicidal assassination'. 'All autoimmune diseases invoke the metaphor of suicide', observes Sarah Manguso in

107 Alice Andrews, *Autoimmunity: Deconstructing Fictions of Illness and the Terrible Future to Come* (PhD thesis: Goldsmiths University of London, 2011), 27; Cohen, 'My Self as an Other', 8; Jacques Derrida, *The Post Card: From Socrates to Freud and Beyond* (1980), trans. Alan Bass (Chicago: University of Chicago Press, 1987), 393; Jacques Derrida, *Rogues: Two Essays on Reason* (2002), trans. Pascale-Anne Brault and Michael Naas (Stanford, CA: Stanford University Press, 2005), 157; Catherine Malabou, *Ontology of the Accident: An Essay on Destructive Plasticity* (2009), trans. Carolyn Shread (Cambridge: Polity, 2012), 18.

108 Mary Felstiner, *Out of Joint: A Private and Public Story of Arthritis* (Lincoln: University of Nebraska Press, 2005), 165.

Two Kinds of Decay.[109] It is not clear that 'metaphor' captures what is really going on, but 'suicide' certainly does.

Immunology barely ever discusses this question of 'suicide' (despite regularly discussing the phenomenon of apoptosis, as we shall shortly discuss). 'The situation appears to be paradoxical. On one hand we have shown that self-recognition is needed, but on the other hand we know that self-reaction can be suicidal', observed Edward Golub in the first edition of his textbook *Immunology*. By the co-authored second edition, the paradox remains, but not the question of suicide.[110] Immunological discourse encourages us to note that the 'paradox' exists but discourages us from thinking it through the lens of suicide. In contrast, psychiatric discourse encourages us to think about suicidal ideation, suicidal intent and suicidal risk; sociologists like to classify suicide along lines such as 'egoistic', 'anomic', 'altruistic', and 'fatalistic'; philosophers like to debate free will and the right to life and death. Yet the autoimmune disease simply cannot figure in any such discussions. It is difficult to use the phrase 'she killed herself' ('killed her *Self*?) about someone who dies from an autoimmune disease or because of an autoimmune disease, since discussions of suicide always point us towards some kind of conscious repudiation of life and contain a voluntarism (of a sort), announcing *Enough!* with a sense of *defiance*, as Virginia Woolf describes one young suicide in *Mrs Dalloway*. 'Death of one's own free choice, death at the proper time, with a clear head and with joyfulness', enabling one to 'die proudly when it is no longer possible to live proudly', as Nietzsche puts it in section 36 of *Twilight of the Idols*. Such a 'free death' as an act of will, refusal, hope, and a huge rebuttal of the idea that 'The Self Must be Defended!' is far from the case with an autoimmune disease which, if considered in terms of suicide, has an air of absurdity about it, a 'decision deciding my life without me'.[111] It has been said that by repudiating a life

109 Alice Andrews, 'Autoimmune Illness as a Death-Drive: An Autobiography of Defence', *Mosaic*, vol. 44, no. 3 (2011), 189–203 (189, emphasis added); Sarah Manguso, *The Two Kinds of Decay* (London: Granta, 2008), 14; Vicki Kirby, 'Autoimmunity: The Political State of Nature', *Parallax*, vol. 23, no. 1 (2017), 46–60.

110 Edward S. Golub. *Immunology: A Synthesis* (Sunderland, MA: Sinauer Associates, 1987), 482, compared with Edward S. Golub and Douglas R. Green, *Immunology: A Synthesis*, 2nd edn. (Sunderland, MA: Sinauer Associates, 1991).

111 Jacques Derrida, *Circumfession: Fifty-Nine Periods and Periphrases*, in Geoffrey Bennington and Jacques Derrida, *Jacques Derrida* (1991), trans. Geoffrey Bennington (Chicago: University of Chicago Press, 1993), 282.

deemed unsatisfactory and unbearable, the suicide throws down a gauntlet to the living, whom they wish to leave; throws down a gauntlet to the dead, whom they wish to rejoin too soon; and throws down a gauntlet to God, since they are God's creation.[112] But an autoimmune disease appears to involve the immune system throwing down the gauntlet *for* you but also *at* you. In his reflection on the fear of breakdown, Winnicott cites a comment from one of his patients (who did go on to kill herself): 'All I ask you to do is to help me to commit suicide for the right reason instead of for the wrong reason.'[113] The autoimmune disease looks and feels like no reason can be given whatsoever, neither right nor wrong. This may well be why the discussion so often falls back on phrases such as 'quasi-suicidal', 'more or less suicidal', or 'semi-intentional' suicide.[114] But its 'more or less', 'quasi' and 'semi-intentional' nature runs the risk of robbing suicide of any meaning and any integrity it may have.

What is at stake in the autoimmune disease, then, is a new hyphenated 'self-' akin to Hamlet's self-slaughter: 'self-murder', '*Selbstmord*', '*meurtre de soi-même*'. This 'horrid crime of destroying one's self', as Samuel Johnson's *A Dictionary of the English Language* (1755) defines self-murder, has historically been seen as the most serious of crimes because, as Blackstone points out in his *Commentaries on the Laws of England* (1765–1769), it is not only the crime of killing the Self – from the Latin *sui* (self) and *caedere* (kill) – but also a *crime against the Sovereign*. Suicide is 'a peculiar species of felony'. For a start, it is a 'felony committed on oneself' or 'a *felo de se*', but the suicide is also guilty of a further 'double offence; one spiritual, in invading the prerogative of the Almighty, . . . the other temporal, against the king, who hath an interest in the preservation of all his subjects'.[115] As such, 'the law has therefore ranked this among the highest crimes', because in destroying one's own body *one destroys the body politic*; destruction of the Self appears as insubordination against the System.

The immunological language we encountered in chapter 1 has suddenly started to take an interesting turn, pointing to a grotesque

112 Gérard Vincent, 'A History of Secrets?', in Antoine Prost and Gérard Vincent (eds.), *A History of Private Life*, vol. V (Cambridge, MA: Belknap Press, 1991), 263.

113 Winnicott, 'Fear of Breakdown', 93.

114 Derrida, *Rogues*, 45; Derrida, 'Autoimmunity', 94; Manguso, *Two Kinds*, 21.

115 William Blackstone, *Commentaries on the Laws of England*, vol. IV (Chicago: University of Chicago Press, 1979), 189.

move on a part of the body against the body, a grotesqueness exacerbated by the fact that its origins lie in the body's hypervigilant security-immunity operation. The terror of immunity cannibalizing its own body, its own Self, is the terror of the body's own security forces. 'I am at war with myself', as Derrida put it towards the end of his life. '*I am* has become *I war*', says Taylor of his autoimmune disease. Taylor again:

> My body is a body politic that is always at war with itself . . . The enemy is not external – a menacing other or invading alien – but internal . . . My body does not know itself, does not recognize itself, and therefore miscalculates by reading the same as an other to be attacked. In this way my body produces the other it then proceeds to destroy and in destroying the 'other' destroys itself.

'How do I know what to fight against?' asks Mary Felstiner. No need to worry – your immune system has decided for you. As Jeanette Winterson says of the character Louise in her novel *Written on the Body* in the passage used as an epigraph to this chapter: here comes your immune system, picking a fight . . . *with you*.[116]

Now, I am aware that to imagine the autoimmune disease in terms of a suicidal death drive might be to take the logic of immunity too far outside immunology's main concerns. That said, recent research into autoimmune disease, inflammation, and suicide suggests the presence of inflammation is present in patients with suicidal behaviour and ideation and shows increased rates of suicidality among those with autoimmune disorders.[117] In a culture in which self-destruction has reached epidemic proportions, this connection is surely significant; I return to this in chapter 5. I am equally aware that to think about the death drive in relation to autoimmune disease might seem too far removed from what Freud was trying to do with that idea. Yet since our speculation has brought us this far, we may as well continue just a little bit further.

---

116 Derrida, *Learning*, 47; Felstiner, *Out of Joint*, 70; Mark C. Taylor, *Field Notes from Elsewhere: Reflections on Dying and Living* (New York: Columbia University Press, 2009), 209–10.

117 L. Brundin, S. Erhardt, E. Y. Bryleva, E. D. Achtyes, and T. T. Postolache, 'The Role of Inflammation in Suicidal Behaviour', *Acta Psychiatrica Scandinavica*, vol. 132 (2015), 192–203.

In *Beyond the Pleasure Principle*, Freud comments that his argument about the death drive involves 'the necessity for borrowing from the science of biology'. This is a good thing, he says, because 'biology is truly a land of unlimited possibilities'.[118] Freud makes this comment despite the fact that in the years leading up to the publication of the book he sought to spell out how his psychological thought was distinct from biology, claiming variously that it is 'necessary to hold aloof from biological considerations during our psycho-analytic work', that psychology should be kept clear from 'biological lines of thought', and that his work on sexuality is characterized by being 'deliberately independent of the findings of biology'.[119] The fact that these three statements insisting on psychoanalysis as a science independent from biology all appear just a few years before the claim that borrowing from biology is necessary points us to the somewhat strained relationship between biology and psychoanalysis in Freud's thought (and in the psychoanalytic movement in general). Joel Kovel points out that Freud sought to ground the concept of *Trieb* materially and grant the body 'a real and disjunctive input with respect to the demands of culture', but this left him trapped in much of the terminology of the sciences whose assumptions he was demolishing (making it difficult, in Kovel's view, to integrate psychoanalytic concepts with a genuinely historical social theory).[120] This issue has been explored at length by Frank Sulloway in his account of Freud as a 'biologist of the mind', who shows the tensions in the ways in which Freud, the onetime biologist, 'strenuously resented any attempt from the direction of biology to rob psychoanalysis of its independent disciplinary status'. But Sulloway also suggests that there remains in Freud's work an *intellectual union of psychology with biology* that has been either ignored or unappreciated in psychoanalysis. And Elizabeth Wilson has made a strong case that the moments of 'biological reduction' in Freud's work often produce his 'most acute formulations about the nature of the body and the character of the psyche'.[121] As late as 1940 in *An Outline of Psycho-Analysis*

118 Freud, *Beyond*, 60.

119 Sigmund Freud, 'The Claims of Psycho-Analysis to Scientific Interest' (1913), *SE13*, 181; 'On Narcissism: An Introduction' (1914), *SE14*, 78–79; *Three Essays on the Theory of Sexuality* (1905), preface to the 3rd edn. (1915), *SE7*, 131.

120 Joel Kovel, *The Radical Spirit: Essays on Psychoanalysis and Society* (London: Free Association Books, 1988), 130.

121 Frank J. Sulloway, *Freud, Biologist of the Mind: Beyond the Psychoanalytic Legend* (London: Fontana, 1979), 26; Henri Ellenberger, *The Discovery of the Unconscious: The History and Evolution of Dynamic Psychiatry* (London: Penguin, 1970), 261, 499, 512–16;

Freud would insist that 'the phenomena with which we are dealing do not belong to psychology alone' but 'have an organic and biological side as well.'[122] 'Psyche is extended; knows nothing about it', Freud observes in a late notebook (in August 1938).[123] Freud 'never forgot that he was a somaticist', comments Wilhelm Reich.[124] *Psyche* is *somatic*.

So, in the light of the speculative nature of *Beyond the Pleasure Principle*, let us speculate a little further about the kind of union of biology and psychology that might be at stake. And while we are at it, we might even speculate about what appears to be Freud's desire to find an almost quasi-biological imperative behind the fragmented and destructive Self. Speculating further still, we might even consider a link or two between Freud and some key immunological themes.

In its biological arguments, *Beyond the Pleasure Principle* reflects the influence of the well-known biologist and neo-Darwinian, August Weismann, whom Freud cites throughout chapter 6 of the book as a way of examining whether the findings of biologists might flatly contradict the idea of a death drive. In so doing, Freud suggests that examining the primitive organization of protozoa might shed some light on the features of higher animals.

> It becomes a matter of complete indifference to us whether natural death can be shown to occur in protozoa or not. The substance which is later recognized as being immortal has not yet become separated in them from the mortal one. The drive forces which seek to conduct life into death may also be operating in protozoa from the first, and yet their effects may be so completely concealed by the life-preserving forces that it may be very hard to find any evidence of their presence . . . But even if protista turned out to be immortal in Weismann's sense, his assertion that death is a late acquisition would apply only to

Elizabeth A. Wilson, *Psychosomatic: Feminism and the Neurological Body* (Durham, NC: Duke University Press, 2004), 3. Interestingly, although Wilson makes her case for a feminist theory through psychopharmacology, neurogastroenterology, hypothalamic structures, and affective neuroscience, she has nothing to say about the autoimmune disease, despite the fact that virtually all autoimmune diseases are suffered more by women than men and despite the emergence in the last few decades of a psychoneuroimmunology (about which we shall say more in chapter 5).

122 Sigmund Freud, *An Outline of Psycho-Analysis* (1940), *SE23*, 195.

123 Sigmund Freud, 'Findings, Ideas, Problems' (1938), *SE23*, 300.

124 Wilhelm Reich, 'Interview: 18 Oct. 1952', in *Reich Speaks of Freud*, 70.

> its *manifest* phenomena and would not make impossible the assumption of processes *tending* towards it.[125]

From this, Freud insists that any expectation that biology would flatly contradict the recognition of a death drive has not been fulfilled. Rather, he finds a 'striking similarity' (or, elsewise, 'unexpected analogy', 'dynamic corollary' or 'significant correspondence') between Weismann's distinction between mortal and immortal parts of the organism and psychoanalysis's distinction between the drives. We can thus apply the libido theory 'to the mutual relationship of cells' and suppose that

> the life instincts or sexual instincts [*Triebe*] which are active in each cell take the other cells as their object, that they partly neutralize the death instincts [*Todestrieb*] . . . in those cells and thus preserve their life; while the other cells do the same for *them*, and still others sacrifice themselves in the performance of this libidinal function.[126]

And, in *The Ego and the Id*, he discusses 'the death drive of the single cell' and how the destructive impulses of that instinct are neutralized by being directed against the external world.[127] So, the 'biological' aspects of Freud's arguments certainly reflect the influence of Weismann. But might they not also reflect, though far less explicitly, the influence of Elie Metchnikoff, in a way that allows us to speculate on the convergence of the immunological and psychoanalytical fields? One way to consider this possibility is through the work of Sabina Spielrein.

In a footnote towards the end of *Beyond the Pleasure Principle*, at the point at which he has flagged the question of sadism, Freud writes that 'a considerable portion of these speculations have been anticipated by Sabina Spielrein in an instructive and interesting paper', and he cites an article by her published in 1912 called 'Destruction as the Cause of Coming into Being'. Despite acknowledging her 'anticipation' of his arguments, Freud adds that Spielrein's own argument 'is unfortunately not entirely clear'. He goes on to misrepresent her argument somewhat. In her 'Destruction' article, Spielrein asks why the sexual drive gives rise

125 Freud, *Beyond*, 49.
126 Freud, *Beyond*, 46, 49–50.
127 Freud, *Ego*, 41.

to negative feelings such as anxiety and disgust as well as positive ones such as pleasure. The fact that in the act of conception 'the unity of each cell is destroyed, and from the product of this destruction, new life originates', means that for such organisms 'the act of creation simultaneously means that they are destroying themselves'. For Spielrein, this implies that

> the drive for self-preservation is a simple drive that consists exclusively of a positive component. Yet the drive for preservation of the species, which must dissolve the old in order to create the new, consists of a positive and a negative component.[128]

This points to a fundamental psychic conflict between the two drives such that 'one feels or finds an enemy inside oneself whose very essence is destruction'. The 'dissolution of the I' in the sexual act, as an act of destruction, thus conjures up the image of death. Spielrein was clearly interested in the ways that such creative destruction is 'psychological' and 'biological', and it may be that Freud was concerned about the more biological side of her argument. The point, however, is that there is no 'death drive' in Spielrein's article. What appears in her article is an argument about the creative power of destruction. A few months prior to the publication of that 1912 paper, however, Spielrein had presented a version of it to the Vienna Psychoanalytic Society. Freud was present at the meeting, and the records of the meeting have him contributing to the discussion. The following day Freud wrote to Carl Jung about the paper and commented that what troubled him most about Spielrein's paper was that the 'little girl' wants to 'subordinate the psychological material to biological criteria', which he found 'no more acceptable than a dependency on philosophy, physiology or brain anatomy'.[129] In this earlier, spoken, version of her argument, Spielrein takes as her 'point of departure the question of whether a normal death instinct exists in man'. And, in letters leading up to the paper and article, she certainly makes it clear that she had a 'death drive' in mind: 'He [Jung] urged me to write

128 Sabina Spielrein, 'Destruction as the Cause of Becoming' (1912), in *The Essential Writings of Sabina Spielrein*, trans. Ruth I. Cape and Raymond Burt (London: Routledge, 2019), 98–99, 120.

129 Freud, letter to Jung, 30 Nov. 1911, in William McGuire (ed.), *The Freud/Jung Letters* (London: Hogarth Press, 1974), 469.

my new study on the death instinct', she wrote to a friend in 1910.[130] Why is all this relevant? Because, in the spoken version of her argument, she cites the immunological work of Metchnikoff as a springboard for her thinking.

In 1903 Metchnikoff published a book called *The Nature of Man*. The book is wide-ranging, but part of its argument is against superstitious fears about death. The 'disharmonies' of nature in which he is interested, and which we discussed in chapter 1, turn out to be focused on the 'fear of annihilation through death'. 'The watchword of all systems of philosophy is to bow to the inevitable, that is to say, to be resigned to the prospect of annihilation'.[131] In the final chapter of the book, Metchnikoff takes up the idea of the cell-soul that we also discussed in chapter 1. Referring, like Freud, to the work of Weismann, Metchnikoff comments on the immortality of unicellular organisms, rooted in the fact that organisms that 'have no highly developed consciousness' can 'renew their lives by repeated divisions with complete regeneration', which is precisely why one can refer to the 'cellular soul'. But the immortality of the soul has no relation to the problem of death. Death requires a different argument entirely.

> The most important question relating to natural death is the following: Is the appearance of natural death in man accompanied by the disappearance of one instinct, the instinct of self preservation, and by the appearance of another instinct, the instinct of death?[132]

Metchnikoff is sensitive to the likely objections to such an idea, yet he maintains his position:

> It may seem altogether surprising and improbable to us that an instinct for death should arise in man, since we are imbued with an instinct of an opposite nature . . . The desire of life and the fear of

130 S. Spielrein, 'Presentation: On Transformation', 29 Nov. 1911, trans. M. Nunberg, in Herman Nunberg and Ernst Federn (eds.), *Minutes of the Vienna Psychoanalytic Society*, vol. III, *1910–1911* (New York: International Universities Press, 1974), 329–35; Letter from around September 1910, cited in Lisa Appignanesi and John Forrester, *Freud's Women* (London: Penguin, 2000), 216.

131 Metchnikoff, *Nature*, 199.

132 Metchnikoff, *Nature*, 278, 287.

> death are manifestations of an instinct deep-rooted in the constitution of man. That instinct is of the same order as the instincts of hunger and thirst, of the need of sleep, of movement and of sexual and maternal love.[133]

Registering the fact that human beings who are 'full of the desire for life believe more easily in eternal life than in the possibility of an instinct of death', Metchnikoff stresses the 'instinct of death' which appears to lie 'deep in the constitution of man'. So deep, in fact, that one might want to describe it as unconscious.

Metchnikoff finishes that book by returning to a point he had made at the beginning of a book from two years previously that was much more narrowly focused on his research on immunity, called *Immunity in Infective Diseases* (1901), where he comments on the ways in which immunity is a theoretical or philosophical issue as much as it is practical, because of 'the marked pessimism developed during the century just closed', which was 'in large measure prompted by the dread of disease and premature death'.[134] In *The Nature of Man*, this dread gets connected to a death drive that comes to the fore 'after a normal life and an old age healthy and prolonged', and he cannot resist connecting the biological drive to the more philosophical and psychoanalytical questions it entails. 'The pessimistic school has often spoken of death as the true goal of human life', Metchnikoff notes, citing philosophers (Schopenhauer) and poets (Baudelaire) as well as biologists (Weismann) in support of this claim. 'The normal end, coming after the appearance of the instinct of death, may truly be regarded as the ultimate goal of human existence.' He finishes the book by commenting on the need to develop a more sophisticated analysis of this drive, in the form of a *thanatology*.[135] The proposed thanatology gets developed in *The Prolongation of Life*, published a few years later, in 1907, and just prior to Spielrein's own work which would lead her to the idea of the death drive.[136] It is perhaps not unreasonable to point out that Metchnikoff attempted suicide twice (in 1873 and then again in 1880)

133 Metchnikoff, *Nature*, 281–82

134 Élie Metchnikoff, *Immunity in Infective Diseases* (1901), trans. Francis G. Binnie (Cambridge: Cambridge University Press, 1905), 1.

135 Metchnikoff, *Nature*, 283, 288, 298.

136 Metchnikoff, *Prolongation*, 126–31, 300, 327.

and, according to his wife, carried on thinking about it until well into his old age.[137]

John Kerr suggests that although 'Metchnikoff goes uncited in Freud's work . . . the latent polemical agenda appears to be remarkably similar'.[138] Metchnikoff's work was widely available in several languages, though it may be that Freud gleaned his ideas about Metchnikoff through Spielrein.[139] Freud's initial hesitancy about the idea of a death drive – he later recalls his own 'defensive attitude when the idea of an instinct of destruction first emerged in psycho-analytic literature, and how long it took before I became receptive to it'[140] – may have led to a certain forgetfulness about sources. Either way, as Derrida once suggested, everything that Freud thought about the death drive really should have led him to ask other questions (which is why Derrida himself returns time and again to the drive, from *Writing and Difference* in 1967 through to his final deconstruction of autoimmunity). But what questions?

One question might be the one we have seen asked by Leader and Corfield: If the death drive is an instinct for destruction turned against the subject's own person, a tendency to self-destruction and thus a drive towards the death of the Self, then is this not the nature of the autoimmune disease? The death drive appears on the one hand then to be a psychological phenomenon but on the other hand appears to be *literally in our cells*.

Cell death had, in fact, been widely studied and discussed throughout the nineteenth century cell theory discussed in the previous chapter. Virchow, for example, writes of necrosis, mortification and degeneration and uses the term 'necrobiosis' to describe '*death* brought on by (altered) *life* – a spontaneous wearing out of living parts – the destruction and annihilation consequent upon life – natural as opposed to violent death (mortification)'.[141] He also uses the term 'degeneration'.

137 Olga Metchnikoff, *Life of Elie Metchnikoff, 1845–1916* (London: Constable and Co., 1921), 77, 80–81, 104, 236.

138 John Kerr, *A Most Dangerous Method: The Story of Jung, Freud, and Sabina Spielrein* (New York: Alfred A. Knopf, 1993), 499.

139 Todd Dufresne, *Tales from the Freudian Crypt: The Death Drive in Text and Context* (Stanford, CA: Stanford University Press, 2000), 22.

140 Freud, *Civilization*, 120.

141 Rudolf Virchow, *Cellular Pathology as Based on Physiological and Pathological Histology: Twenty Lectures* (1858), trans. Frank Chance (London: John Churchill, 1860), 318.

Through the nineteenth century and into the twentieth century, cell death was understood either as part of the normal developmental process, with the death of the cell as a progressive phenomenon, or in terms of injury caused by forces external to the cell, an acute pathological and 'uncontrolled' death resulting from a violent cell injury, which eventually became known as *necrosis*. Much later, in the early 1970s and thus during immunology's 'golden age', John Kerr, Andrew Wyllie, and Alastair Currie suggested that normal cell death appears quite different from cell death caused by injury. When injured cells get inflamed and burst, they spill over into neighbouring cells and inflame the tissue; the process is uncontrolled. In contrast, when normal cells die, they shrink into a corpse, the remains of which are subject to phagocytosis; the process is a mode of *controlled cell death*. Kerr, Wyllie, and Currie sought to capture this death with the term *apoptosis*, a word adopted from the Greek term for the 'falling off' of petals from flowers or leaves from trees. They suggested that apoptosis takes place not only in the process of developmental cells, as in embryology, but in all cells and in all kinds of animals: it is a general mechanism of 'controlled cell deletion' or 'programmed cell death'. This is the basis for the major distinction that developed in the field between apoptosis and necrosis.[142] Despite the gentleness suggested by the 'falling off' of petals, however, apoptotic cell death is in fact truly *spectacular*. 'One moment a cell looks peaceful and happy, the next it enters a violent programme of cytoplasmic blebbing worthy of the death throes of an actor in a B-movie.'[143] The result is that most discussions of cell death involve an imaginative array of narratives and images concerning a systemic and systematic machinery of death: cell murder, cell death genes, cell death pathways, sacrifice, cellular fratricide, euthanasia, grim reapers, and cellular cannibalism. Most of all, because apoptosis appears to be a kind of programme for self-destruction operating within the system, the process is often described as a *controlled suicide programme* or a *cell suicide machinery*. In chapter 1 we considered the cell as biological machine and political machine. It turns out to also be a suicide machine.

142 J. F. R. Kerr et al., 'Apoptosis: A Basic Biological Phenomenon with Wide-Ranging Implications in Tissue Kinetics', *British Journal of Cancer*, 26 (1972), 239–57 (241).

143 William C. Earnshaw, 'A Cellular Poison Cupboard', *Nature*, 397 (1999), 387–89 (387).

Apoptosis is understood as contributing to many pathologies, including tumours, immunodeficiency, and autoimmune diseases. It is significant, I think, that the most thorough and creative treatment of this idea of a controlled suicide machinery has been undertaken by an immunologist, Jean Claude Ameisen. In *La sculpture du vivant: le suicide cellulaire ou la mort créatrice* (1999) Ameisen treats cellular self-destruction as the heart of the living organism. The very notion of programmed cell death understands the initiation of the cell's 'execution', and hence the decision to live or die, to depend on the cell itself. Death is prescribed in the very heart of the cell, as a potential to be realized. For Ameisen, this has been present from the very origin of the very first cell. In contrast to those who posit in the history of life a primordial time in which each living cell possessed the promise of immortality, Ameisen offers a different scenario entirely, one which sees 'the ability to self-destruct in the heart of the very first cells'. But he also argues that the process can be seen in embryonic development. In the sanctuary of the womb, largely protected from disease and accident, 'mysterious events' occur in the embryo. Parts dissolve and disappear even as they are being built, organs are shaped, fingers are sculpted by having spaces created between them. Embryonic apoptosis is like 'the carving of a sculpture'. But this active sculpting means that the cells that compose the tissues and organs of the embryo are the site of death on a massive scale. In effect, death is at the heart of the process of active self-organization. The same tools that creatively 'sculpt the face of cellular life' also possess the potential to 'sculpt the face of death'. The cell simply stops affirming its belonging to the community of the living. Apoptosis is thus a feature of cellular life and an indicator of the 'shifting frontier . . . between the kingdom of life and the kingdom of death'. Indeed, Ameisen suggests that what this process reveals is that 'life cannot be easily separated from death'. Life is in fact *inhabited* by death.[144] Life death.

The information, signals, and communication that lead to the death – the *programming* of the death – mean that the death being sculpted is *suicidal*. Sensitive to the 'ambiguities' surrounding the notion of suicide, Ameisen insists that recognizing 'the ability to self-destruct in the heart of the very first cells' means that rather than 'imagining the existence of

144 Jean Claude Ameisen, *La sculpture du vivant: Le suicide cellulaire ou la mort créatrice* (Paris: Seuil, 1999), 38, 40, 51, 67, 69, 104, 312, 316.

a universe before suicide', we should recognize instead that suicide has been with us 'from the birth of life'. It is the same cellular mechanisms of execution and protection, the architects of cellular survival and duplication, that control the suicide.[145] The cells of the embryo are ready to destroy themselves from the beginning. A self-destructive but genetically controlled and programmed cell-suicide machinery means death is the default option for the cell.[146] From the birth of each cell, the cell is only an instant away from death. Only the body's protective system allows the cell to continue.[147] Suicide is always ready to be triggered.

All told then, the cell, seen by many as the protagonist of life, turns out in the end to be the protagonist of suicidal death. 'In the animal world, death by suicide is the norm', observes Martin Raff; suicide is 'the inevitable complement to cell division'.[148] The drive towards death is inside our very cellular being; the cell exhibits a death drive. If 'biology has . . . entered an existentialist phase', as one expert on cell death puts it, this is no doubt because biology has stumbled upon something important: 'Suicide determines self.'[149] You are, literally, a suicide machine.

To the extent that cells can be said to constitute a 'social' organism in the manner outlined by cell theory, this suicidal death is *socially controlled*. In apoptosis, a dying cell loses its adherence for neighbouring cells and quickly becomes subject to phagocytosis. The rapid, solitary and incident-free death usually causes no injury, inflammation or scarring, as surrounding cells fill the space left free by the dead cells and leave no trace of the work of self-destruction.[150] Communication comes in from *the System* instructing the cell to die, and the cell follows orders, shutting down its own metabolism and awaiting disposal. That is, as a universal feature inherent to multicellular organisms, apoptosis relies on cells being so committed to the stability of the organism that they are willing to sacrifice themselves by actively carrying out the process of cell

145 Ameisen, *La sculpture*, 67, 312, 316.

146 Jean Claude Ameisen, 'The Origin of Programmed Cell Death', *Science*, vol. 272, no. 5266 (1996), 1278–79.

147 Ameisen, *La sculpture*, 102–04, 285.

148 Martin C. Raff, 'Death Wish', *Sciences* (July–August 1996), 36–40 (36).

149 Respectively: Gerry Melino, 'The Meaning of Death', *Cell Death and Differentiation*, vol. 9, no. 4 (2002), 347–48 (347); J. P. Medema and A. K. Simon, 'Suicide Determines Self', *Cell Death and Differentiation*, vol. 9, no. 4 (2002), 364–66.

150 Ameisen, *La sculpture*, 57.

death. Cells appear to conform to the 'social' expectations demanded of them by their body politic. 'Cellular neighborhoods seem to be more conformist than any middle-class suburb', observes Raff. A cell seems to need signals from other cells in the System to proliferate and survive, and 'in their absence the cell kills itself by activating an intrinsic suicide programme'. In other words, cell survival and cell death appear subject to the equivalent of some kind of 'social control'.[151] Ameisen points out that the permanent coupling of the fate of each cell to the nature of the interactions it can establish with the collective body to which it belongs represents one dimension of the social control of cell survival and cell death. If, however, we take into account what we have observed in this chapter and the previous one, that 'each individual cell may be considered as a complex entity, a "society" by itself, a mingling of heterogeneous organelles and components that behaves as a whole', then what we are talking about is a kind of collective self-destruction of the cell as a society, driven by signals originating from inside the cell-society itself.[152]

This complicates our previous discussions, since we have noted that much of the thinking around the cell likes to think of it in terms of sovereign power, with those cells working within the immune process as committed to the security of the body politic. Yet it now appears that this sovereign power and security agent par excellence is a protagonist of death. You fight for and exercise the right to self-defence and security, but also kill yourself. This is the very reason why Ameisen connects the cellular suicide machinery to autoimmune disease. For Ameisen, any lymphocyte incapable of interacting with its own body or Self, and hence unable to perform the role of defending the body, will after the birth of the child be incapable of interacting with the general activities of the cells guarding the body. In effect, such lymphocytes are no longer able to protect the body.

> Any lymphocyte whose receptor interacts too well with one of the assemblages that constitute the self, risks attacking the very body to which it belongs and destroying tissues or an organ. This is a

151 Raff, 'Death Wish', 39; Martin C. Raff, 'Social Controls on Cell Survival and Cell Death', *Nature*, 356 (1992), 397–400 (397).

152 Jean Claude Ameisen, 'On the Origin, Evolution, and Nature of Programmed Cell Death: A Timeline of Four Billion Years', *Cell Death and Differentiation*, vol. 9, no. 4 (2002), 367–93 (367).

> lymphocyte that reveals the potentially dangerous nature of the lymphocyte, provoking a disease we call 'autoimmune'.[153]

Unable to distinguish Self from non-Self, such lymphocytes survive by turning on the body itself and producing autoimmune diseases. The organism itself would ordinarily destroy such lymphocytes through a signal that suicide is their next step; in effect, the suicide of the cells is part and parcel of the immune process. When this fails, the immune system engages in a mortal combat not with something *other*, not something *nonself*, not some *external enemy* in the form of a microbe or infectious disease, but *with its own body*.

> The coexistence of our body and our immune system is a dynamic balance, a permanent fight . . . The maintenance of our integrity and identity is the result of a complex relation of forces, an armed peace interrupted by sudden fighting.[154]

Once again, then, 'immunity' points to a combat between the body and its own security. It points to the body's security turning against the body, security's violence turned inward.

The death drive has long been a concept through which critical theorists have sought to think with Freud. During the 1950s and onwards, critical theorists began to use the death drive to think in different ways about history and repression, exploring the political implications of Freud's argument. The Frankfurt School in particular, in varied ways and to varying degrees, thought that Freud clearly foresaw the rise and nature of destructive forces such as fascism, even if he only foresaw them through the lens of the psychological. After all, the question of how it is that human beings can be so self-destructive has an obvious social application. What if the death drive is manifested in history as well as the body? If the death drive reveals part of our truth as cellular selves and systems, might it also reveal some truth about our *social system* and the *sovereign power* which presides over it? The problem posed for society and culture by such questions plagued Freud until the end, as he moved from implicit to far more explicitly stated political

153 Ameisen, *La sculpture*, 82.
154 Ameisen, *La sculpture*, 222.

ideas. It has long been easy to read *Beyond the Pleasure Principle* as a product of but also a reflection on a period of mass destruction and violence in European civilization, and in the decade between the publication of that book in 1920 and *Civilization and Its Discontents* in 1930, Freud increasingly uses the destructiveness of the death drive to grapple with the destruction of the lives of others and the kind of violence carried out by states. In 'The Economic Problem of Masochism' (1924) he suggests that a *cultural suppression of the death drive* 'holds back a large part of the subject's destructive instinctual components from being exercised in life',[155] and in *Civilization and Its Discontents* he writes that 'the fateful question for the human species seems to me to be whether and to what extent their cultural development will succeed in mastering the disturbance of their communal life by the human drive of aggression and self-destruction'.[156] The 'overpowering' and 'erotic mastery' over an object is connected to a drive for mastery (*Bemächtigungstrieb*) more generally. Hence, *Civilization and Its Discontents* is far more concerned with an aggressive destructiveness in the death drive. Civilization tends towards both *sovereign mastery* and *self-destruction*,[157] a line of thought he continues in his exchange of letters with Albert Einstein in 1932 on the question of 'Why War?' But Freud ultimately lacked a social theory dynamic enough to do much more with these ideas; such is the task of critical theory.

Might the suicidal death drive operate not only in cells and selves but also in systems and even the System? Might the vulnerability of the Self be writ large in the vulnerability of the System? Is not the *fear of breakdown* a political as well as personal fear? This would then help us push further the idea of the autoimmune disease as terror: a terror not only of one's own suicidal death drive but also the suicidal death drive of the security state. We will pursue this towards the end of chapter 5 and then into chapter 6. To get there, however, we need to engage in a rather long detour through the immune system *as a system*, and that detour is a long one since it begs a yet broader question: What is this thing called System?

155 Freud, 'Economic Problem', 170.
156 Freud, *Civilization*, 145.
157 Freud, *Civilization*, 121–22.

3

Imagining an Immune System: Politics of Systems I

> *And if the system in question were the collective as such? What relations do we really have with each other? How do we live together? What really is this system that collapses at the slightest noise?*
>
> Michel Serres, *The Parasite* (1980)

What do we imagine when we imagine an immune *system*? What do we imagine when we imagine any system? What does it mean to be governed by systems, through systems, in systems or by *the System*? How do we even begin to live in an age of system? What *is* a system?

The topic of the annual *Cold Spring Harbor Symposia on Quantitative Biology* in 1967 was 'antibodies'. Many of the other *Symposia* had a profound impact in their fields and this one was no different, quickly coming to be viewed as a watershed moment in immunology. In his history of immunology, Arthur Silverstein goes so far as to suggest that from this point 'all of the important conceptual questions of the past 80 years appeared for the first time to be answerable'.[1] We will return to the *Symposia* at various points below, but I mention it here simply because of its timing in conceptual history, for it was around 1967 that the idea of an 'immune system' was invented. Anne Marie Moulin's extensive survey of the titles and tables of contents of the main books

1 Arthur M. Silverstein, *A History of Immunology* (San Diego, CA: Academic Press, 1989), 143.

published on immunity in the 1950s reveals no hint of an 'immune system' as a key idea or framework in that period.[2] From the mid-1960s, however, in the context of the 'new immunology' that was emerging in the middle of that decade, the idea of an 'immune system' began to appear. Robert A. Good, a leading American researcher, began to show that the lymphocytes integrated the whole process (some went further and spoke of a 'dictatorship of the lymphocyte', as leading Soviet immunologist Rem Petrov put it). Good published an article in 1967 on the 'systematic' nature of the action of lymphocytes and called it 'Disorders of the Immune System'. The idea of an immune system then becomes a trend in the early 1970s (the period during which immunology was becoming properly 'institutionalised' as a discipline). In his closing speech at the 1967 *Symposia*, Niels Jerne sought to reconcile those who believed that immunity depends on antibodies and those who believed it depends more on specialized cells, and in so doing he used the word 'system' to capture both. He then started to describe lymphocytes as the immune system and eventually presented a popular outline of the immune process in the article published in *Scientific American* in 1973, already cited in chapter 1, with the title 'The Immune System'.[3] The British Library catalogue lists the first book with 'immune system' in the title as the one based on a Symposium on Molecular Biology held at UCLA in 1974 and published later that year, edited by Eli Sercarz, Alan Williamson and C. Fred Fox (*The Immune System: Genes, Receptors, Signals*, 1974). The year later sees the publication of both M. J. Hobart's *The Immune System: A Course on the Molecular and Cellular Basis of Immunity* and a volume edited by Erwin Neter, *The Immune System and Infectious Diseases*, based on yet another international conference on immunology in 1974, this time in Buffalo, New

2 Anne-Marie Moulin, *Le dernier langage de la médecine: Histoire de l'immunologie de Pasteur au sida* (Paris: Presses Universitaires de France, 1991), 341; Anne-Marie Moulin, 'Immunology Old and New: The Beginning and the End', in Pauline M. H. Mazumdar (ed.), *Immunology 1930–1980: Essays on the History of Immunology* (Toronto: Wall and Thompson, 1989), 293.

3 Robert A. Good, 'Disorders of the Immune System', *Hospital Practice*, vol. 2, no. 1 (1967), 38–53; Rem Petrov, *Me or Not Me: Immunological Mobiles* (1983), trans. G. Yu. Degtyaryova (Moscow: Mir Publishers, 1987), 99; Niels Kaj Jerne, 'Summary: Waiting for the End', *Cold Spring Harbor Symposia on Quantitative Biology*, vol. 36 (1967), 591–603; Niels Kaj Jerne, 'The Immune System', *Scientific American*, vol. 229, no. 1 (July 1973), 52–60.

York. The number of publications with 'immune system' in the title rises exponentially through the 1970s and thereafter. During the same period, the content of major textbooks would also shift to reflect the rise of 'system'. For example, the first edition of Stewart Sell's, *Immunology, Immunopathology and Immunity* (1972), which would become a leading textbook in the field for decades, begins on the question of immunity, immune reactivity, immune mechanisms, and immune deficiency, but has nothing to say about the 'immune system' other than a brief mention of the 'lymphoid system'. In contrast, later editions open by defining immunology as 'the study of the *system* through which we identify infectious agents as different from ourselves and defend or protect ourselves against their damaging effects' and include chapters based on the idea of system.

In effect, the idea of an immune *system* gradually became a core concept in immunology's theoretical framework and the popular imagination. Indeed, as regards the discussion in the previous chapter, the idea of an immune *System* came to be a defining aspect of the idea of an immune *Self*.

This, in turn, contributed to the expansion of immunology in the 1970s. Following the very first International Congress of Immunology in August 1971 – 'Independence Day for immunology' (Rem Petrov) – more and more immunology journals entered the field, with some forty-seven new journals launched through the 1970s and into the 1980s. Petrov reports a similar development for the same period in the Soviet Union. In a way, the very idea that there is such a thing as the immune *system* helped create modern immunology, not least because 'system' implies the need for a body of disciplinary knowledge to study it.[4]

So, the idea of the *immune system* was born, nurtured and rose to maturity between roughly 1967 and 1973. Why?

From within immunology, the answer tends to run as follows: the emergence of the idea of the 'immune system' captures the fact that the 1960s was a significant and productive decade for immunological research. The period witnessed the emergence of immunology out of the

4 Petrov, *Me or Not Me*, 260, 352; Anne Marie Moulin, 'The Immune System: A Key Concept for the History of Immunology', *History and Philosophy of the Life Sciences*, vol. 11 (1989), 221–36; Pauline M. H. Mazumdar, 'Immunology', in Kenneth F. Kiple (ed.), *The Cambridge World History of Human Disease* (Cambridge: Cambridge University Press, 1993), 137.

'dark ages' and into a new period of flourishing and creative work, which was then consolidated in the 1970s. So, one answer as to why 'immune system' came to the fore is that the term coincides with the establishment of the disciplinary field of immunology in that period. Yet maybe there is more to be said. For, although the notion of system is now so established that it is hard for us to believe that it has a history, the fact that it does have a history, and that this history is philosophically relevant and politically revealing, means that we might wish to situate the 'immune system' within it. What we therefore need to do is to situate the idea of the immune system within the much wider intellectual context of what we might call *the age of system*.

Our world is dominated by the idea of System: *a* system, system*s*, systems *of* systems, *the System*. We speak of markets and politics, production and consumption, migration and territoriality, finance and pensions, philosophy and culture, the telephone and the television, the internet and the satellite, roads and railways, air traffic control and shipping lanes, and virtually everything else, in terms of systems. The list could go on and include the body and its various 'systems'. Try saying 'nervous' without saying 'system'. Try saying 'immune' without saying 'system'. The word 'system' seems to flow naturally and seems able to attach itself to everything. The idea of the immune system thereby points to the ways in which 'system' dominates our cultural imagination and our political thinking. This domination comes at a price, as we learn from a young age that there is a logic to systems and to the System, rooted in the lived reality of rules, hierarchies, processes, and regulations through which systems operate, but also the lived reality of the systems within us that we are told constitute us as bodies. This ubiquity of system makes 'system' feel 'contentless', as Robert Flood suggests. And yet 'system' is packed with assumptions and tropes which are deeply political.[5]

To understand what might hang on the fact that the process of immunity is presented to us as a system, then, and to understand what might hang on this politically, we must turn to the growth of systems theory in general. By the time that the immune system was being born, systems theory had become one of the most pervasive ideas within intellectual

5 Robert L. Flood, *Liberating Systems Theory* (New York: Plenum Press, 1990), 71. Also Robert Lilienfeld, *The Rise of Systems Theory: An Ideological Analysis* (New York: John Wiley, 1978).

discourse. In 1966, leading systems theorist Stafford Beer could write that 'the idea of *system* has emerged as all-important', and in 1968 another leading system theorist, Ludwig von Bertalanffy, could write that 'if someone were to analyse current notions and fashionable catch-words he would find "systems" high on the list'.[6] And, in 1967, just a few months before the *Cold Spring Harbor Symposia*, the Special Subcommittee on the Utilization of Scientific Manpower in the US suggested that 'the so-called "systems approach"' and 'systems management techniques' would be key to the future of government. The Likert management system approach had taken hold within management theory, following the work of Rensis Likert, and based on the idea that there are essentially four management systems: Exploitative Authoritative (System I), Benevolent Authoritative (System II), Consultative (System III), and Participative (System IV). The Special Committee on Manpower held that systems theory offers a set of tools for understanding and managing urban transportation, urban housing, urban renewal, urban development, education, health, welfare, and law enforcement.[7]

At the very least, the immune 'system' needs to be understood within this wider context. Doing so will be the focus of this chapter. The two chapters to follow, however, will broaden the range further by pointing out that twentieth-century systems theory itself emerges from a long history of thinking about systems. To put the argument in its broadest terms: the idea of something constituting a system had existed since at least the early seventeenth century, often being used to describe the movement of objects and bodies. System could be found in the natural sciences (such as Galileo's *Dialogue concerning the Two Chief World Systems* in 1632 and book III of Newton's *Principia* in 1687, called *The System of the World*), in political philosophy (such as Hobbes's 'system' of bodies and the body politic in *Leviathan*), in political economy (such as the mercantile system or Adam Smith's system of liberty), and in biology (such as the circulation system). This flurry of work in the

6 Stafford Beer, *Decision and Control: The Meaning of Operational Research and Management Cybernetics* (London: John Wiley and Sons, 1966), 241; Ludwig von Bertalanffy, *General Systems Theory: Foundations, Development, Applications* (1968) (London: Allen Lane, 1971), 1.

7 *Hearings before the Special Subcommittee on the Utilization of Scientific Manpower*, Ninetieth Congress, First Session, 24–27 January and 29–30 March 1967, 18–19.

seventeenth and then eighteenth centuries was the first period in a three-hundred-year frenzy of systems (to tweak an observation of Theodor Adorno's). We will return to the earlier period of this frenzy in the following chapter, but in this chapter, we focus on the later part of the frenzy, the second half of the twentieth century, when something self-consciously describing itself as *systems theory* came to dominate intellectual life, becoming central to the biological and social sciences and, concomitantly, to the political administration of capital. And this centrality connects the politics of immunity to the wider politics of modernity. To unpack this, however, will require covering a lot of ground, moving back and forth across the three-hundred-year frenzy, in order to unravel not only what it means to imagine *immunity* as a system but also what it means to imagine *anything* as a system, including other images through which the body and security are imagined. The questions asked at the beginning of this chapter spiral out into a series of questions about what it means to imagine organization and organism, illness and health, order and chaos, terms that are widely used to understand the Self and the System, the body and the body politic.

Remaining in this chapter firmly in the twentieth and twenty-first centuries, I will first connect the immune system to the idea of systems in general, to systems theory and hence the politics of systems theory. This will expand the politics of immunity that we have been developing in previous chapters. I then use a critique of Niklas Luhmann's sociological systems theory as a springboard back, in chapter 4, to the earlier period of the three-hundred-year frenzy, in order to unearth some of the deep-rooted assumptions that operate when we think about system, including harmony and balance, internal laws, and self-regulation. In chapter 5, I ask a very different kind of question, about the nervous system as a political issue, in order to bring the politics of immunity back into the picture through the question of nerves, the nervous Self and the nervous state. All of which is designed to connect us back through a circuitous route to the idea of security once more and will allow us to then pick up on some of the issues left to one side in chapter 2.

This chapter and the two that follow it are therefore dealing with two broad issues. On the one hand, the historical rise of the idea of self-policing systems encouraged the belief that the System works best when the humans within it recognize themselves as part of the system and believe in it to the extent that they accept their own embeddedness in it.

They come to believe that the System is self-regulating and hence cannot be interfered with, never mind controlled. The formula runs as follows: You are at liberty to question the System, but never forget that your security and salvation lies within it. This has the effect of making it appear as though power does not reside in human beings. The System seeks to undermine solidarity and replace human bonds with the process of the System itself. The System wants to align the Subject with the System in such a way that the former is subsumed under the latter; it wants to be *the Subject* itself.

On the other hand, one of the ways in which we imagine systems is through the lens of control and the extent of our powerlessness when systems go wrong. Is the system beyond control? Is it out of control? Is it becoming chaotic? There is nothing that is 'naturally' order or chaos, as we shall discuss in the next chapter, and for the most part, when what appears to be a system works well (for us) or looks orderly (to us) then it feels like one of nature's well-ordered systems. When it appears to have gone wrong, it appears the epitome of chaos. We are faced day after day with processes and institutions of varying degrees of power that are presented to us as systems, systems so complex that they often appear to us and feel to us as though they are chaos but which we are told time and again are orderly; unless, that is, we are told that the system is in danger of collapsing into chaos. Either way, the story remains the same: *order must be restored.*

The Mystique of the System

Twentieth-century systems theory has deep roots in Ludwig von Bertalanffy's work on biological organisms in the late 1920s and early 1930s, in which he sought to show how a system maintains its stability when in constant exchange with its environment. Systems theory then received a major push through the RAND Corporation in the late 1940s.[8] RAND was formed at the end of World War II to develop a 'science of war' by calculating the efficiency of various weapons

8 Ludwig von Bertalanffy, *Modern Theories of Development: An Introduction to Theoretical Biology* (1928), trans. J. H. Woodger (New York: Harper Torchbooks, 1966), 7, 177–80, 187.

systems and developing mathematical models to assess conflicts, potential conflicts, and the cost-effectiveness of various weapons. RAND quickly branched out into the new and burgeoning field of 'national security', applying games theory, mathematical modelling, logistical simulations, and probability theory to the field, all of which became core to systems analysis. The extent to which modern statecraft and the political administration of capitalist modernity operates through modes of quantification, information, codification and standardization can be seen operating here, in the origin of systems theory.

Although systems analysis emerged from wartime operations research, it differed in both its creativity and its claim to universality. Here is RAND's own gloss on this:

> The term 'systems analysis' generally includes broader and more difficult problems than those traditionally covered by the terms 'operations analysis' or 'operations research'. The latter terms are often applied to studies of existing systems, or to some tractable facets of a single system, designed to uncover more effective ways to perform specific missions. Systems analysis, on the other hand, refers to the far more complex problem of choice among alternative future systems, where the degrees of freedom and the uncertainties are large, where the difficulty lies as much in deciding what ought to be done as in how to do it.[9]

What is to be done? The question was designed to elicit creative answers. Part of this requirement for creativity lay in the changing nature of weaponry. Operations research analysts during World War II were working with real data drawn from real combat situations, but the newer systems analysis had to be based heavily on speculative thinking, not least because the ultimate weapon of the period was now atomic. They were primarily engaged in analysing weapons of the future, which they did through technological games and assumptions about the security of capitalist states. This involved elaborate modelling and game-theoretical calculations to estimate the accuracy of weapons and likely battle losses, which were then projected onto scenarios of future war. Systems analysts were, in effect, charged with creating imaginary scenarios and 'thinking the unthinkable'.

9 RAND Corporation, *The RAND Corporation: The First Fifteen Years* (Santa Monica, CA: RAND, 1963), 27–28.

Systems analysis came to form the centrepiece of RAND's methodological innovations through the early Cold War, with RAND research becoming almost synonymous with systems analysis. The approach received a further push in 1961, through President Kennedy's appointment of Robert McNamara as secretary of defense, who established an Office of Systems Analysis within the Department of Defense, and brought to staff it a team of 'whiz kids' from RAND and Ford Motor Company. McNamara then remained secretary of defense through most of the Vietnam War before becoming president of the World Bank. Given the task of bringing to the management of 'national security' the methods he had used in industry, especially at Ford, McNamara was core to developing thinking about security as a form of political administration and through systems of systems.[10] Hence the 'revolution in military affairs' announced by General Westmoreland in 1969, known for its arguments about an 'electronic battlefield', was in fact a systems revolution, envisaging 'systems that heretofore have been unknown'.[11] 'System' became a dream of total organization: not just military systems, weapons systems, defence systems, and early-warning systems, but also control systems, expert systems, production systems, communication systems, integrated systems, centralised systems. This extended to biological as well as political systems, as we shall see.

In 1950, the US Air Defense Systems Engineering Committee published a report which included an analysis of the committee's name. The report declares itself 'reasonably certain of the meaning of the words "Air Defense"', but 'system' was said to require explanation.

> Let us look into the meaning of the word 'system'. The word itself is very general; *Webster's* [*Dictionary*] gives fifteen different meanings for 'system'. There are, for instance: the 'solar system' and the 'nervous system', in which the word pertains to special arrangements of matter;

10 RAND, *First Fifteen Years*, 32, 41–46; David Hounshell, 'The Cold War, RAND, and the Generation of Knowledge, 1946–1962', *Historical Studies in the Physical and Biological Sciences*, vol. 27, no. 2 (1997), 237–67 (245); Paul N. Edwards, *The Closed World: Computers and the Politics of Discourse in Cold War America* (Cambridge, MA: MIT, 1996), 113, 125–29.

11 'Address by General W. C. Westmoreland' to the Association of the US Army, Washington, DC, 14 October 1969, 'Appendix A' in Paul Dickson, *The Electronic Battlefield* (Bloomington: Indiana University Press, 1976), 219.

there are also systems of philosophy, systems for winning with horses, and political systems; there are the isolated systems of thermodynamics, the New York Central System and various zoological systems.

From this, the report makes quite a leap:

The Air Defense System has points in common with many of these different kinds of systems. But it is also a member of a particular category of systems: the category of organisms. This word, still according to *Webster*, means 'a structure composed of distinct parts so constituted that the functioning of the parts and their relation to one another is governed by their relation to the whole.' The stress is not only on pattern and arrangement, but on these also as determined by function, an attribute desired in the Air Defense System.

This, in turn, requires further explanation:

What then are organisms? They are of three kinds: animate organisms which comprise animals and groups of animals, including men; partly animate organisms which involve animals together with inanimate devices such as is the Air Defense System; and inanimate organisms such as vending machines. All these organisms possess in common sensory components, communication facilities, data analysing devices, centers of judgement, directors of action, and effectors, or executing agencies.

Organisms also have the power of development and growth and the possibility of decay and death. Moreover, they require to be supplied with material. Since armies are organisms, it is not surprising that these functions parallel the divisions of a general staff . . .

Nearly all organisms can sense not only the outside world, but also their own activities. It is often the case that some of the component parts of a complicated organism are themselves complete organisms.

It is the function of an organism to interact with and alter the activities of other organisms, generally to achieve some defined purpose, but not always with any particular other organism or group of organisms.[12]

12 Air Defense System Engineering Committee [ADSEC], *The Air Defense System*, final report, 24 October 1950.

To think of war as a system, then, is to think of the system as an animate techno-organism.

Much of the impetus behind this systems turn came from thinking about war technologies such as cockpits, missiles, bomb control units, fire control systems, anti-aircraft technology, and, most generally of all, what Peter Galison calls a 'mechanized Enemy Other'.[13] Between 1942 and 1960, a series of conferences took place named after their sponsoring institution, the John Macy Foundation, bringing together a multi-disciplinary range of experts and researchers, including Norbert Wiener, Claude Shannon, John von Neumann, Warren McCulloch, Gregory Bateson, Margaret Mead, Arturo Rosenblueth, Julian Bigelow, Paul Lazarsfeld, William Ross Ashby, and William Grey Walter. The 'Macy Conferences' sought to consider the possibility of a unified science integrating biological and technical ideas about control; a similar series of conferences took place in the UK from 1949 to 1958 under the auspices of the Ratio Club. To succeed in this endeavour, the participants combined a theory of information, a model of neural functioning that treated neurons as information-processing systems, and computers working with binary code. Wiener and von Neumann were also working on developing computing machines powerful enough to perform the calculations required for the Manhattan Project to produce the atomic bomb, under the general direction of Robert Oppenheimer.

Those meeting at the Macy Conferences conceived of their work as a completely new way of looking at human beings, a 'cybernetic' approach rooted in a general theory of 'control and communication in the animal and the machine', as suggested in the subtitle of Wiener's *Cybernetics* (first published in 1948 and then in a longer second edition in 1961). On this view, in contrast to the age of clockwork mechanisms (seventeenth and early eighteenth centuries) and the age of the steam engine (later eighteenth and nineteenth centuries), the present age was understood as an age of communication and control. This logic of communication and control runs through animal and machine and reaches from the cellular and molecular level of the body through to social systems themselves.

13 Norbert Wiener, *Cybernetics; or, Control and Communication in the Animal and the Machine*, 2nd edn. (Cambridge, MA: MIT Press, 1961), 3, 4, 14, 15; Peter Galison, 'The Ontology of the Enemy: Norbert Wiener and the Cybernetic Vision', *Critical Inquiry*, vol. 21, no. 1 (1994), 228–66 (231).

The mechanisms through which machines multiply are the same as the physiological processes undergone by living organisms. When machines are thought of as systems, they can be thought of as organisms; when organisms are thought of as systems, they can be thought of as machines. 'Organism' is thus not so much a biological term as it is a term which connotes organization and order. This vision of human-machine integration was thought to require new modes of analysis.

> The newer study of automata, whether in the metal or in the flesh, is a branch of communication engineering, and its cardinal notions are those of the message, amount of disturbance or 'noise' – a term taken over from the telephone engineer – quantity of information, coding technique, and so on.[14]

Machines are, in effect, organic systems, but organic systems such as human beings can also in turn be considered as machines.

> When human atoms are knit into an organization in which they are used . . . as cogs and levers and rods, it matters little that their raw material is flesh and blood. *What is used as an element in a machine, is in fact an element in the machine.*[15]

What matters is that they are all *systems*.

The idea that humans are like machines or even *are* machines has a long history in mechanical philosophies, as noted in the introduction. But many of those earlier debates were also driven by what might also distinguish humans from machines, whether that be called 'soul', 'animus', 'vitality', 'vital substance', 'spirit', 'penetrating force', or 'formative power'. The new ideas concerning the man-machine in post-war systems theory, emerging from a wider intellectual flurry of interest in technology – in works such as Siegfried Giedion's *Mechanization Takes Command* (1948), Jacques Ellul's *The Technological Society* (1954), and Gilbert Simondon's *On the Mode of Existence of Technical Objects* (1958) – do away with the distinction entirely, rendering obsolete the earlier

14 Wiener, *Cybernetics*, 41–42.

15 Norbert Wiener, *The Human Use of Human Beings: Cybernetics and Society* (1950) (Boston: Da Capo Press, 1954), 185.

concepts of man and machine along with key debates such as the distinction between the natural and the technological or the organic and inorganic. Systems thinking developed a theory encompassing all systems, drawing a 'functional equivalence' between, say, computers and the human brain or even, in a telling example, the prison and the steam engine.[16] 'Everything we learn of organisms leads us to conclude not merely that they are analogous to machines but that they *are* machines', wrote Warren McCulloch.[17]

In this vision, the critical test of whether a machine can 'think' should not succumb to the outmoded debate about whether or not the machine has (or is) a brain, because 'the brain is not a thinking machine'. Rather, the brain is an *acting* machine. The brain gets information and then acts on or with it. The important question is not what a thing *is* but what it *does*. This is a question concerning its qualities *as a system* and how it is locked into itself *as a system*.[18] Hence, to think of the 'mechanization' of the human is to really be thinking about its 'systematization', in the sense that its 'life' is its system and its system is its 'life'. How is the man-machine organized as a system? How is it controlled as a system? How does it adapt in the face of feedback entering the system? How does it function as a system within and around other systems? Living organisms are like automata in having inputs and outputs and thus being self-reproducing, and this is true of 'the System' itself.[19] Because human input is simply an input like any other input, feeding information into the System, human beings become technical components; technically significant components, perhaps, but not ontologically significant. Their role is 'maintaining the system that performs the work'.[20]

16 Stafford Beer, *Management Science: The Business Use of Operations Research* (London: Aldus Books, 1967), 146.

17 Warren S. McCulloch, '*Mysterium Iniquitatis* of Sinful Man Aspiring into the Face of God', in McCulloch, *Embodiments of Mind* (Cambridge, MA: MIT Press, 1965), 163.

18 W. Ross Ashby, *An Introduction to Cybernetics* (London: Chapman and Hall, 1957), 1; *Design for a Brain: The Origin of Adaptive Behaviour* (1952) (London: John Wiley, 2nd edn. 1960), 36, 40.

19 John von Neumann, 'The General and Logical Theory of Automata' (1948), in Lloyd A. Jeffress (ed.), *Cerebral Mechanisms in Behavior: The Hixon Symposium* (New York: Hafner, 1951), 2; John von Neumann, *Theory of Self-Reproducing Automata* (Urbana: University of Illinois Press, 1966), 71.

20 Herbert A. Simon, *The New Science of Management Decision* (New York: Harper and Row, 1960), 35, 38.

In a famous interview with *Der Spiegel* in 1966, Martin Heidegger was asked about what might be taking the place of philosophy. He replied: cybernetics. Such an answer was plausible only because cybernetics had become synonymous with systems theory. The Macy Conferences had in fact been conferences on systems, with titles such as 'Circular Causal Systems and Feedback Mechanisms in Biological and Social Systems', only becoming known as cybernetics conferences after the publication of Wiener's *Cybernetics* in 1948; groups such as the Systems Science and Cybernetics group were being formed; the *Journal of Cybernetics* would eventually become *Cybernetics and Systems: An International Journal*; and key figures such as Bertalanffy came to the view that cybernetics, as a general theory of systems, was in fact part of systems science in general.[21] For systems theory, what is at stake is less the kind of issues that came to interest those obsessed with the 'cyborg', a term that appeared in an article by Manfred E. Clynes and Nathan S. Klines in the September 1960 issue of *Astronautics*, and much more the ability to understand what Bertalanffy describes as the *world as organization*. The idea of organization applies from a single cell through to the universe as a whole, including its technological forms. Organization concerns the 'cell-body-personality-nature-culture-society system', observed Anthony F. C. Wallace in applying these ideas within anthropology.[22] Or, as we might say: from Cell to Self to Sovereign. Through a complex of techniques, technologies, practices, fictions, fantasies, and ideas, *system* came to permeate the workings of power and ideology, and *systems theory* sought to be a universal theory of everything: for systems theory, *system* = *life*, and the human world must be imagined and managed as a systematised world. Part of system's productive power lay in the idea that it could indeed have a universal applicability. 'A man, a people, or an epoch are among other things also systems.'[23] Death is

21 Bertalanffy, *General Systems*, 13, 15, 43, 96, 157–58; Gordon Pask, *An Approach to Cybernetics* (London: Hutchinson, 1961), 22–23; Heinz von Foerster and Berhnard Poerksen, *Understanding Systems: Conversations on Epistemology and Ethics* (2001), trans. Karen Leube (New York: Kluwer, 2002), 136; Claus Pias, 'The Age of Cybernetics', in Claus Pias (ed.), *Cybernetics: The Macy Conferences 1946–1953 – The Complete Transactions* (Zurich-Berlin: Diaphanes, 2015), 11–26.

22 Anthony F. C. Wallace, 'Revitalization Movements', *American Anthropologist*, vol. 58, no. 2 (1956), 264–81 (266).

23 Karl W. Deutsch, *The Nerves of Government: Models of Political Communication and Control* (New York: Free Press, 1963), 140.

simply the transformation of one system's material into another; 'when a living creature dies, it loses its "system-ness"'.[24]

We can examine this power and applicability through four different lenses – security, urban policy, public policy, and the world system – separated out here for purely analytical reasons in order to help situate the emergence and rise of the idea of the immune system.

We are now familiar with security systems and the security system. Prior to 1950, however, there were very few books published with 'security system' or 'system of security' in their titles. Yet the idea of 'national security' that emerged in the years following World War II very quickly came to be understood through the lens of systems theory. The British Library catalogue contains 152 books with 'security system' in the title. Of those, only fourteen were published prior to 1955, and those books were overwhelmingly concerned with social security, a term which had developed slightly earlier than 'national security'. The year 1955 is significant because it sees the publication of Charles P. Curtis's *The Oppenheimer Case: The Trial of a Security System*, following which one gets a flurry of publications with 'security system' in the title, after which the term becomes more and more commonplace.

The subtitle of Curtis's book is significant. Oppenheimer had been director of the laboratory at Los Alamos which developed the technology of the atomic bomb, yet a series of events led to Oppenheimer being brought before the Personnel Security Board at the Atomic Energy Commission (AEC). This was a period in which thousands of American citizens were being tested for their loyalty to the security state that was by then becoming so dominant, with many losing their jobs for being considered disloyal or not loyal enough. The exact details of the reasons why Oppenheimer had to appear before such a board are not relevant to us here, but the Oppenheimer case represents a watershed moment for national security policy, because although Oppenheimer's interrogators employ the concept 'national security', that very concept was increasingly supplanted by 'security system'. The board articulated the whole question of Oppenheimer's (dis)loyalty in terms of the security system and the idea that what was at stake was not simply loyalty to the state or

24 Donella H. Meadows, *Thinking in Systems: A Primer* (White River Junction, VT: Chelsea Green Publishing, 2008), 12.

the country but loyalty to that very security system. The board held that Oppenheimer's failure was his lack of loyalty to the system itself.

> The record of his actions reveals a frequent and deliberate disregard of those security regulations which restrict a man's associations . . . The requirement that a man in this position should relinquish the right to the complete freedom of association that would be his in other circumstances is altogether a reasonable and necessary requirement. The exact observance of this requirement is in all cases essential to the integrity of the security system.

Once the (security) system exists, it is *the System that matters*. This much is evident from the 'Statement' on the Oppenheimer case issued by the AEC on 29 June 1954. One section is titled 'Security in 1954' and deals with the need to construct a 'thoroughly articulated security system that will be effective in protecting strength and yet maintain the basic fabric of our liberties'. Such a system must be maintained, should not be challenged, and must itself be secured as a system.

To make this case, the AEC has recourse to the trope of the healthy body: a 'healthy security system' is one protected from any 'security risk', but the health of this system is under threat not only from outside dangers but also from within. 'The "security risk" concept has evolved in recent years as a part of our search for a security system which will add to the protection of the country.' This poses a problem, of course, since 'one inherent difficulty is that every human being is to some degree a security risk'. 'So long as there are normal human feelings like pain, or emotions like love of family, everyone is to some degree vulnerable to influence, and thus a potential risk in some degree to our . . . security system.' The system is under permanent siege. Once the security system exists as a system, individuals must recognize that they are a component of it, that they are an organic part of the overall organization of the system as a system and that this is the basis of their subservience within it and to it.

This is the aspect of the security system which the board held 'has had insufficient public attention'. What is this aspect? First, the system 'must involve a subordination of personal judgment'; second, the system 'must entail a wholehearted commitment to the preservation of the security system'; third, the system must ensure 'the avoidance of conduct tending to confuse or obstruct'; and fourth, more than anything, the system must prioritise '*the protection and support of the entire system itself*'.

> Those who stand within the security system *are not free to refuse their cooperation with the workings of the system*, much less to confuse or obstruct them . . . If this manner of cooperation is not forthcoming, *the security system itself, and therefore the interests of the United States which it protects, inevitably suffer* . . . To permit a man in a position of the highest trust to set himself above any of the laws of Security would be to *invite the destruction of the whole security system.*[25]

It has been said that Oppenheimer was 'curiously passive throughout the proceedings', even thanking the board for its patience and consideration despite the fact that the board had, in the name of security, completely humiliated him.[26] One explanation for his measured complicity with the process might be that as a 'specific intellectual' – that is, an intellectual who has 'a direct and localized relation to scientific knowledge and institutions' and who could bring into play 'his specific position in the order of knowledge' (to cite Foucault, whose example of the specific intellectual is, in fact, Oppenheimer) – he was hounded because of the very specific knowledge at his disposal.[27] Then again, the explanation for his measured complicity might be the fact that Oppenheimer, like his accusers, accepted the legitimacy of the security system itself. They accepted its legitimacy *as a system* and perhaps even as *the System*. His passivity and compliance are perhaps evidence of the System's success *as a system*. In this regard, the references to a 'healthy security system' become more pertinent as an early indicator of what would later become known as 'vital systems security': the idea that the security system has *vital* needs connects *security* with *life* itself. (In this context, the later stress in Colonel John Warden's *universal system model* on the organic nature of the system,

25 United States Atomic Energy Commission, *In the Matter of J. Robert Oppenheimer: Transcript of Hearing before Personnel Security Board and Texts of Principal Documents and Letters* (Cambridge, MA: MIT Press, 1971), 999, 1013, 1061; United States Atomic Energy Commission, *Statement by the Atomic Energy Commission*, 29 June 1954, emphasis added.

26 Allen Weinstein, 'The Symbolism of Subversion: Notes on Some Cold War Icons', *Journal of American Studies*, vol. 6, no. 2 (1972), 165–79 (174).

27 Michel Foucault, 'Truth and Power' (1976), in *Power/Knowledge: Selected Interviews and Other Writings 1972–1977* (Brighton: Harvester Press, 1980), 127–28.

discussed in the introduction, becomes even more significant, especially the key importance attached to the subsystem which protects it, namely the immune system.)

In December 1950, *Life* magazine ran a long presentation and discussion of what it called the 'Wiener defense plan' in which the problem of population dispersal during a nuclear attack became the solution to the problems of the city more generally. If, for example, a well-developed system of highways would enable the prompt dispersal of the population during war, would not such a system improve cities in general? The rhetorical question presupposed that we would all agree that highways are the key 'communication mechanism' in a city, which is itself 'primarily a communications center', Wiener suggested.[28] With the idea that the city could be envisaged as a system like all other systems, as organism and machine, so systems analysis could be brought to bear on the so-called 'urban crisis', a belief that intensified as the crisis became more and more of an issue through the 1960s. The key lay in the framing of the urban crisis as a national security crisis. Time and again, through the 1960s, one gets formulations along the lines that the guerrilla warfare on the city streets in the capitalist West was connected to the guerrilla warfare in the non-capitalist East; that the problems facing cities could be considered more lethal than the nuclear bomb; that national security now depended less on weaponry and more on managing the crisis in the cities; that disturbances in the cities were a form of civil war; that cities were a breeding ground for communism; that 'the enemy' could take many forms and assume many guises and that most of these appeared in the form of urban problems such as poverty, crime or drugs.

In this context, security experts were appointed as advisers to city governments, as the techniques and technologies developed under the rubric of the security system became the cornerstone of the political administration of American cities. The concepts of 'civil system' and 'urban system' were developed as police tools for the political administration of the security system as a whole, most obviously in the various 'war on X' programmes. Policies were implemented through the Community Renewal Program, itself a product of the spread of systems analysis. Behind such programmes was a body of university research

28 [No author], 'How U.S. Cities Can Prepare for Atomic War: M.I.T.', *Life*, 18 Dec. 1950, 77–86 (85).

emerging from units such as the 'Urban Systems Laboratory' (MIT), 'Systems Analysis and Urban Planning' (New School for Social Research), and 'Cybernetics and Urban Analysis' (University of Southern California).[29] Civil system, urban system, security system: each had become an analogue of the other, each reinforcing the idea of the social field as war zone and each an instantiation of the extent to which systems analysis had become a technique of police power. The broad objective of systems theory is to make *any* and *every* system operate to a model of efficiency laid down by the administrative forms of the state. In a period in which the City appeared to the ruling class to mirror the world in being increasingly ungovernable and in which workers and other political movements appeared more and more able and willing to challenge the system, so systems theory was to be deployed as a security measure.

For systems theory, civil society is a *social system* and can thus be studied, understood and administered like any other system. This should not be too surprising, given that the original systems thinkers within the military were often social scientists and, even more often, economists. Indeed, one of the fundamental propositions of systems analysis is that there is no distinct field of military knowledge as such, but rather that the military just happened to be the first real application of systems theory. In the words of RAND member Arthur E. Raymond, in a presentation of 'Project RAND' in 1947, the organization was concerned 'with systems and ways of doing things, rather than particular instrumentalities, particular weapons'. And, he added,

> we are concerned not merely with the physical aspects of these systems but with the human behavior side as well. Questions of psychology, of economics, of the various social sciences, so-called, are not omitted because we all feel that they are extremely important in the conduct of warfare.[30]

29 Jennifer S. Light, *From Warfare to Welfare: Defense Intellectuals and Urban Problems in Cold War America* (Baltimore: Johns Hopkins University Press, 2003), 51, 57.

30 Arthur E. Raymond, 'Presentation on Project RAND', RAND Document D-188, 13 August 1947, cited in David R. Jardini, 'Out of the Blue Yonder: The Transfer of Systems Thinking from the Pentagon to the Great Society, 1961–1965', in Agatha C. Hughes and Thomas P. Hughes (eds.), *Systems, Experts, and Computers* (Cambridge, MA: MIT Press, 2000), 315.

As security research incorporated more sophisticated socio-economic approaches, so the approach came to be applied more broadly to the social and economic policy fields, as public health specialists, educational researchers, sociologists, political scientists, lawyers, and a plethora of other specialists adopted systems thinking. Personnel from RAND, the Pentagon, security agencies, and security think tanks entered into the field of social reform programmes. By 1972, almost half of RAND's research was devoted to 'welfare' issues.[31] Yet 'welfare' was now so closely aligned with 'warfare' that they were understood together as a security system. When President Johnson declared a 'war on poverty' in 1964, he was thus doing much more than using a phrase with which we have since become awfully familiar. He was, in fact, facilitating the drive towards the diffusion of systems analysis into the policy agenda more generally, as every policy question came to be thought of as war – war on drugs, war on crime, war on poverty – and national security merged with social security. The logic of security was thereby stretched across the social field, and social policy was increasingly defined through the lens of systems analysis and in the name of security. If *the System* was to be defended, it was to be defended through the systems approach.

In this context, as academics, intellectuals, think tanks, and policy circles were eased and teased into thinking through systems, systems theory was used to signal the end of ideology. 'System' was intended to be a 'category for destroying categories',[32] and there is a sense in which systems theory wanted to be the transdisciplinary moment of the post-war era – although, along with much post-war thought, in seeking to perform this role it still proved itself to be a deeply ideological enterprise. One way this universal language was developed was through the internationalization of systems theory. The International Institute for Applied Systems Analysis (IIASA) was formally founded in 1972, following its very slow birth through the 1960s. The founding parties were the US, USSR, and ten other countries from across the Eastern and Western blocs. The Soviet Union already had something akin to 'systems

31 Ida R. Hoos, *Systems Analysis in Public Policy: A Critique* (Berkeley: University of California Press, 1972), 47–48; Roger E. Levien, 'RAND, IIASA, and the Conduct of Systems Analysis', in Hughes and Hughes (eds.), *Systems, Experts, and Computers*, 433–61.

32 Stafford Beer, *Cybernetics and Management*, 2nd edn. (London: English Universities Press, 1967), 27.

analysis' (*sistemnyi analiz*) in the form of Aleksandr Bogdanov's early work on 'tektology' (taken from the Greek word 'tekton', meaning builder). Despite Lenin's early condemnation of Bogdanov's work as 'sheer nonsense' and a 'reactionary muddle', and the dismissal of cybernetics as an idealistic reactionary pseudoscience and bourgeois deviation from dialectical materialism, the USSR gradually came around to adopting its own version of systems theory, making it amenable to an organization such as the IIASA, which was eventually nicknamed 'the East-West Institute' and 'the East-West RAND'.[33] Systems theory went some way to 'opening up the cold war', to use Eglė Rindzevičiūtė's phrase.[34] Those working in systems theory came to consider it beyond the major ideological fault line of the second half of the twentieth century. More generally, this was also the period in which the idea of the world itself as the ultimate system came to the fore. An emerging 'new ecology' treated the whole earth as an 'ecosystem', a concept developed by Eugene P. Odum and his colleagues and then 'systemized' in the 1970s, not least by the publication of the Club of Rome report *Limits to Growth* (1972). The Club of Rome's report was itself inspired by Jay Forrester's original systems approach to US defence and his application of this to urban systems (*Urban Dynamics*, 1969) and then world systems (*World Dynamics*, 1971).[35] These connections then came together in the ecosystems philosophy developed by writers such as James Lovelock, Fritjof Capra, Lynn Margulis, and Dorion Sagan.

On the other hand, this transdisciplinary agenda entailed a massive transformation of the research agenda across all academic disciplines, connected more broadly to the structuralist revolution in the social sciences, the game-theoretic revolution in economics, the behaviouralist revolution in political science, and the cognitive revolution in

33 A. Bogdanov, *Essays in Tektology* (1921), trans. George Gorelik (Seaside, CA: Intersystems Publications, 1980); Simon Kassel, *Soviet Cybernetics Research: A Preliminary Study of Organizations and Personalities* (Santa Monica, CA: RAND, 1971); Slava Gerovitch, *From Newspeak to Cyberspeak: A History of Soviet Cybernetics* (Cambridge, MA: MIT Press, 2002).

34 Eglė Rindzevičiūtė, *The Power of Systems: How Policy Sciences Opened Up the Cold War World* (Ithaca, NY: Cornell University Press, 2016).

35 Jay W. Forrester, 'System Dynamics: A Personal View of the First Fifty Years', *System Dynamics Review*, vol. 23, no. 2–3 (2007), 345–58; Donella H. Meadows and Dennis Meadows, 'The History and Conclusions of *The Limits to Growth*', *System Dynamics Review*, vol. 23, no. 2–3 (2007), 191–97.

psychology. Major conferences on systems theory saw economists, analytical philosophers, engineers, biologists, and operations specialists all rubbing shoulders, trying to replicate the kind of interdisciplinary participation and intellectual reach of the Macy Conferences. New academic titles were created such as 'Professor of General Systems' (Stafford Beer's title at the Open University in the UK). But the impact was also within the disciplines themselves. Hunter Heyck's analysis of the leading American academic journals in politics, sociology, anthropology, economics, and psychology shows that talk of 'systems' was not especially common before 1950, started increasing from 1950 onwards, and by the late 1960s had become one of the most common terms in the social sciences. Academic articles that employed the term 'system' in a deliberate and coherent way rose from less than 7 percent in 1930 to just over 60 percent by 1970.[36]

By 1970, then, 'system' was carrying a huge conceptual weight within the social sciences. In sociology, for example, one can trace the rise of systems theory through the titles of some of Talcott Parsons's key works, which move, in effect, from 'structure' to 'system': his interest in *The Structure of Social Action*, the title of his 1937 book, shifts to *The System of Modern Societies*, the title of his book published in 1971.[37] A sample of titles of key texts from within the social sciences between 1965 and 1975 makes clear what had happened: *A Systems Analysis of Political Life* (by David Easton, 1965); *A Sociology of Human Systems* (Joseph Monane, 1967); *Sociology and Modern Systems Theory* (Walter Buckley, 1967); *Systems Analysis and the Political Process* (James Schlesinger, 1967, published by RAND); *Systems of Political Science* (Oran Young, 1968); *Modern Systems Research for the Behavioral Scientist* (Walter Buckley, 1968); *General Systems Theory and Psychiatry* (Frederick Duhl, William Gray and Nicholas Rizzo, 1969); *Industrial Relations Systems* (John Dunlop, 1970); *Information, Systems, and Psychoanalysis* (Emanuel Peterfreund and Jacob T. Schwartz, 1971); *Physical Geography: A Systems Approach* (Richard J. Chorley and Barbara A. Kennedy, 1971); *System and Structure: Essays in Communication and Exchange* (Anthony

36 Hunter Heyck, *Age of System: Understanding the Development of Modern Social Science* (Baltimore: Johns Hopkins University Press, 2015).

37 See Talcott Parsons, 'On Building Social System Theory: A Personal History', *Daedalus*, vol. 99, no. 4 (1970), 826–81.

Wilden, 1972); *Introduction to Systems Philosophy: Toward a New Paradigm of Contemporary Thought* (Ervin Laszlo, 1972); *The World System: Models, Norms, Applications* (also by Laszlo, 1973); *A General Systems Philosophy for the Social and Behavioral Sciences* (John W. Sutherland, 1973); *Perspectives on General System Theory: Scientific-Philosophical Studies* (Bertalanffy again, 1975). What also emerges is cross-disciplinary work such as the world systems analysis of Immanuel Wallerstein in the early 1970s, the first volume of which, *The Modern World-System*, appeared in 1974, designed to break with both 'sovereign state' and 'national society' as the key unity of analysis and not entirely unrelated to the development of the aforementioned 'international system'. In other words, the scope and impact of the concept of 'system' was extensive, stretching across the social sciences, to the extent that most social scientists would have agreed with Niklas Luhmann, to whose work we turn below, in his claim that 'a society is not possible other than as a system'.[38]

All told, the 'age of system' was thought to need systems theory, considered by its proponents to be a 'universal' theory. As David Noble puts it, 'Everything and everyone, after all . . . could be viewed as a part ("component") related to other parts in a large whole ("system"), and thus as amenable to systems analysis, which by the 1960s had attained a force and an aura all its own.'[39] It extended, for example, to artistic production, with exhibitions on 'Systemic Painting' (at the Guggenheim Museum, New York, in 1966); Jack Burnham's work on 'Systems Aesthetics'; the rise of 'cybernated art' in the work of artists such as Nam June Paik and Roy Ascott; and the birth of the 'Systems Group' of British artists in 1969. Straddling all these fields and more – straddling *all* fields – what was announced was nothing less than a universal systems theory as a *new paradigmatic form of knowledge* for all the sciences and a fundamental change in the narration of modernity. Highlighting issues that we will take up in the chapter that follows, systems theory is, ultimately, *a view of the whole world* with several overlapping precepts: a stress on the *whole* rather than the parts; a stress on *process* rather than structure; the insistence that

38 Niklas Luhmann, *Introduction to Systems Theory* (2002), trans. Peter Gilgen (Cambridge: Polity Press, 2013), 9.

39 David F. Noble, *Forces of Production: A Social History of Industrial Automation* (New York: Alfred A. Knopf, 1984), 54.

systems are *self-regulating*, maintaining themselves (*homeostasis*) through continuous *exchange of information* – that is, *communication*; that this exchange of information constitutes *feedback*; that systems have *interconnectivity* but vary in their *complexity* and thus their ability to maintain a homeostatic equilibrium; the idea of self-organization (*autopoiesis*), in the sense that the system is a self-organizing or self-regulating process. Even the Self was now a System, as Gregory Bateson put it in his essay 'A Cybernetics of "Self"', a text that has become a locus classicus in the field.[40] And, if the Self is a System, then so too is the body.

Bodies as Systems

'Medicine discovered systems theory rather late', note William Arney and Bernard Bergen.[41] Nonetheless, when it did, the impact of the discovery was huge. In 1943, a paper by Warren S. McCulloch and Walter Pitts, 'A Logical Calculus of the Ideas Immanent in Nervous Activity', treated the responses of neurons as equivalent to propositions, such that 'physiological relations existing among nervous activities correspond . . . to relations among the propositions'. If propositions are treated purely formally, following the work of Rudolf Carnap, Bertrand Russell and Alfred North Whitehead, then neurons could be seen as obeying similar mathematical rules as the inputs and outputs of signals. John von Neumann was one of many convinced of the paper's importance and sought to apply mathematical logic work to the idea of self-reproducing machines as living systems. Taking Max Delbrück's work on the phage as the simplest system, von Neumann believed a virus or bacteriophage could be studied as a self-reproductive information-processing organism. On this basis, notes Lily Kay, von Neumann broke the problem of living organisms down into two parts: first, the functioning of their elementary units, and second, the overall system through which the units are assembled.[42] The functioning of the system was

40 Gregory Bateson, *Steps to an Ecology of Mind* (Chicago: University of Chicago Press, 1972), 331–33.

41 William Ray Arney and Bernard J. Bergen, *Medicine and the Management of Living: Taming the Last Great Beast* (Chicago: University of Chicago Press, 1984), 72.

42 Warren S. McCulloch and Walter Pitts, 'A Logical Calculus of the Ideas Immanent in Nervous Activity', *Bulletin of Mathematical Biophysics*, vol. 5, no. 4 (1943), 115–33

understood as a problem distinct from the knowledge of their elementary units. Von Neumann used this to grapple with the ways the central nervous system organizes the information produced by the sense organs, but the general point about wholes and units was encouraged by the extensive use of biological ideas within systems thinking. Here is David Easton writing in 1957 about political systems:

> The way in which the body works represents responses to needs that are generated by the very organization of its anatomy and functions; but in large part, in order to understand both the structure and the working of the body, we must also be very sensitive to the inputs from the environment.

Easton goes on to add that 'in the same way [as the body works], the behavior of every political system is to some degree imposed upon it by the kind of system it is, that is, by its own structure and internal need'.[43] Political systems are *systems* in the same way that the *body* is a system; the body politic is, first and foremost, a body. Talcott Parsons likewise argues that 'developments in biological theory and in the social sciences . . . have created firm grounds for accepting the fundamental continuity of society and culture as part of a more general theory of the evolution of living systems'.[44]

So on the one hand, what one finds in systems theory is nothing less than a 'systematization' of the image of the body as an organization. 'One can look at a cell or an atom . . . as systems', notes Ervin Laszlo in *The Systems View of the World* (1972), just as one can the family, community, nation, and economy. Glossing Laszlo's argument a year later, Howard Brody, in an article in the journal *Perspectives in Biology and Medicine*, suggested that 'systems theory . . . has long appeared to hold great promise for a comprehensive overview of natural phenomena'. He goes on to argue that analysing living things as natural systems has

---

(117); von Neumann, 'General and Logical', 21–24; Lily E. Kay, *Who Wrote the Book of Life? A History of the Genetic Code* (Stanford, CA: Stanford University Press, 2000), 105–06, 109.

43 David Easton, 'An Approach to the Analysis of Political Systems', *World Politics*, vol. 9, no. 3 (1957), 383–400 (386).

44 Talcott Parsons, *The System of Modern Societies* (Englewood Cliffs, NJ: Prentice-Hall, 1970), 2.

implications for the concepts 'health' and 'disease', with the former defined as 'the harmonious interaction of all hierarchical components' and the latter as 'the result of a force which perturbs or disrupts hierarchical structure'. But, more broadly and ambitiously, the systems approach is said to provide a 'solid philosophical foundation' to understand the whole natural hierarchy from subatomic particles through to the biosphere.[45]

By the early 1970s, then, systems thinking had made it into biology. 'Biologists no longer study life today', observed Nobel Prize winner François Jacob in *La logique du vivant* (1970). 'Instead, they investigate the structure of living systems.' (The eventual English translation of his book should really have been *The Logic of the Living* rather than *The Logic of Life*, to better capture the living 'systemness' he was trying to describe.)[46] Biology textbooks would soon be repeating the point that what exists and can be studied is living systems. From this, two things will be of interest to us: first, once system is imagined as organism and organism imagined as a system of systems, what will finally emerge is the 'immune system'; and second, and picking up on the arguments in chapter 1, such a move facilitates a conception of the body as a security system and hence a politics of immunity as security. We will have to get there in a roundabout way, through the concept of information.

'What is man, or any material object, but a sum of information?' The question posed by Stanisław Lem in 1964 could only be rhetorical because, for Lem, 'everything is information'.[47] This was not an especially controversial claim. We have already noted Ashby's comment that the brain is an acting machine, in the sense that it gets information and then it does something about it. What was crucial to systems analysis was the idea that a key measure of the degree of organization and order in any system is the quantity of information and the successful

45 Ervin Laszlo, *The Systems View of the World: The Natural Philosophy of the New Developments in the Sciences* (Oxford: Basil Blackwell, 1972), 14, 23; Howard Brody, 'The Systems View of Man: Implications for Medicine, Science, and Ethics', *Perspectives in Biology and Medicine*, vol. 17, no. 1 (1973), 71–92 (71, 75, 77, 82).

46 François Jacob, *The Logic of Life: A History of Heredity* (1970), trans. Betty E. Spillman, in François Jacob, *The Logic of Life* and *The Possible and the Actual* (London: Penguin, 1989), 299–300.

47 Also see Stanisław Lem, *Summa Technologiae* (1964), trans. Joanna Zylinska (Minneapolis: University of Minnesota Press, 2013), 221.

communication of this information within it. Every system produces and processes information and 'feedback' to produce a controlled order or 'homeostasis'. A system is a system of information. Yet, at the same time, the concept of 'information' underwent a radical transformation, to the extent that it too was imagined as having a life of its own: 'Information should develop from information, just as organisms develop from organisms', Lem suggested, adding that 'Its bits should fertilize one another; they should cross-breed.'[48] These ideas can be traced back to work conducted on telegraphic transmission by Bell Labs, which was also involved in research into strategic communications. In 1948, the year of publication of Wiener's *Cybernetics*, Bell announced a new electronic semiconductor called a 'transistor', and the corporation's in-house journal, the *Bell System Technical Journal*, published an article by Claude Shannon called 'A Mathematical Theory of Communication'. The article was developed into a book of the same title the following year, with a substantial introductory essay by Warren Weaver. Shannon and Weaver's work outlined a theory of communication based on a definition of 'information' that made it capable of being quantified. Information is not to be confused with meaning and is used in a way that should not be confused with its ordinary usage. A message that might be heavily loaded with meaning and a message which is pure nonsense are equivalent as regards information. Information must be divorced from meaning in order to be understood as a quantitative measure of communication. In other words, 'information' is not concerned with whether what people tell each other makes any sense but, rather, is the communication of what Shannon called 'binary digits' ('bits') down a single wire. As a unit, the 'bit' enables codification such that *meaning* disappears and gets replaced by the idea of a formal processing of information as a mechanism for controlling systems: 0 or 1. A system is thus bits of information sustained through time, and systems theory is a logic of information without meaning.[49]

It is difficult now, accustomed as we are to living in an 'Information Society', to imagine just how new this idea of information was. When, in

48 Lem, *Summa Technologiae*, 251.

49 Claude E. Shannon, 'A Mathematical Theory of Communication', *Bell System Technical Journal*, vol. 27, no. 3 (1948), 379–423, and vol. 27, no. 4 (1948), 623–56; Claude E. Shannon and Warren Weaver, *The Mathematical Theory of Information* (Urbana: University of Illinois Press, 1949).

1952, Encyclopaedia Britannica published *The Great Ideas: A Syntopicon of Great Books of the Western World*, covering 'Angel' to 'Love' in volume I and from 'Man' to 'World' in volume II, there was no entry for 'information'. Such an absence now is unthinkable. It is unthinkable because the belief in systems and hence the world *as* information came to shape everything from biology to the social sciences and from post-war art movements to popular culture. Not only did engineers start talking about information as something technical, quantifiable, and measurable, but large swathes of intellectual discourse also adopted the same epistemological assumption that systems are systems of information. It 'is *information* and not *legislation* that changes us', systems thinkers such as Beer liked to say. Information is '*a difference which makes a difference*', as Bateson put it.[50]

The material base of this idea of information is clear, not least in the drive for 'communication and control' – that is, the control of communication – by the combined forces of a leading capitalist organization (the Bell Telephone Company) and the leading security state (the USA). The vision of information fits perfectly into the capitalist conception of production and order in the twentieth century. It does so in two broad ways, one inside the workplace and one outside.

First, 'information' about production techniques was once held largely by workers, learnt through 'word of mouth' and hence possessed as 'local' or 'traditional' knowledge rather than intellectual property. In the twentieth century, employers sought to appropriate this information as part of the 'deskilling' process of 'scientific management'. Frederick Winslow Taylor commented in his book *The Principles of Scientific Management* that workmen in many of the trades employed in an industrial establishment 'have had their knowledge handed down to them by word of mouth, through the many years in which their trade has been developed'. In such a scenario, knowledge of the various skills used in the production has failed to be 'codified or systematically analysed or described'. Taylor recommended that management appropriate the 'information' held by workers. The argument was double-sided: on the one hand, so long as information was held by the workers, they possessed a certain kind of power (they 'carry the firm's information base', as

50 Stafford Beer, 'The Liberty Machine', *Futures*, vol. 3, no. 4 (1971), 338–48 (347); Bateson, *Steps*, 459, emphasis added.

Kenneth Arrow put it much later); on the other hand, by appropriating the information held in common by the workers, management would, in turn, be transforming it into the intellectual property of capital. Dispossessing workers of their knowledge was crucial to the development of twentieth-century capitalism and the meaning of the firm as owner of information about the production process. Class warfare *in* the information age, to use Michael Perelman's term, would be class warfare *over* information.[51]

The second way information has been central to capital lies in the concept of a market order being developed by thinkers such as F. A. Hayek. Hayek articulates a defence of the market conceived of as a system in which 'the relevant information is communicated to all'. The problem with any planned economy, for Hayek, lies in the fact that no system could ever possess or manage information well enough to enact such a plan. In contrast, information in the market system is communicated throughout the system. The price system, for example, is 'a mechanism for communicating information'. As Hayek put it when receiving the Nobel Prize for such ideas, the market is the most 'efficient mechanism for digesting dispersed information than any that man has deliberately designed'. This mechanism means that individual participants need to only know the bits of information relevant to their choices and actions.[52] This concept of information thus helped shape a new conception of rationality rooted in logistics, game theory, and quantitative methods, reducing rationality to rules or rational choice.

Both inside and outside the workplace, then, the capitalist system – that is, the body of capital – came to be conceptualized as a system for the communication of information, the firm conceptualized as the legitimate owner of information, and the human *conceptualized as an information-processing body*. So central to capitalism was this 'cult of information', to borrow the title of Theodore Roszak's book on the

51 Frederick Winslow Taylor, *The Principles of Scientific Management* (New York: Harper and Brothers, 1911), 31–32, 103; Kenneth J. Arrow, 'The Economics of Information: An Exposition', *Empirica*, vol. 23, no. 2 (1996), 119–28 (126); Michael Perelman, *Class Warfare in the Information Age* (New York: St. Martin's Press, 1998).

52 Friedrich A. Hayek, *Individualism and Economic Order* (Chicago: University of Chicago Press, 1948), 86–87; 'The Pretence of Knowledge', Nobel Prize lecture, 11 December 1974; *Law, Legislation and Liberty: A New Statement of the Liberal Principles of Justice and Political Economy*, vol. II (London: Routledge, 1982), 71, 178.

subject, that Nobel Prizes would eventually be awarded to some of the cult's leading figures: Kenneth Arrow for work on the economics of information (1972), Hayek for work on the market as an information processor (1974), and Herbert Simon for work on information and decision-making (1978). The idea of a critique of information as part of the wider critique of political economy, and thus a critique of systems theory and even perhaps a critique of imagining the body in this way, became increasingly difficult. After all, who could be against information? But the question 'Who could be against information?' is a version of a much larger question: Who could be against the smooth running of systems and the System? Information is always understood as integrated in systems, to the extent that information is regarded as impossible outside of a system.

'You must get acquainted with this symbolic system, if you want to gain entrance to entire orders of reality', Jacques Lacan observed about information systems in the mid-1950s.[53] His comment was partly designed to draw attention to the world of the machine: information gets 'jammed' when it gets corrupted, becomes 'noise', and entropy ensues; the human subject is a jammed machine. These are terms which we shall take up in the chapter to follow. But Lacan was also picking up on the idea of the body as a system for the processing of information and the ways in which the logic of information had made its way into biological thinking. An organism is held together by the acquisition, use, and transmission of information, Wiener had suggested in *Cybernetics* in 1948. But the major turning point was molecular biology's paradigm shift that took place around 1953.

Why 1953? 'Molecular biology' was a term coined by Warren Weaver in 1938 during his time as director of the Rockefeller Foundation's natural science division and intended to capture the essence of the foundation's programme – namely, its emphasis on the ultimate minuteness of biological entities but also the significance of this in the social universe as well as the biological. The year 1953 saw the publication of new work on the structure of deoxyribonucleic acid (DNA) by J. D. Watson, F. H. C. Crick, and Rosalind Franklin. They developed the idea of 'genetical

53 Jacques Lacan, *The Seminar of Jacques Lacan*, bk. II, *The Ego in Freud's Theory and in the Technique of Psychoanalysis 1954–1955* (1978), trans. Sylvana Tomaselli (New York: W. W. Norton and Co., 1988), 83.

information' said to be residing exclusively in the 'base' of each nucleotide. Just before the publication of the first article in April 1953, by Watson and Crick (Franklin having been shunted aside and left uncredited), Watson had co-written a letter to *Nature* suggesting that one solution to the 'confusing situation' created by the 'rather chaotic growth in technical vocabulary' of bacterial genetics could be 'the use of the term "inter-bacterial information"', and the letter ended by suggesting 'the possible future importance of cybernetics at the bacterial level'. By the time of their second main article in *Nature*, published the following month, the information they were talking about had become *code*. Taking up Erwin Schrödinger's suggestion in *What Is Life?* (1944) that a relatively small number of atoms might provide a 'code-script' similar to the Morse code, Watson and Crick proposed that it seemed very likely 'that the precise sequence of bases is the code which carries the genetical information'.[54]

The outcome of their work was the double helix of DNA conceived of as an information system. 'Information means . . . the precise determination of sequence, either of bases in the nucleic acid or of amino acid residues in the protein.' This claim becomes the 'central dogma' of heredity. As stated by Crick in his 1958 overview of the previous decade's research in this field, the central dogma states that

> once 'information' has passed into protein *it cannot get out again*. In more detail, the transfer of information from nucleic acid to nucleic acid, or from nucleic acid to protein may be possible, but transfer from protein to protein, or from protein to nucleic acid is impossible. Information means here the *precise* determination of sequence, either of bases in the nucleic acid or of amino acid residues in the protein.[55]

54 Boris Ephrussi, Urs Leopold, J. D. Watson, and J. J. Weigle, 'Terminology in Bacterial Genetics', *Nature*, vol. 171 (18 April 1953), 701; J. D. Watson and F. H. C. Crick, 'Genetical Implications of the Structure of Deoxyribonucleic Acid', *Nature*, vol. 171, no. 4361 (1953), 964–67; J. D. Watson and F. H. C. Crick, 'Molecular Structure of Nucleic Acids: A Structure for Deoxyribose Nucleic Acid', *Nature*, vol. 171, no. 4356 (1953), 737–38; Erwin Schrödinger, *What Is Life?* (1944), in *What Is Life? with 'Mind and Matter' and 'Autobiographical Sketches'* (Cambridge: Cambridge University Press, 1992), 20–22, 49.

55 Francis Crick, 'On Protein Synthesis', *Symposium of the Society for Experimental Biology*, 12 (1958), 138–63 (153).

In 1953, then, *genes became information*.[56] There were no 'genetic messages' in the 1930s, and genes did not 'transfer information' before the 1950s; they simply possessed biochemical specificities. But by the end of the 1950s, genetics had become a form of information science. A popular textbook such as Steven Rose's *The Chemistry of Life*, first published in 1966, could claim that 'it is now possible to see why both the geneticists and the cyberneticians (information scientists) are interested in protein synthesis', for what is at stake is '*information transfer*: the converting of information from one form to another'.[57] A new conception or redefinition of 'life' had been produced, organized around 'breaking the Code'. As J. D. Bernal put it, 'Life is beginning to cease to be a mystery and becoming practically a cryptogram, a puzzle, a code that can be broken.'[58] This was a shift that helped construct 'DNA' as a key cultural icon, as a secular replacement for the soul, and as the basis for the further colonization of the body by capital. In terms of the discussion in the previous chapters, we might say that the DNA in the cell was now considered the DNA of the Self.

This shift was not without its critics at the time. For example, leading biologist Peter Medawar warned that 'we have to be on our guard against treating information content as a measure of biological organisation', though he conceded that 'the ideas and terminology of information theory would not have caught on as they have done unless they were serving some very useful purpose'. Colin Cherry made a similar argument in his major text *On Human Communication* (1957).[59] But the momentum towards the 'Information Age' was too powerful, and so the characteristics of living organisms that had historically been employed to define 'life', such as growth, development, and reproduction, were gradually displaced by a complex of ideas about systems of information: the nervous system became a system for digital information-processing, the cell became informed matter, organisms became bundles of infor-

56 Matthew Cobb, '1953: When Genes Became Information', *Cell*, no. 153 (2013), 503–506.

57 Steven Rose, *The Chemistry of Life* (Harmondsworth: Penguin, 1970), 161–62. The passage was removed from later editions of the book.

58 J. D. Bernal, 'Definitions of Life', *New Scientist* 33 (5 Jan. 1967), 12–14 (13).

59 Peter Medawar, 'Herbert Spencer and the Law of General Evolution' (1963), in Peter Medawar, *Pluto's Republic* (Oxford: Oxford University Press, 1982), 225–26; Colin Cherry, *On Human Communication: A Review, a Survey, and a Criticism* (Cambridge, MA: MIT Press, 1957), 40, 68.

mation and even viruses came to be defined as information-delivery systems. Priscilla Wald has pointed out that S. E. Luria's major textbook *General Virology*, published in 1953, makes nothing of information or code, but by the second edition, published in 1967 (co-authored with James Darnell), the chapter on the origin and nature of viral reproduction had been rewritten to treat viruses as information-delivery systems.[60]

'Messages, information, programs, code, instructions, decoding: these are the new concepts of the life sciences', noted Georges Canguilhem in 1966.[61] By 1970, the two books by Nobel Prize winners François Jacob (*Logic of Life*) and Jacques Monod (*Chance and Necessity*) treated organisms as 'a triple flow of matter, energy and information', as Jacob put it. Biologists now study 'the algorithms of the living world', and heredity must be understood 'in terms of information, messages and code' that coordinate 'the activities of the system'. 'Information . . . is at one and the same time what is measured, what is transmitted and what is transformed. Every interaction between the members of an organization can accordingly be considered as a problem of communication.' This applies just as much to a human society as to a living organism or to an automatic device. It also applies to the Self: 'The intention of a psyche has been replaced by the translation of a message.'[62] Monod makes the same points. So too do most other biologists of the period. In a popular overview of current ideas in biology called *The Life Science* (1978), Peter Medawar and Jean Medawar begin by pointing out that the idea of information crops up repeatedly in the book because no book on biological ideas can be without it. The organism as a *system* simply *is* information, because 'information embodies, expresses, and often specifies order', as they put it in their book *Aristotle to Zoos: A Philosophical Dictionary of Biology*. Information *informs* – that is, *forms* – the material

60 Priscilla Wald, *Contagious: Cultures, Carriers, and the Outbreak Narrative* (Durham, NC: Duke University Press, 2008), 181, comparing chapter 18 of S. E. Luria, *General Virology* (New York: John Wiley and Sons, 1953), with chapter 19 of S. E. Luria and James Darnell Jr., *General Virology*, 2nd edn. (New York: John Wiley and Sons, 1967), especially 442–45.

61 Georges Canguilhem, 'The Concept of Life' (1966/1968), trans. Arthur Goldhammer, in *A Vital Rationalist: Selected Writings from Georges Canguilhem*, ed. Francois Delaporte (New York: Zone Books, 2000), 316.

62 Jacob, *Logic*, 1–2, 95, 251, 299–300.

world.[63] In effect, life became information. The extent to which this was so can be seen in the way that information gradually displaced the idea of biological *specificity*. Watson and Crick's articles illustrate this to some extent, but so too do comments made by biologists in the period: 'Specificity . . . or "information" as it is called nowadays', Joshua Lederberg commented in 1956, despite his own 'philosophical reservation' about treating them as synonyms; 'we use the term "information" in the sense of molecular specificity, that which distinguishes one protein from another', noted George Gamow, Alexander Rich, and Martynas Yčas in an article published in the same year as Lederberg's.[64]

Susan Oyama asks: If there is 'information' in things such as chromosomes or the cells of the immune system, then what is it information about, exactly? Or as N. Katherine Hayles puts it, 'If our body surfaces are membranes through which information flows, who are we?'[65] Better still: *What* are we? Molecular biologists struggling to build a new biology distinct from and even in opposition to older biological frameworks sought to eradicate from their language traditional ideas of 'function', 'purpose', 'organization', and 'harmony', focusing instead on what Haraway describes as '*the translation of the world into a problem of coding*'. But, as Haraway and others also point out, the emergence of information, communication, and coding within biology was intimately connected to the problem of coding in the Cold War's project of command-control-communication-intelligence (C^3I).[66] The more the

---

63 Jacques Monod, *Chance and Necessity: An Essay on the Natural Philosophy of Modern Biology* (1970), trans., Austryn Wainhouse (London: Penguin, 1997), 38, 62–63, 78, 80; P. B. Medawar and J. S. Medawar, *The Life Science: Current Ideas of Biology* (London: Paladin, 1978), 13–14; P. B. Medawar and J. S. Medawar, *Aristotle to Zoos: A Philosophical Dictionary of Biology* (London: Weidenfeld and Nicolson, 1984), 205.

64 Joshua Lederberg, 'Comments on the Gene-Enzyme Relationship', in Oliver H. Gaebler (ed.), *Enzymes: Units of Biological Structure and Function* (New York: Academic Press, 1956), 167; George Gamow, Alexander Rich, and Martynas Yčas, 'The Problem of Information Transfer from the Nucleic Acids to Proteins', in John H. Lawrence and Cornelius A. Tobias (eds.), *Advances in Biological and Medical Physics*, vol. IV (New York: Academic Press, 1956), 40, 66.

65 Susan Oyama, *The Ontogeny of Information: Developmental System and Evolution*, 2nd edn. (Durham, NC: Duke University Press, 2000), 6; N. Katherine Hayles, *How We Became Posthuman: Virtual Bodies in Cybernetics, Literature, and Information* (Chicago: University of Chicago Press, 1999), 109.

66 Donna J. Haraway, *Simians, Cyborgs, and Women: The Reinvention of Nature* (London: Free Association, 1991), 164; Scott L. Montgomery, *The Scientific Voice* (New York: Guilford Press, 1996), 144–53.

biological imagination turns from the physiology of organic form to systems theory, so more emphasis gets placed on the idea of the body as an information-processing mechanism. The perpetual strategic battles and security operations between Self and non-Self, discussed at length in the previous chapter, get recast as a set of operations for the production and communication of information, data, and code. And yet, the powers of war and police were themselves simultaneously becoming 'informationalized', thereby allowing the body and its information to become imagined in new security ways. For all the radicalism involved in reconceptualizing the body in this new and novel way, we were still to be considered as objects of security.

The question 'Why 1953?', then, is obvious, and yet it points us in still more interesting directions. The year that Watson and Crick used the word 'code' when identifying the double helix was also significant as a key year in the Cold War. Stalin's death opened new possibilities for East-West relations, the end of the Korean War appeared to signify a shift in the global order of violence, the extension of the loyalty programme in the US extended the logic of security into realms previously unimaginable, and, most pertinently here, new methods and technologies in cryptanalysis transformed information theory. Watson and Crick's use of the word 'code' was significant, then, because it also seemed to somehow connect their discovery to the *encoding* of security information and the whole problem of secret-code breaking. Some of the thinkers in question, such as Shannon working at Bell, considered their work on communication and cryptology to be one and the same. The wider context of Cold War biology has been well documented. When Gamow, Rich, and Yčas claim, in their 1956 article on information transfer from nucleic acids to proteins, that their research for the article faced difficulties 'similar to those encountered by an Armed Force's Intelligence Office trying to break an enemy code', the comment is telling.[67] Much of the biological research of the time was funded by the national security state, and this facilitated the adoption of the language of the Cold War into biology. After all, to reformulate the question just asked, if the genetic code is a code, then what exactly is it code for?

67 Gamow, Rich, and Yčas, 'Problem of Information', 40, 66.

Humberto Maturana suggested in 1970 that the whole language of 'code' in biology is untenable.[68] But it is only untenable if one treats the discourse in entirely its own terms. If, however, one considers it in the context of the problem of coding faced by the burgeoning security state, then it begins to make more sense, since it can then be understood as part of the culture of security. When the language of 'information' was used consciously as a means of framing the problem of the genetic 'code', and biological thought in general was reshaped by the same language, it resonated with the new language within the security state, which had by then become a state obsessed with information: collecting information, processing information, communicating information, classifying information, protecting information, and, of course, coding information. Coded information as life's secret force paralleled coded information as the groundwork of security. The idea of a 'code' helped bring together ideology and biology, oscillating between foundational ideas of security and intelligence ('breaking the code') and foundational ideas of medical knowledge ('breaking the genetic code'). This might also explain why the trope of the *secret* was so important to biological thought at this time. Watson and Crick's description of their work as a search for the secret of the gene employs a language that was more than a little grandiose and definitely provocative compared to most biological discussion, yet it paralleled the secret of death that was at that point at the heart of security politics, as Evelyn Fox Keller puts it.[69] The 'secret of life' paralleled by the 'secret of death'. *Life death* once again, only this time as a secret code.

The *OED* has the term 'classified information' appearing in English in the 1950s. As Wald writes, the idea that information might be 'classified' made it appear to be the material substance of the state's security from enemies.[70] At the same time, the idea that a body held information containing the body's secrets reinforced the idea that it was the security

68 Humberto R. Maturana, *Biology of Cognition* (1970), in Humberto R. Maturana and Francisco J. Varela, *Autopoiesis and Cognition: The Realization of the Living* (Dordrecht: D. Reidel Co., 1980), 6, 53.

69 J. D. Watson, 'Growing Up in the Phage Group' (1966), in John Cairns, Gunther S. Stent, and James D. Watson (eds.), *Phage and the Origins of Molecular Biology: The Centennial Edition* (Cold Spring Harbor, NY: Cold Spring Harbor Laboratory Press, 2007), 239; Evelyn Fox Keller, *Secrets of Life, Secrets of Death: Essays on Language, Gender and Science* (New York: Routledge, 1992), 42, 106–07.

70 Wald, *Contagious*, 183.

of the body politic that was somehow at stake. Nature, body, and state could all now be conceived in terms of 'information security' and life imagined within security's regimes of signification. All told, biology's paradigmatic shift to information theory and systems analysis connected biology to discourses of communication and control in the security state. This imagination of a biological *information* and *messages*, *programmes* and *instructions*, *code* and *decoding* was to become the biggest fantasy and most fundamental of semiotic systems of Western thought thereafter.[71] A new knowledge-power nexus had formed: life was now information and system and thus an information system. The body as a system of systems could now be imagined through the lens of information and the translation of messages.

All of which takes us back to politics of immunity and the idea of the immune system.

As we saw in the previous chapter, when gradually introducing the idea of the Self into immunology's lexicon in the late 1940s, Frank Macfarlane Burnet sought a language for immunology that resonated with our cultural understanding of ourselves. The book he co-authored with Frank Fenner, *The Production of Antibodies*, appeared at the same time that the idea of 'bits' of 'information' was becoming paradigmatic, and immunologists in general started imagining the immune process (not yet called 'immune system') through the lens of information, communication, command, and control. In *Enzyme, Antigen and Virus* (1956), Burnet presented an overview of the recent developments in research into immunity, and the book concludes with a section on information theory. There, Burnet connects the immune process to the work on information systems and a possible 'communications theory of the cell'. 'Since 1945 there has been a widespread recognition amongst scientists and the public generally of the importance of the principles which have emerged from experience in the development of electronic communications and control systems.' In this light, he announces that his book 'was originally conceived as an attempt to develop something analogous to a communications theory that would be applicable to the concepts of general biology'. He admits, however, to some difficulty in developing such a theory: 'Only the most generalized sketch of an outline has yet

71 Jean Baudrillard, *Simulations*, trans. Paul Foss, Paul Patton, and Philip Beitchman (New York: Semiotext[e], 1983), 106–09.

been given of how information theory at the strict level can be applied to biology.' Nonetheless, the concept of 'communication through molecular pattern' was intended to become part of 'the normal tools of every biological investigator and theorist'.[72]

The same year as *Enzyme, Antigen and Virus* was published, Shannon warned that information theory was in danger of becoming too much of a 'bandwagon' and oversold as another fashionable field alongside computing, cybernetics, and automation.[73] Despite such warnings, Burnet and many others continued to pursue the idea. Writing about autoimmune diseases in the *British Medical Journal* in 1959, Burnet suggested that we might think of them in 'modern terms':

> It is a 'central dogma' of modern biochemical genetics that only genetic processes can modify the information in a genetic system. There is no presently conceivable way by which information from the environment can be incorporated into the nucleic acid of the chromosomes. The flow of information is always from the genetic to the biochemical, never in the opposite direction. The chain of command is pictured as deoxyribonucleic acid–ribonucleic acid protein (D.N.A.-R.N.A.-protein), and on this dogma there is no way in which an antigen can meaningfully influence the pattern of D.N.A.[74]

In *The Integrity of the Body*, published in 1962 and targeted at a more general readership, Burnet sought to bring immunology into more general line with these ideas and refers again to the dogma concerning 'information flow'. He had followed closely the developments in genetics in the previous decade, and as we have seen in a passage already cited in chapter 2, in general terms the idea now was that 'it is one of the concise statements of modern immunology that the body will accept as itself only what is genetically indistinguishable from the part replaced'. Thus, for tissue to be rejected it must be recognizably different, and the difference lies in genetic information that the tissue provides. 'During

72 F. Macfarlane Burnet, *Enzyme, Antigen and Virus: A Study of Macromolecular Patter in Action* (Cambridge: Cambridge University Press, 1956), 163–75.

73 Claude Elwood Shannon, 'The Bandwagon' (1956), in Shannon, *Collected Papers*, ed. N. J. A. Sloane and Aaron D. Wyner (Piscataway, NJ: IEEE Press, 1993), 462.

74 Sir Macfarlane Burnet, 'Auto-Immune Disease I: Modern Immunological Concepts, *British Medical Journal*, vol. 2, no. 5153 (1959), 645–50 (649).

embryonic life the body *acquires or generates the information* which allows it to differentiate immunologically between what is self and what is not self.' With this link established, Burnet treated the immune process as a system of information transfer along the lines of genetics and used the language of computing to try and express this.[75]

In other words, immunology proved to be a conducive environment for 'information' to flourish, and, for immunology, information appeared to be an idea full of potential. Before long, the immune system was being conceptualized as a telecommunications network, sending signals and communicating information. We will see in chapter 5 that this was also a way of imagining the immune system in the very language that had long been used to describe the nervous system, enabling the two systems to be brought together. The point here is that immunity imagined as information-processing encouraged the idea of the Self as a collection of data. The immunological dichotomy of Self and non-Self could be reinterpreted through the ideas of information and code: the antigen was now a *code* for 'foreignness' or 'non-Self', and the antibody carried the right *information* or *pattern of information* for Self and System to deal with that antigen.[76] All of which effected a shift in immunological discourse and its popular understandings. The idea of biological *specificity*, for example, had for a long time been at the heart of immunology, but just as *specificity* was being displaced by *information* elsewhere, as we have seen, so it was also gradually displaced in immunology, as the problem of how an antigen is recognized became conceptualized as an 'informational problem'. The view emerged that 'an antigen . . . could itself provide the information which directed the synthesis of an antibody molecule'.[77] In a chapter called 'Information' in a popular book for a general readership called *The Lives of a Cell* (1974), leading American biologist Lewis Thomas wrote that 'it seems to be in the nature of

75 Sir Macfarlane Burnet, *The Integrity of the Body: A Discussion of Modern Immunological Ideas* (Cambridge, MA: Harvard University Press, 1962), 13, 94–96, 103, emphasis added; *Changing Patterns: An Atypical Autobiography* (London: Heinemann, 1968), 192; *Self and Not-Self: Cellular Immunology*, bk. I (Victoria: Melbourne University Press, 1969), 24.

76 Alfred I. Tauber, 'Historical and Philosophical Perspectives Concerning Immune Cognition', *Journal of the History of Biology*, vol. 30, no. 3 (1997), 419–40; Alfred I. Tauber, *Immunity: The Evolution of an Idea* (Oxford: Oxford University Press, 2017), 142.

77 Medawar and Medawar, *Life Science*, 101.

biologic information that it not only stores itself up as energy but also instigates a search for more'. Hence lymphocytes 'carry specific information in their surface receptors, presented in the form of a question: is there, anywhere out there, my particular molecule configuration?' In particular, 'lymphocytes are apparently informed about everything foreign around them'.[78] This was the basis for the emergence of what some call 'immuno-informatics' or 'immuno-semiotics', treating the immune system as a system of signs. Major international conferences began to take place on the semiotics of cellular communication in the immune system, including one organized under the auspices of NATO, held in 1986, including among its participants leading cultural theorists such as Umberto Eco.[79]

In the process of thinking about immunity as a process of *information*, then, immunity was also becoming increasingly *imagined as a system*, and vice versa. It is this that gives immunity its 'systemness'. Moulin suggests that the immune system 'emerged out of a two-fold need to explain both the organization of the various cells and the means of regulation' and identifies the main features of the immune system as follows:

> (i) immunity is a permanent feature of the entire body; (ii) the representation of immunity requires anatomical and histological knowledge of its parts (on originating organs, for example); (iii) knowledge of immunity, beyond this morphological description, refers to a logical category, the so-called immunocompetence of cells, whatever be their localization; (iv) all immunological phenomena can be described and explained in terms of the immune system.[80]

What was emerging from all of this was a dogma, centred on a theory of antibody formation and then clonal selection, which would be as central to immunology as the dogma concerning DNA within genetics; 'Burnet's Dogma', as Jerne dubbed it, in his closing speech at the 1967 *Cold Spring Harbor Symposia*. And core to this dogma was the concept of system

78 Lewis Thomas, *The Lives of a Cell* (1974) (London: Futura, 1976), 109–10.

79 Eli E. Sercarz et al. (eds.), *The Semiotics of Cellular Communication in the Immune System: Proceedings of the NATO Advanced Workshop, Il Ciocco, Lucca, 9–12 Sept. 1986* (Berlin: Springer-Verlag, 1988).

80 Moulin, 'Immune System', 227, 230.

which had, by the time of immunology's 'Golden Age', come to permeate almost all thinking about immunity.

In one sense, what was at stake was the scientific legitimation that immunology could receive by adopting systems analysis and its key concepts, tropes, and images and thus in a sense *adapting itself* to systems analysis. Yet in another sense what this also meant was that immunity was imagined through the key tropes and images through which systems in general were imagined. And, as should by now be clear, such claims about systems are deeply political.

To press this point home, we can turn our attention to the work of one of the leading 'systems theorists' of the late twentieth century, Niklas Luhmann, whose first publications appeared in the 1960s and who, through the 1970s and onwards, articulated an advanced sociological systems theory. Luhmann's work has been widely and positively received in law, the social sciences, and humanities, and much of it underpins the immunitary turn within biopolitics by thinkers such as Esposito. But I want to address Luhmann's work here for how it thinks about immunity in relation to system. This will then open the space in the following chapter for an exploration of eighteenth-century conceptions of order, balance, and harmony and the ways in which these conceptions both relied on and reinforced ideas about physiological systems and social systems, from the body to the body politic.

The Mystique of the Immune Social System

Luhmann was a major figure in what is broadly known as 'second-order systems theory', along with others such as Heinz von Foerster, Stafford Beer, and Gregory Bateson, who each in various ways pursued the idea that Society and Self are not possible other than as a system. Luhmann was also a conduit for the development of both 'autopoiesis' and 'immunity' as sociological categories. In the 1970s, the idea of the 'immune system' became the foundation of the 'immune network',[81] with networks imagined as *systems that constitute themselves*. This idea was

81 Niels Jerne, 'Towards a Network Theory of the Immune System', *Annales d'immunologie*, 125C (1974), 373–89; G. W. Hoffmann, 'A Network Theory of the Immune System', *European Journal of Immunology*, vol. 5 (1975), 638–47.

developed by Francisco Varela, Humberto Maturana, and their various co-authors as an argument about autopoiesis. From the Greek *auto*, meaning 'self', and *poiesis*, meaning 'making', 'autopoiesis' generates the idea of *self-making* systems: autonomous and operationally closed, self-regulating but responsive to the environment and thereby able to maintain and regenerate themselves. We can view the cell as the basic autopoietic system, one which, by virtue of the network of components that regenerate it as a system, constitutes the mechanism that generates the identity of the living. The autopoietic system maintains stability through a circular series of interactions and exchanges with its environment, enabling it to retain its identity and thereby reproduce itself as a system. This is especially the case for the immune system qua system.[82] For Varela, the system's 'network constitution' involves 'a global co-operation which spontaneously emerges as the states of all participating components become mutually satisfactory'.[83] This idea about co-operation was designed as part of Varela's rebuttal of the militarized model which dominated immunology, as we saw in chapter 1, but the point here is that in an autopoietic system 'there is no need for a central processing unit to guide the entire operation, and the external impacts do not turn the system's axle'. External stimuli can modify the local environment, 'but the system's operation and properties will always emerge from global co-operativity'.[84] As Varela and Maturana put it in their later book *The Tree of Knowledge*, 'the most striking feature of an autopoietic system is that it pulls itself up by its own bootstraps'.[85]

The achievement of Varela and his various co-authors was to consolidate the idea of the immune system as a *system* on the basis that its stability rests on a process of self-organization and self-determination. Luhmann gave these ideas a decidedly sociological twist. For Luhmann, the idea that the system *generates itself* and *is its own work* applies to

82 Maturana and Varela, *Autopoiesis*, 59–134; Franciso J. Varela, *Principles of Biological Autonomy* (New York: North Holland, 1979), 211–37.

83 N. M. Vaz and F. J. Varela, 'Self and Non-Sense: An Organism-Centered Approach to Immunology', *Medical Hypotheses*, vol. 4, no. 3 (1978), 231–67 (232); Varela, *Principles of Biological Autonomy*, 212.

84 Francisco J. Varela and Antonio Coutinho, 'Second Generation Immune Networks', *Immunology Today*, vol. 12, no. 5 (1991), 159–66 (159–60).

85 Humberto R. Maturana and Francisco J. Varela, *The Tree of Knowledge: The Biological Roots of Human Understanding* (1987), trans. Robert Paolucci (Boston: Shambala Publications, 1992), 46.

society. 'A social system comes into being whenever an autopoietic connection of communications occurs and distinguishes itself against an environment by restricting the appropriate communications.'[86] Just as a system of molecules produces other molecules such that the system continues, so communications produces yet more communications.

Luhmann's generalization of the idea of autopoiesis to social and psychic systems and his treatment of modernity as a system of systems was intended to revolutionize what he describes as the classical epistemological position on the subject-object distinction, a distinction he regards as 'meaningless'.

> In my approach to systems theory, you will see that I try to leave this subject-object distinction behind and replace it with the distinction between, on the one hand, the operation that a system actually performs when it performs it, and, on the other hand, the observation, be it seen by this system or be it by another system.[87]

Concomitantly, 'rationality' refers to the system, not to the subject. Hence, classical sociological 'action theory' must likewise now apply the concept of action to the *system itself*.[88] In effect, the concept of the rational acting subject becomes dispensable.

This tallies with the idea that once we define human and machine alike as systems, the concept of 'man' can be renounced and replaced instead with 'personality systems', 'psychic systems', and 'communication systems', all of which are far more observable, recordable, and predictable.[89] This approach allows an 'objective' and hence scientific

86 Niklas Luhmann, *Ecological Communication* (1986), trans. John Bednarz (Chicago: University of Chicago Press, 1989), 145–46.

87 Niklas Luhmann, 'Self-Organization and Autopoiesis' (1991–1992), in Bruce Clarke and Mark B. N. Hansen (eds.), *Emergence and Embodiment: New Essays on Second-Order Systems Theory* (Durham, NC: Duke University Press, 2009), 146.

88 Niklas Luhmann, *Social Systems* (1984), trans. John Bendarz, Jr., with Dirk Baecker (Stanford, CA: Stanford University Press, 1995); *A Sociological Theory of Law* (1972), trans. Elizabeth King-Utz and Martin Albrow (London: Routledge, 1985); *Introduction to Systems Theory* (2002), trans. Peter Gilgen (Cambridge: Polity, 2013); *The Differentiation of Society*, trans. Stephen Holmes and Charles Lamore (New York: Columbia University Press, 1982).

89 Niklas Luhmann, 'The Cognitive Program of Constructivism and the Reality That Remains Unknown' (1990), in Luhmann, *Theories of Distinction: Redescribing the Description of Modernity*, trans. Joseph O'Neil et al. (Stanford, CA: Stanford University

understanding of the system, its autopoiesis and its homeostasis, rather than any non-mathematical and unquantifiable 'subjective' judgements. ('Homeostasis' was a term developed within systems theory to connote the stability and equilibrium of a body, based on the work of Walter B. Cannon, who in *The Wisdom of the Body*, first published in 1932 and then expanded in a new edition in 1963, applied the idea to social bodies such as industrial and domestic organizations, as well as the body politic.)

For Luhmann 'the old philosophy of the Subject' was a philosophy caught up in an ontology that is now obsolete, and is to be abandoned along with notions such as 'man' or 'humankind'. Working instead with systems and their environments avoids the political trap into which the old philosophy of the Subject falls, namely of positing a Collective Subject along the lines of Hegel's state or Marx's revolutionary class (the philosophical tradition associated with 'Old Europe'). It is the System rather than any kind of Subject that engages in action. No Subject is needed to transform history because 'the system creates its own history'.[90]

As well as distancing himself from any kind of collective political project, this is Luhmann's way of saying that there is no standpoint from which the social totality can be even observed, let alone grasped as a totality. For this reason, *critique* is neither viable nor desirable as an intellectual and political project, and is in fact misguided and unnecessary. Social theory must start with the fact that society is a self-describing system. Such 'self-description' is an 'all-encompassing system of communication' and '*excludes everything else*'.[91] A society's operations *are* its communications. Building on Shannon and Weaver's account of

---

Press, 2002), 147; *Social Systems*, 255–77; *Love as Passion: The Codification of Intimacy* (1982), trans. Jeremy Gaines and Doris L. Jones (Stanford, CA: Stanford University Press, 1986), 15–16, 23; 'Self-Organization and Autopoiesis', 146, 148; 'The Paradoxy of Observing Systems', *Cultural Critique*, 31, 1995, 37–55.

90 Niklas Luhmann, '"What Is the Case?" and "What Lies Behind It?": The Two Sociologies and the Theory of Society' (1993), trans. Stephen Fuchs, *Sociological Theory*, vol. 12, no. 2 (1994), 126–39 (134–35); *Love as Passion*, 4; *Social Systems*, 442; 'Self-Organization and Autopoiesis', 143; 'The Control of Intransparency', *Systems Research and Behavioral Science*, vol. 14 (1997), 359–71 (367).

91 Niklas Luhmann, 'Deconstruction as Second-Order Observing' (1993), in *Theories of Distinction*, 106; *Ecological Communication*, 145–46; *Observations on Modernity*, trans. William Whobrey (Stanford, CA: Stanford University Press, 1998), 108.

information, Luhmann argues that communication is something performed by systems rather than humans, who must, in turn, adjust to that communication. Communication is possible because there is a communication system and not because there are subjects who communicate. The outcome of this is a sociological theory in which human beings are remarkably absent.[92] All we are left with is the idea of a world of instrumental systemic rationality. The obsolescence of philosophical arguments about reason as the basis of political claims about historical change leaves us with the rationality of systems, with which human desires and hopes are to be aligned. Whereas, in the past, 'man' has been first, in the future, the System must be first.

What might it mean to stop or change or control the system? For Luhmann, the question comes from a position of ignorance. The 'poiesis' in 'autopoiesis' may point to a 'making' or even 'production' of society, but this should in no way be taken to imply *control*. The lesson of the system, and thus the lesson of systems theory, is that we cannot and should not seek to control things. Control is an attribute of the System, comes from within the System, and operates as a series of movements and flows throughout. Managers are part of the system they manage and are thus managed, regulation is itself regulated, controls are themselves controlled. For these reasons, the concept of control requires redefinition, to better capture the way autopoietic and homeostatic systems regulate themselves. In this context, individuals (but not collectives) have a choice. An individual can 'choose to escape by declaring that society, not he, is sick. The repertoire that is then available ranges from anarchism through terrorism to resignation.' These are the actions of people 'driven literally to an extreme', but they are 'not what one finds in real life', for the System survives by bringing uncontrolled Subjects into line. For the most part, individuals will be a *homme-copie*, aping the actions of others within the system and thereby allowing the System to perpetuate itself. There is little point trying to change this: 'Protest against this is as futile as protest against domination.'[93] Forget protest,

92 Alan Wolfe, 'Sociological Theory in the Absence of People: The Limits of Luhmann's Systems Theory', *Cardozo Law Review*, vol. 13, (1992), 1729–43; John Mingers, *Self-Producing Systems: Implications and Applications of Autopoiesis* (New York: Plenum Press, 1995), 141–50.

93 Luhmann, 'Self-Organization and Autopoiesis', 151; *Introduction to Systems Theory*, 5, 119; *Social Systems*, 271, 483; 'Control of Intransparency', 367.

forget changing history, just observe and adjust to what has been *realized as the System*. To be sure, what has been realized as the System might generate within us the *worst fears*, but it *cannot be rejected*.[94] Herein lies the ultimate achievement of systems theory. The critique of political economy? Futile. The critique of alienation? Futile. Protest? Futile. Critique and protest are merely part of the System's information flows, processed by the System to help it adjust and stabilize. The System is meant to accommodate all difference such that opposition and protest become part of that very accommodation, bringing a close fit between the requirements of the System and individual desires.[95] In systems terms, what Luhmann really ought to have said is that protest is *noise*, merely a form of feedback which the System uses to autopoietically *rebalance* itself, turning disturbance and stability back into order. The rebalancing allows the System to incorporate its own deficiencies, its own contradictions, its own lacks, and the noise that affects it. The System uses the noise to maintain the silence.

This is systems theory as nothing other than affirmation of the system of bourgeois modernity and its modes of policing and administration. On the one hand, the twentieth-century project of 'scientific management' began by arguing precisely the point made by Luhmann and systems theory in general – namely that 'man' must be replaced by 'system'.[96] That was the starting point of Frederick Winslow Taylor's *Principles of Scientific Management*, for example, which paved the way for targets, management by numbers, and key performance indicators in order to train us into working flexibly, within the system and transform us in line with the system's 'mission', 'vision', and 'strategic goals'. On the other hand, the image was of systems that autopoietically change to stay the same. Politics becomes an affirmation of the System *and nothing else*. Luhmann's position highlights the fact that for all its purported radicalism in being a universal theory, systems theory deep down glorifies what exists for fear of its inevitable destruction. The System becomes its own ideology, as Habermas puts it, which is then

94 Niklas Luhmann, 'I See Something You Don't See' (1990), in *Theories of Distinction*, 193; *Love as Passion*, 172–78; *Theory of Society*, vol. II (1997), trans. Rhodes Barrett (Stanford, CA: Stanford University Press, 2013), 183–225.

95 Luhmann, *Introduction to Systems Theory*, 27.

96 Taylor, *Principles*, 7.

perpetuated by systems theory itself.[97] The System works in such a way *as if* it cannot be confronted, *as if* the political administration of society has been fully consolidated and *as if* our relations with one another can only ever be simply one subsystem among many. The System works *as if* the security of the System is all that matters.

Jack Burnham once commented that systems theory might ultimately be just 'another attempt by science to resist the emotional pain and ambiguity that remain an unavoidable aspect of life'.[98] It does so by encouraging us to give ourselves over to systems and to lose ourselves in the System. It is thus unsurprising that critics see in Luhmann's work an ideology for a certain kind of politics: neoconservative (at least in the terms laid out by Habermas), lacking any kind of emancipatory impetus, espousing a grim technocratic functionalism and a thinly disguised apology for the status quo, or even a fundamental inhumanity, in which all that matters is that the social order continues to be a system of social order, generating the necessary 'yes' to the social system that it needs to survive. Of course, in one sense, Luhmann and the systems theorists are on to something, in that the System does indeed present itself as an independent substance, endowed with a motion of its own and passing through a life-process of its own. That said, the point is to understand that this is how the System, like capital, *presents itself*. It is how capital and the system want to *appear*. Luhmann accepts the appearance and offers us a sociological argument to accommodate ourselves to it.

All of which brings us to Luhmann's argument that the System *immunizes itself against contradiction and change*. Herein lies the politics of immunity in Luhmann's work. Protest is a *social infection*, which is why the System must be immunized against it. The necessary 'yes' to society can be obtained through a reformulation of the opposition between modern capitalism and the totality of the social movements stimulated by it. Such reformulation requires a *socio-legal metabiology with immunity as its key security mechanism*.

97 Jürgen Habermas, 'Technology and Science as "Ideology"' (1968), trans. Jeremy J. Shapiro, in *Toward a Rational Society* (Cambridge: Polity, 1987), 102–07. He reiterates the point in *The Theory of Communicative Action* (1981), and *Between Facts and Norms* (1992).

98 Jack Burnham, *Great Western Salt Works: Essays on the Meaning of Post-Formalist Art* (New York: George Braziller, 1974), 11.

Writing the bulk of his work during immunology's 'golden age', Luhmann suggests that we distinguish between the 'structure of expectations' on the one side and 'society's immune system' on the other. This imagines a social immunology through which the System polices and secures itself. From this perspective, social conflict and law are part of the immune process. Social conflicts are things against which a social system needs to be immune, but in fact they themselves can serve as processes of immunization: they alert the social system to whatever might threaten it and then allow those threats to be recruited by the system. This is an integral part of society's immune system. 'The immune system protects' by processing the information that comes in the form of 'alarm signals' within the system. Conflict, contradiction, and crisis 'function as an alarm in the society's immune system' and can thus be imagined as problems of *system integration*. An immune system must be able to 'correct deviations' and re-establish the status quo ante. Yet the immune system must also be able to accept some changes to enable it to preserve the structures under attack. In that sense, the social system's immune system protects the system *through* conflict rather than against it: the system does not immunize itself *against the no* but *with the help of the no*; it does not protect itself *against changes* but *with the help of changes*. The immune system protects the system's autopoietic self-reproduction as a system and thus preserves social order as an order.[99] Immunity: security.

The fact that the immune system is essentially a security operation is why law is so central to Luhmann's account. For Luhmann, the task of a 'sociology of law' is to simply observe and describe the legal system as a system, but not to connect the legal system with the social and material foundations of that order. Describing the work of law reveals that 'the legal system serves as society's immune system'. How? Through 'the security attained by law'. 'Law must fulfil the function of an immune system', and it must do so by adopting and adapting conflicts and contradictions through a process of 'social immunization'. In this sense, the legal system functions 'immunologically'.[100] In *Law as a Social System* (1993) Luhmann offers a 'juridical immunology' to try to understand 'the autopoietic reproduction of the immune system of law'. In this

99 Luhmann, *Social Systems*, 69–72, 398, 404.

100 Luhmann, *Sociological Theory of Law*, 287; *Social Systems*, 373–76, 387.

system, law's 'immunity structures . . . prevent ongoing disappointment from resulting in the annulment of the structures'.[101]

> It remains open, uncertain and unpredictable when and for what reason someone will opt for a conflicting course and will oppose a potential norm with another potential norm. And there is, as is generally the case in immunology, no concrete, ready answer for such incidents. The legal system makes no prognosis about when conflicts will happen, and what the particular situation will be, who will be involved and how strong their involvement will be. The mechanisms of the legal system are geared to operate 'without fear or favour'. And they need time to prepare their immune response . . . One can also talk of an immunization system with respect to the way in which solutions, once found, reduce the probability of new 'infections'.[102]

Luhmann stresses that the idea of law as the immune system of society is part and parcel of a system's autopoiesis and is not meant metaphorically. As an immune system, law 'gets along without knowledge of its environment' and simply deals with the symptoms of the unknown causes of the disturbances, without being 'directed by the ideal of a rational practice'. 'The immune system enables society to cope with the structural risk of the continuous production of conflicts.' On this basis, the law's 'formulation of rules is like the generation of antibodies for specific cases'. The legal 'immune system stores its own system history'.[103] In similar fashion, Esposito comments that law performs an 'immune function' for community. Even before being codified, 'law is necessary to the very life of the community' through its 'immunizing role'. 'Just as the immune system functions for the human organism, law ensures the survival of the community in a life-threatening situation. It protects and prolongs life, snatching it from the jaws of death.' Building on Luhmann's arguments, Esposito suggests that 'we can no longer say that the immune

101 Niklas Luhmann, *Law as a Social System* (1993), trans. Klaus A. Ziegert (Oxford: Oxford University Press, 2004), 384, 477.

102 Luhmann, *Law*, 171.

103 Luhmann, *Law*, 476–77. Also see the applications and extension of this in Gunther Teubner (ed.), *Autopoietic Law: A New Approach to Law and Society* (Berlin: Walter de Gruyter, 1988); Gunther Teubner, *Law as an Autopoietic System* (Oxford: Blackwell, 1993).

system is a function of law' but, rather, that 'law is a function of the immune system'.[104]

The legal system's sensitivity as a system allows it to adjust in order to be able to offer security to the social system as a whole by incorporating and learning from threats, in classic autopoietic and immunologic fashion. Indeed, this ability to adjust is perhaps the defining characteristic of modernity: 'Certain historical tendencies stand out, indicating that since the early modern period, and especially since the eighteenth century, endeavors to secure a social immunology have intensified.' This is why Luhmann insists, when discussing protest, disobedience, and the 'no', that the system immunizes itself with help of the no. It is the immunological incorporation and accommodation of conflict that enables autopoietic reproduction.[105] Or, in Esposito's terms, the neutralization of protest or conflict does not eliminate it, but, rather, works 'for its incorporation in the immunized organism as an antigen at once necessary to the continuous formation of antibodies'.[106] Processes such as deflation and inflation are 'reactions of the immune system', as are reforms brought about by religious institutions in response to social change.[107] The System renders itself secure by immunizing itself. And, moreover, nothing goes wrong with this process: immunity is *always* security.

Situating Luhmann's work within the development of immunology as well as systems theory allows us to see why autopoiesis was such an important concept to Luhmann, whose account of social immunization was far more 'Varelean' than 'Burnetian'. With this in mind, it is not unreasonable to also note that Varela was deeply concerned about the extension of autopoiesis to the fields of social analysis, and the reasons for his concern are telling. At one point in *Autopoiesis and Cognition* (1972), co-authored with Maturana, after a discussion of autopoiesis in biological phenomena, they ask, 'What about human societies, are they,

---

104 Roberto Esposito, *Immunitas: The Protection and Negation of Life* (2002), trans. Zakiya Hanafi (Cambridge: Polity, 2011), 21, 50.

105 Luhmann, *Social Systems*, 370, 382–84, 394–97, 402.

106 Roberto Esposito, *Bíos: Biopolitics and Philosophy* (2004), trans. Timothy Campbell (Minneapolis: University of Minnesota Press, 2008), 61–62.

107 Niklas Luhmann, *Organization and Decision* (2011), trans. Rhodes Barrett (Cambridge: Cambridge University Press, 2018), 336; *Risk: A Sociological Theory* (1991), trans. Rhodes Barrett (Berlin: Walter de Gruyter, 1993), 177; *A Systems Theory of Religion* (2000), trans. David A. Brenner and Adrian Hermann (Stanford, CA: Stanford University Press, 2013), 128.

as systems of coupled human beings, also biological systems?' They openly acknowledge that between them they 'do not fully agree on an answer to the question posed by the biological character of human societies', and so they agree to postpone the discussion.[108] In work that followed, however, both Maturana and Varela fought against the kind of extension or transposition of autopoiesis such as undertaken by Luhmann. Varela several times from the late 1970s onwards expressed a fear that autopoiesis was being stretched to the point where it was losing its explanatory power.

> Frankly, I do not see how the definition of autopoiesis can be directly transposed to a variety of other situations, social systems for example . . . The notion of autopoiesis becomes a metaphor and loses its power. This is what has happened, in my view, with the attempts to apply autopoiesis directly to social systems . . . Autopoiesis seems mostly adequate to the domain of cells and animals.[109]

In a 1978 lecture on the Chilean Civil War, Varela suggests that such transpositions have a long historical dark side, underpinning the civil war and being found in the rise of fascism, all of which he connects to the violent polarising epistemology employed during those periods.[110] Maturana made his own position clear in a series of interviews conducted by Bernhard Poerksen. Asked about the sociological extension of the concept of autopoiesis and specifically Luhmann's use of the term, Maturana responded by suggesting that such an extension involves a real danger:

> Autopoiesis as a biological phenomenon involves a network of molecules that produces molecules. Molecules produce molecules, form themselves into other molecules, and may be divided into molecules. Niklas Luhmann, however, does not proceed from molecules producing molecules; for him, everything revolves around communication

108 Maturana and Varela, *Autopoiesis*, 118.

109 Francisco J. Varela, 'Describing the Logic of the Living: The Adequacy and Limitations of the Idea of Autopoiesis', in Milan Zeleny (ed.), *Autopoiesis: A Theory of Living Organization* (New York: Elsevier North Holland, 1981), 38.

110 Francisco Varela, 'Reflections on the Chilean Civil War', *Lindisfarne Letter*, 8 (1979), 13–19.

> producing communications . . . The decision to replace molecules by communications places communications at the centre and excludes the human beings actually communicating. The human beings are excluded and even considered irrelevant; they serve only as the background.

Pressed further, he goes on to make much the same political point as Varela:

> Just imagine for a moment a social system that is . . . functioning autopoietically. It would be an autopoietic system of the third order, itself composed of autopoietic systems of the second order. This would entail that every single process taking place within this system would necessarily be subservient to the maintenance of the autopoiesis of the whole. Consequently, the individuals with all their peculiarities and diverse forms of self-presentation would vanish. They would have to subordinate themselves to the maintenance of autopoiesis. Their fate is of no further relevance. They must conform in order to preserve the identity of the system. This kind of negation of the individual is among the characteristics of totalitarian systems.[111]

Varela once commented in conversation that Luhmann was 'the worst thing to have happened to him'.[112] Maturana seems to be suggesting that Luhmann's vision of the System could turn out to be the worst thing to happen to all of us. In the service of the security of the social order, the politics of immunity is a politics of subordination.

Whatever the rights and wrongs of the kind of extension of autopoiesis to non-biological systems, what is certainly significant about the kind of argument espoused by Luhmann is the fact that his socio-legal systems theory depends ultimately on the idea of immunity. The social system must have an immune subsystem and must do so for the security of the system. This is what allows its autopoietic self-regulation. The mystique of the immune system becomes part of the mystique of

---

111 Humberto R. Maturana/Bernhard Poerksen, *From Being to Doing: The Origins of the Biology of Cognition* (2002), trans. Wolfram Karl Koeck and Alison Rosemary Koeck (Heidelberg: Carl-Auer Verlag, 2004), 104–09.

112 The comment was made to Mark Hansen, cited in 'System-Environment Hybrids', in Clarke and Hansen (eds.), *Emergence and Embodiment*, 131.

autopoiesis, and both immunity and autopoiesis become categories through which the social system is understood, ultimately giving rise to the most powerful mystique of all: the mystique of a system that is so self-regulating that we should subordinate ourselves to it.

Where does all this leave us? A simple answer cannot be given. A complex answer will take us through the chapters to follow.

For now, let us note a comment made by Kenneth E. Boulding in a foreword to a collection of essays on autopoiesis published in 1981. Boulding notes that, although the concept of autopoiesis is a recent one, the idea existed in embryonic form in the eighteenth century, in Adam Smith's idea of an invisible hand. F. A. Hayek made more or less the same point, seeing the roots of twentieth-century systems theory's notions of autopoiesis, homeostasis, and self-organization in eighteenth-century liberal ideas concerning a system of natural liberty ordered by an invisible hand.[113] Their observation is telling, connecting as it does a concept at the heart of one way of imagining organismic self-regulation to three centuries of imagining capitalist order. It was another way of pointing to the three-hundred-year frenzy of systems thinking, but also spelling out for us in decidedly political terms that this tradition contains some remarkably liberal beliefs. Imagining a system in this way plays on a whole series of political arguments about order as well as system, and especially orders of 'natural liberty': how they are made (or not made), how they are managed (or not managed) and how they are policed (or not policed).

One of the problems that is left undealt with, however, is the System's *openness*. Conventional physics likes to deal with closed systems, in which 'no material enters or leaves', as Bertalanffy puts it. But, as Bertalanffy points out, the characteristic state of the living organism is of being an open system, where the boundaries are not well defined, the system is in constant exchange with its environment, and there is an import and export of material.[114] Or, as Jacob puts it, the living organism

113 Kenneth E. Boulding, 'Foreword', in Zeleny (ed.), *Autopoiesis*, xii; F. A. Hayek, *The Fatal Conceit: The Errors of Socialism* (London: Routledge, 1988), 9, 148; F. A. von Hayek, 'The Results of Human Action but Not of Human Design', in *Studies in Philosophy, Politics and Economics* (London: Routledge, 1967), 96–105; F. A. Hayek, *Law, Legislation and Liberty*, vol. I (London: Routledge, 1982), 20.

114 Bertalanffy, *General Systems*, 38, 128; Ludwig von Bertalanffy, 'The Theory of Open Systems in Physics and Biology', *Science*, vol. 111, no. 2872 (1950), 23–29 (23).

cannot be a closed system, because 'it cannot stop absorbing food, ejecting waste-matter, or being constantly traversed by a current of matter and energy from outside'.[115] It is abundantly clear that, politically speaking, talking about completely closed systems is pointless. The systems that are of any interest are thus always open and complex, always subject to flows of matter and energy rather than in a steady-state equilibrium. This openness points to the other, the foreign, the outside, the dangerous, the non-Self, the threatening, but also to internal cracks, fissures, and contradictions. 'In man, there's already a crack, a profound perturbation of the regulation of life', observes Lacan in writing about the death drive. Rule number one, then, according to Haraway: always look for the cracks.[116] In contrast to the system of genetic information transfer, which might be understood as 'totally, intensely conservative, locked into itself, utterly impervious to any "hints" from the outside world',[117] the immune system simply cannot be imagined in the same way. Immunity must be imagined as open. It is this openness that is precisely what enables us to look for immunity's cracks, fissures, and contradictions and which makes it so politically interesting as an idea because bodies are also so obviously open, also liable to cracks and fissures, also rent apart by contradictory forces. Even a body (politic) – that is, a society – imagined as immune.

The system of interest, then, is stable yet vulnerable, balanced yet with a potential for chaos, ordered but tending towards disorder. The system is liable to crack, even to break down. What is this system that collapses at the slightest noise? What is the noise that makes it collapse? How does the system manage its own noise, its own knowledge of its own potential chaos and collapse? How does it manage its fear of cracking, of breaking down? Do these things not make it nervous?

115 Jacob, *Logic*, 253.

116 Lacan, *Seminar, II*, 37; Donna Haraway in 'Nature, Politics, and Possibilities: A Debate and Discussion with David Harvey and Donna Haraway', *Environment and Planning D*, vol. 13, no. 5 (1995), 507–27 (514).

117 Monod, *Chance*, 110.

4

Order, Energy, Entropy, Bodies: Politics of Systems II

> *Callisto had learned a mnemonic device for remembering the Laws of Thermodynamics: you can't win, things are going to get worse before they get better, who says they're going to get better? . . . That spindly maze of equations became, for him, a vision of ultimate, cosmic heat-death . . . He found in entropy or the measure of disorganization for a closed system an adequate metaphor to apply to certain phenomena in his own world. He saw, for example, the younger generation responding to Madison Avenue with the same spleen his own had once reserved for Wall Street: and in American 'consumerism' discovered a similar tendency from the least to the most probable, from differentiation to sameness, from ordered individuality to a kind of chaos. He found himself, in short, restating Gibbs' prediction in social terms, and envisioned a heat-death for his culture in which ideas, like heat-energy, would no longer be transferred, since each point in it would ultimately have the same quantity of energy; and intellectual motion would, accordingly, cease.*
>
> Thomas Pynchon, 'Entropy' (1960)

There is a story told of a device invented for Archduchess Maria Theresa by Wolfgang von Kempelen in the 1770s. The device was an automated puppet in Turkish clothing, sat in front of a wooden table with a chess-board on top. A series of cogs, gears, and clockwork devices made it appear as though the machine could play chess, and the challenge was for people to try to beat this 'Mechanical Turk', who turned out to be remarkably adept at the game. The trick was that the contraption also held a skilful but hidden chess player who controlled the Turk's movements from inside the cabinet. The device is often used to capture the main contours of the debate about mechanical devices and the nature of life in the eighteenth century: Is there a hidden life, soul or vital substance within the mechanism? A ghost in the machine or demon in the wiring? But one might equally use the device to think about the concept of system. The terms 'machine' and 'system' were more or less synonymous in the eighteenth century, and so questions concerning the self-regulation, balance, control, and preservation of machines were simultaneously questions about systems. The system appears to be a life of its own, but is there something hidden?

At the same time, the ease and regularity with which the system appears able to regulate itself – 'like clockwork', as they said then and as we still say now – also posed another set of questions: A system might appear to be self-regulating, but does not its perpetual movement create the possibility of it going wrong? This tension was touched on early in the eighteenth century by Leibniz in a wry comment about how we feel when we think of our bodies as machine-systems:

> In German, the word for the balance of a clock is *Unruhe* – which also means disquiet; and one can take that for a model of how it is in our bodies, which can never be perfectly at their ease. For if one's body were at ease, some new effect of objects – some small change in the sense-organs, and in the viscera and bodily cavities – would at once alter the balance and compel those parts of the body to exert some tiny effort to get back into the best state possible; with the result that there is a perpetual conflict which makes up, so to speak, the disquiet of our clock; so that this appellation [*Unruhe*] is rather to my liking.[1]

The system's movement, then, is also a form of unrest and hence a kind of disquiet. For Leibniz, this *Unruhe*, with its connotations of restlessness, agitation, noise, and disturbance, is a model for our body always in

1 G. W. Leibniz, *New Essays on Human Understanding* (1704), trans. Peter Remnant and Jonathan Bennett (Cambridge: Cambridge University Press, 1996), 167.

motion, always having its 'balance' shifted and thus never perfectly at ease.

What, then, of other systems, such as the market system or political system? After all, as much as 'system' and 'machine' were treated as synonymous in the eighteenth century, both terms were also regarded as synonymous with 'economy' and 'constitution' and closely related in turn to 'institution' and 'organization'. These terms circulated around each other, reinforced one another, and played off one another. That they did so by moving between the body and the body politic, and between the state and the market, is a reminder that the image of *system* carried a lot of philosophical, historical, and political weight long before twentieth-century systems theory got hold of it. And, if 'system' carries a lot of weight on its own, then a *self-regulating system* carries even more. We are dealing, then, with one of the central fictions of systems and the System: self-regulation.

Immunological texts are replete with comments such as 'regulation truly serves as the nexus of understanding the immune system'.[2] But what does this mean? For a start, what does it mean now that we know that the immune system can go so badly wrong that it can destroy the thing it is meant to protect? But also, what does it mean given that the sentence could be reformulated with any 'X' standing in for 'immune': 'regulation truly serves as the nexus of understanding the capitalist system', to give just one example. With that in mind, this chapter puts immunity on the back burner while we further unravel the idea of system. One reason we do so is because in the chapter to follow we turn our attention to the immune system's close cousin, the nervous system. This attention will help develop an argument about the politics of nerves and nervousness that will allow us to bring immunity back to the front burner. This will then allow us to complete some of the arguments begun in chapters 2 and 3. A further reason to unravel the idea of system is because of how powerful the idea is, as we saw in the previous chapter. But, as we also started to note there, although a system can be described as self-regulating, homeostatic, autopoietic, and balanced, a system is always most interesting for its cracks, faults, fissures, and failures. What

2 Scott H. Podolsky and Alfred I. Tauber, *The Generation of Diversity: Clonal Selection Theory and the Rise of Molecular Immunology* (Cambridge, MA: Harvard University Press, 1997), 327.

interests us is a system's ability to splinter, collapse, and decay. Its faults are viewed as discord and destruction, while threats to its balance are understood as enemies of the system. Using the previous chapter's discussion of late twentieth-century systems theory as a springboard, this chapter will explore these issues by moving back to the earlier period in the three-hundred-year frenzy of systems; the period, that is, during which bourgeois thought sought to find system in the set of processes that constitute the heart of capitalist modernity. This will allow us to connect liberal assumptions about orderly systems and bodies in eighteenth-century Enlightenment liberalism with a similar set of assumptions central to systems theory in the second half of the twentieth century.

There is a kind of hidden mystery in Enlightenment liberalism, just as there is in systems thinking in general, and it is a mystery that remains with us and permeates the wider culture. It is a mystery that dominates the way we think about systems and the System, often leaving us enraptured – *mystified* even – by the 'webs of systematicity' within which we live.[3] As Clifford Siskin puts it, we experience system and 'the System' as working too well but also not well enough: you can't beat the system, we are told time and again, and yet we are also often ready to blame it for our woes, or for just breaking down. The power of 'system', Siskin suggests, 'is indexed by our capacity to find it everywhere and blame it for everything'. The more embedded we become in systems, the more we feel that we are being used by them; the more powerful the System becomes, the less we feel we can do anything to change it. 'System' has become one of the primary means of totalizing and rationalizing our experience of the social, political, and biological world.[4] As we saw in the previous chapter, this is a deeply ideological process. Amid such webs of systematicity, what happens to the human? How do we live with, in, and through a world of systems?

'The systems world may be for humans, *but* it may not be', William Ray Arney once wrote. His point was that the idea of systems creates a certain kind of equivalence between human systems and all other

3 To use a phrase from a discussion between Donna Haraway and David Harvey, in 'Nature, Politics, and Possibilities: A Debate and Discussion with David Harvey and Donna Haraway', *Environment and Planning D*, vol. 13, no. 5 (1995), 507–27.

4 Clifford Siskin, *System: The Shaping of Modern Knowledge* (Cambridge, MA: MIT Press, 2016), 154.

systems, to the extent that there is 'no priority of human living over any other sub-system within the global system'. From a systems perspective, the subsystem of 'living human beings' is conceptually equivalent to the subsystem of 'waste management'. This generates a feeling of powerlessness.[5] One of the consequences of thinking of the body as a system or system of systems is a feeling of empowerment – empowered as a complete system, empowered by the thought that everything works systematically – undermined by the feeling of being merely a system of systems. This simultaneous feeling of empowerment and powerlessness has implications for how we think about bodies, as Emily Martin has argued.[6] This is especially the case when the system goes wrong and spirals out of control, and even more the case when the system goes wrong and turns against its own body. Tacking back and forth between conceptions of system in the eighteenth century and conceptions of system in the twentieth will allow us to register an important dimension of how bodies are imagined: the body natural and the body politic, of course, but also the body of capital. After all, if anything appears to have a *life of its own*, it is capital, and one of the crucial periods of its seemingly *natural growth* was in the eighteenth century, the very period in which it became 'systematized', the first century of the three-hundred-year frenzy of systems and the very period alluded to in Boulding and Hayek's comments cited towards the end of the previous chapter.

Invisible Hands, Invisible Systems, Invisible Governors

In *Truth and Method* (1960), Hans-Georg Gadamer observed that the rise to prominence of the idea of system in the eighteenth century was in part an outcome of the new natural sciences forcing philosophy into 'systematization' as a means of harmonizing the old and new.[7] The attempt by philosophers to 'systematize' the world produced more and

5 William Ray Arney, *Experts in the Age of Systems* (Albuquerque: University of New Mexico Press, 1991), 58.

6 Emily Martin, *Flexible Bodies: Tracking Immunity in American Culture – From the Days of Polio to the Age of AIDS* (Boston: Beacon Press, 1994), 122, 134–35.

7 Hans-Georg Gadamer, *Truth and Method* (1960), trans. W. Glen-Doepel (London: Sheed and Ward, 1979), 516.

more works along the lines of the Earl of Shaftesbury, in his *Characteristicks of Men, Manners, Opinions, Times* (first published in 1711), who sketched out a system of bodies.

> Now, if the whole System of Animals, together with that of Vegetables, and all other things in this inferior World, be properly comprehended in *one System* of a Globe or Earth: And if, again, this *Globe* or *Earth* it-self appears to have a real Dependence on something still beyond; as, for example, either on its Sun, the Galaxy, or its Fellow-Planets; then is it in reality a PART only of some other System. And if it be allow'd, that there is in like manner a SYSTEM *of all Things, and a Universal Nature*; there can be no particular Being or System which is not either good or ill in that *general one* of the *Universe*.
>
> Therefore if any Being be *wholly* and *really ILL*, it must be ill with respect to the Universal System; and then the System of the Universe is ill, or imperfect.[8]

For Shaftesbury, this Universal System was a way of understanding 'the publick Good', which was also identified as the 'Good of the System', and the whole '*Animal-Order* or *OEconomy*'.

We shall return to the idea of 'animal-order' as 'oeconomy' shortly, in our discussion of the work of Adam Smith. For now, let us note that other thinkers created system after system, for example *A System of Natural Philosophy* (Thomas Rutherford, 1748), *Traité des systèmes* (Étienne Bonnot de Condillac, 1749), and *A System of Moral Philosophy* (Francis Hutcheson, 1755). And books started to appear on, for example, *System for the Compleat Interior Management and Oeconomy of a Battalion of Infantry* (1768), and *A System of the Law of Marine Insurances* (1787). As can be seen from these titles, 'system' was being applied to technologies and processes, objects and concepts, metaphysics and divinity, physics and geography, and even to systems themselves ('systems of systems'). Rousseau describes in his *Confessions* the many walks on which he reflected on his 'system', and Adam Ferguson would declare towards the end of the century that 'the love of science and the

8 Shaftesbury, *Characteristicks of Men, Manners, Opinions, Times* (1711), vol. II, bk. I, pt. II, sect. I.

love of system are the same'.[9] By the end of the eighteenth century, then, the intellectual universe of the West had become saturated with systems. In 1700, less than 5 percent of the publications in England had 'system' in their pages; a century later the figure had risen to just over 40 percent. A new word, 'systematize', emerged from the 1760s onwards.[10]

On the one side, system had become part of the *universe of knowledge*; on the other side, system was part of the *physical universe itself*. On the one side, systems of minds; on the other side, systems of bodies. With its suggestion of the world as a world of systems and the implication that this world could be known, the idea of 'system' had become a major trope of liberal Enlightenment: human beings had to learn to think in systems and to live in systems. My interest here lies in the liberal politics that lies within this idea of 'system' and how that gave rise to assumptions about regulation and order. For eighteenth-century liberalism, systems were understood as being capable of maintaining their equilibrium and balancing the various forces within, without the need for a higher authority. In other words, the good system was self-regulating. Otto Mayr writes that 'the concept of a self-regulating system was a subtle and abstract notion, a totally artificial mental construct'. Several decades of discussions about the idea, combined with the fact that it was assimilated in a manner that was not always clearly articulated and was perhaps even unconscious, gradually allowed the concept to enter popular usage and be applied to the most varied social and physical phenomena.[11] Herein lies the significance of Smith's work. Commenting early in his career on the successful English ability to invent and discover many

9 Jean-Jacques Rousseau, *The Confessions* (1782), trans. J. M. Cohen (Harmondsworth: Penguin, 1953), 343, 456; Adam Ferguson, *Principles of Moral and Political Science*, vol. I (Edinburgh: Strahan and Cadell, 1792), 278.

10 Ernst Cassirer, *The Philosophy of the Enlightenment* (1932), trans. Fritz C. A. Koelln and James P. Pettegrove (Princeton, NJ: Princeton University Press, 1951), 6–8; Siskin, *System*, 18, 29, 34–35, 56, 110–11, 126, 143, 150–52; Walter J. Ong, 'System, Space, and Intellect in Renaissance Symbolism', *Bibliothèque d'Humanisme et Renaissance*, vol. 18, no. 2 (1956), 222–39; Michel-Pierre Lerner, 'The Origin and Meaning of "World System"', *Journal for the History of Astronomy*, vol. 36, no. 4 (2005), 407–41; Richard A. Barney and Warren Montag, 'Introduction: Systems of Life, or Bioeconomic Politics', in Richard A. Barney and Warren Montag (eds.), *Systems of Life: Biopolitics, Economics, and Literature on the Cusp of Modernity* (New York: Fordham University Press, 2019), 8.

11 Otto Mayr, *Authority, Liberty and Automatic Machinery in Early Modern Europe* (Baltimore: Johns Hopkins University Press, 1986), 155.

things yet their distinct lack of success in 'arranging and methodizing their discoveries', Smith observes that 'there is no tolerable system of natural philosophy in the English language'.[12] It would be no exaggeration to say that Smith's life work was to produce just such a system.

In his early *Lectures*, Smith praises Descartes for inventing a system 'connecting all together by the same Chain' that could be 'universally received by all the Learned in Europe'. This is praise indeed, given Smith's view that the system in question contains 'not . . . a word of truth'.[13] Not a word of truth, but worthy of praise for *being a system*. In 'The History of Astronomy', Smith makes a similar point about the importance of system:

> Philosophy is the science of the connecting principles of nature . . . Philosophy, by representing the invisible chains which bind together all these disjointed objects, endeavours to introduce order into this chaos of jarring and discordant appearances, to allay this tumult of the imagination, and to restore it, when it surveys the great revolutions of the universe, to that tone of tranquillity and composure, which is both most agreeable in itself, and most suitable to its nature.

On this basis, philosophy is 'the most sublime of all the agreeable arts', the main task of which is to imagine and examine 'all the different systems of nature'.[14] In one sense, then, Smith is suggesting that what is important is *thinking through systems*, and, moreover, thinking through grand systems that simplify and popularize the other minor systems.

The full title of Smith's essay that is known as 'The History of Astronomy' is *The Principles which lead and direct Philosophical Enquiries; illustrated by the History of Astronomy*. The very same format is used in the title of two other early essays: *The Principles which lead and direct Philosophical Enquiries; illustrated by The History of the Ancient Physics*, and *The Principles which lead and direct Philosophical Enquiries; illustrated by The History of the Ancient Logic and Metaphysics*. These two essays are

---

12 Adam Smith, 'Letter to the Edinburgh Review' (1755), in Smith, *Essays on Philosophical Subjects*, eds. W. L. D. Wightman and J. C. Bryce (Indianapolis: Liberty Fund, 1982), 245.

13 Adam Smith, *Lectures on Rhetoric and Belles Lettres*, ed. J. C. Bryce (Indianapolis: Liberty Fund, 1985), 145–46.

14 Adam Smith, 'The History of Astronomy' (c. 1748–1758), in *Essays*, 45–46.

generally known by the titles 'History of Ancient Physics' and 'History of Ancient Logic and Metaphysics', but placing the full titles of the three essays together makes clear that they were an attempt to build a single argument, along the lines of a *Principles which lead and direct Philosophical Enquiry, Parts I, II, and III*. This suggestion would be supported by the fact that although Smith destroyed many of his manuscripts and papers prior to his death, he wanted those three essays to be kept by friends and considered for possible publication, perhaps as three chapters of a single book. If so, the book might well have ended up with the title *Principles of Philosophical Enquiry*. The point here is that such principles would appear to rest first and foremost on the ability to identify 'the order, harmony, and coherence' of nothing less than 'the Universal System', one part of which would turn out to be the 'system of natural liberty'.[15]

Smith closes the essay on astronomy by stressing the coherence of 'the Newtonian system' over that of Descartes. Newton's system, according to Smith, 'prevails over all opposition, and has advanced to the acquisition of the most universal empire that was ever established in philosophy'.

> And even we, while we have been endeavouring to represent all philosophical systems as mere inventions of the imagination, to connect together the otherwise disjointed and discordant phaenomena of nature, have insensibly been drawn in, to make use of language expressing the connecting principles of this one, as if they were the real chains which Nature makes use of to bind together her several operations. Can we wonder then, that it should have gained the general and complete approbation of mankind, and that it should now be considered, not as an attempt to connect in the imagination the phaenomena of the Heavens, but as the greatest discovery that ever was made by man, the discovery of an immense chain of the most important and sublime truths, all closely connected together, by one capital fact, of the reality of which we have daily experience.[16]

By treating systems as imaginative devices allowing the human mind to connect seemingly disjointed and discordant phenomena, Smith's

15 Adam Smith, 'The History of the Ancient Physics', in *Essays*, 116.
16 Smith, 'Astronomy', 103–05.

contribution to our understanding of system is a contribution in the realm of epistemology: privileging system at the level of *knowledge* means privileging the philosopher's ability to *invent systems in the imagination*. 'System' is a human endeavour 'to introduce order and coherence into the mind's conception' of what might otherwise appear to be a 'chaos of dissimilar and disjointed appearances'. System is thus a method to render 'the great theatre of nature [as] a coherent spectacle to the imagination'.[17] At the same time, however, the intention is to discover and understand the systems that really do exist as an 'immense chain' in the material world. The 'imaginary machine invented to connect together in the fancy those different movements and effects' works because the movements and effects are the reality. Orderliness stems from the 'connecting chains' that generate and regulate the 'regular and harmonious movement of the system, the machine or oeconomy'. Systems 'in many respects resemble machines', and each machine is 'a little system'. Moreover, 'the fitness of any system or machine to produce the end for which it was intended, bestows a certain propriety and beauty upon the whole'.[18] The beauty of the system is precisely its ability to *keep performing as a system*.

The task of the moral philosopher, then, is the production of systems of thought to help us understand not just physics and astronomy but also social order, public welfare, and good government. To this end, *system* and *police* go hand in hand. As he puts it in *The Theory of Moral Sentiments*, just after introducing the concept of an invisible hand:

> The same love of system, the same regard to the beauty of order, of art and contrivance, frequently serves to recommend those institutions which tend to promote the public welfare. When a patriot exerts himself for the improvement of any part of the public police, his conduct does not always arise from pure sympathy with the happiness of those who are to reap the benefit of it . . . The perfection of police, the extension of trade and manufactures, are noble and magnificent objects. The contemplation of them pleases us, and we are interested in whatever can tend to advance them. They make part of the great

17 Smith, 'Physics', 107.

18 Adam Smith, *The Theory of Moral Sentiments* (1759), eds. D. D. Raphael and A. L. Macfie (Indianapolis: Liberty Fund, 1982), 179–83.

> system of government, and the wheels of the political machine seem to move with more harmony and ease by means of them. We take pleasure in beholding the perfection of so beautiful and grand a system, and we are uneasy till we remove any obstruction that can in the least disturb or encumber the regularity of its motions. All constitutions of government, however, are valued only in proportion as they tend to promote the happiness of those who live under them. This is their sole use and end. From a certain spirit of system, however, from a certain love of art and contrivance, we sometimes seem to value the means more than the end, and to be eager to promote the happiness of our fellow-creatures, rather from a view to perfect and improve a certain beautiful and orderly system, than from any immediate sense or feeling of what they either suffer or enjoy.[19]

Now, Smith will not fully work out how the police machine and the system of natural liberty can be reconciled until much later in his work, as I have discussed at length in *A Critical Theory of Police Power*. The point here is that the spirit and love of system on the part of the human agents appears to reside in the desire to identify the harmony and regularity of the systems in the world, most obviously and necessarily the system of natural liberty. What is at stake is therefore not only the extent to which 'system' itself could be imagined and become 'systematic knowledge', but also what it means for a system to exhibit the kind of harmony and regularity in which liberty can best be exercised.

Smith's *Theory of Moral Sentiments* is best known for its account of sympathy, which is the basis of the 'regular and harmonious movement of the system'.[20] The final part of that book is on systems of moral philosophy that help establish the principles which ought to be the foundation of law.

> Human society, when we contemplate it in a certain abstract and philosophical light, appears like a great, an immense machine, whose regular and harmonious movements produce a thousand agreeable effects. As in any other beautiful and noble machine that was the production of human art, whatever tended to render its movements

19 Smith, *Theory*, 185.
20 Smith, *Theory*, 183.

> more smooth and easy, would derive a beauty from this effect, and, on the contrary, whatever tended to obstruct them would displease upon that account: so virtue, which is, as it were, the fine polish to the wheels of society, necessarily pleases; while vice, like the vile rust, which makes them jar and grate upon one another, is as necessarily offensive. This account, therefore, of the origin of approbation and disapprobation, so far as it derives them from a regard to the order of society, runs into that principle which gives beauty to utility, and which I have explained upon a former occasion; and it is from thence that this system derives all that appearance of probability which it possesses.[21]

He also suggests that 'amidst the turbulence and disorder of faction, a certain spirit of system is apt to mix itself with that public spirit which is founded upon the love of humanity'.[22] System brings unity, not faction; system is natural; system defeats disorder.

All told, then, the main task of the moral philosopher becomes the study of systems of civil government and their varied constitutions, commerce and defence.[23] Smith concludes *The Theory of Moral Sentiments* by suggesting that in a future work he will attempt precisely such a comprehensive analysis. That work was to be nothing less than Smith's 'Master System' (to use Siskin's phrase), called *An Inquiry into the Nature and Causes of the Wealth of Nations* (1776).

'When I first saw the plan and superstructure of your very ingenious and very learned Treatise', wrote Thomas Pownall to Smith on the publication of *The Wealth of Nations*, 'it gave me a compleat idea of that system, that might fix some first principles in the most important of sciences, the knowledge of the human community.' Smith's *Wealth of Nations* was, according to Pownall, '*principia* to the knowledge of politick operations'.[24] The reference to Newton's system (the *Principia*) was indicative of the extent to which *Wealth of Nations* is centrally organized around the same principle of system. The words 'system', 'systems' and

21 Smith, *Theory*, 316.

22 Smith, *Theory*, 232.

23 Smith, *Theory*, 186.

24 'A Letter from Governor Pownall to Adam Smith', 25 Sept. 1776, in *The Correspondence of Adam Smith*, eds. Ernest Campbell Mossner and Ian Simpson Ross (Indianapolis: Liberty Press, 1987), 337.

'systematical' together make close to two hundred appearances in *Wealth of Nations*.[25] There are no chapters in *Wealth of Nations* with the kind of titles found in standard economics texts, such as 'The Optimal Allocation of Resources' or 'The Efficiency of Free Trade'. There are, however, chapters on 'Systems of Political Economy' and 'The Principle of the Commercial, or Mercantile, System'. The reason, suggests Athol Fitzgibbons, is because Smith's primary purpose was less to challenge bad economic theories and more to challenge bad philosophical systems.[26] And he was to challenge them with his own Master System: the system of natural liberty.

Book IV of *The Wealth of Nations* is devoted to 'Systems of Political Economy'. There, Smith discusses commercial, mercantile, and agricultural systems, but he does so largely as a way into an argument about 'the simple system of natural liberty'. This system of natural liberty consists of a system of moral sentiments, a system of jurisprudence, and a system of political economy; it is a system of systems. Unlike 'systems either of preference or of restraint', the system of natural liberty is one in which 'the sovereign is completely discharged from a duty . . . of superintending the industry of private people'. Once this is achieved, the system of natural liberty 'establishes itself of its own accord'.[27] Despite this argument being famous for supposedly containing Smith's claims for 'free trade' or 'laissez-faire', Smith rarely uses these terms, preferring instead the 'system of natural liberty' or the 'system of perfect liberty', precisely because such a designation implies not just a set of economic arrangements but also much wider arrangements – much more *systemic* arrangements – concerning morals, laws, and constitutions.

The *naturalness* of the system is important here. Smith tells us that there is a *natural* price to which the prices of all commodities continually gravitate.[28] As the idea of *gravitating* prices suggests, the political economy of the system of natural liberty follows the model of the

25 Fred R. Glahe, *Adam Smith's "An Inquiry into the Nature and Causes of the Wealth of Nations": A Concordance* (Lanham, MD: Rowman and Littlefield, 1993), 520.

26 Athol Fitzgibbons, *Adam Smith's System of Liberty, Wealth, and Virtue: The Moral and Political Foundations of* The Wealth of Nations (Oxford: Oxford University Press, 1995), 172.

27 Adam Smith, *Inquiry into the Nature and Causes of the Wealth of Nations* (1776), eds. R. H. Campbell, A. S. Skinner, and W. B. Todd (Indianapolis: Liberty Fund, 1979), 687.

28 Smith, *Wealth*, 75.

sciences in discovering the system of laws behind the behaviour of natural objects. Smith describes such regulation when discussing, most importantly, the mechanism of the market. In the lingua franca of twentieth-century systems theory, the market generates information, most obviously about prices, which is used by actors to behave in the most rational of ways when making economic decisions. 'Every man . . . lives by exchanging, or becomes in some measure a merchant.'[29] In this way, the 'stability and permanency of the whole system' described in *The Theory of Moral Sentiments*, and the 'great connecting principle' described in the essay on astronomy, take on a far more political dimension, becoming in *The Wealth of Nations* the basis for discharging the sovereign body from any duty of managing the industry of private people or of directing such industry towards the interest of the society as a whole. The statesman who seeks to direct the desire for private gain into socially useful ends 'would not only load himself with a most unnecessary attention, but assume an authority which could safely be trusted, not only to no single person, but to no council or senate whatever'. Since every person becomes, in some measure, a merchant, the 'natural effort of every individual to better his own condition' is sufficiently 'capable of carrying on the society of wealth and prosperity'. At the heart of this system of natural liberty, the principles of moral sentiment and political economy are conjoined in such a way as to facilitate the self-regulation of commercial society, a society where things are 'left to follow their natural course'.[30] Out of this there emerges the self-regulation of the System as a whole.

Now, although this process of self-regulation is a natural process, about which we will shortly have more to say, the naturalness is somewhat occluded by Smith's invocation of the trope for which he has become most famous, and which is in turn one of the most notable images of the body in political thought: the invisible hand. Individuals neither intend to promote the public interest nor know how much they are promoting it. They intend only their own gain, and in this they are 'led by an invisible hand to promote an end which was no part of [their] intention'.[31] The invisible hand is identified with nothing in particular:

29 Smith, *Wealth*, 37.
30 Smith, *Wealth*, 116.
31 Smith, *Wealth*, 456.

not the sovereign, not the state, not the government. In the *Theory of Moral Sentiments*, he comments on the wealthy being 'led by an invisible hand' to distribute the necessaries of life to the poor and thereby 'advance the interest of the society and . . . the multiplication of the species',[32] a process which in turn generates a 'beautiful and grand system' of government. The 'invisibility' of the process would appear to be fundamental to the stability of the system.

The seemingly immaterial nature of a *body* that is *invisible* allows the system itself to appear to perform the action of the controlling subject. The invisible hand is a power inherent in the system, working only for the good of the System. The subject at the heart of Smith's system is subsumed under *the System*, which is somehow managed by 'a hand with no subject', as Catherine Packham puts it, in effect dematerializing any notion of power within the system, or at least rendering any such power invisible, and thereby dispensing with any question of a controlling function.[33] In one sense, the invisible hand is the process of self-regulation itself, and the moral economy and vital principle of sympathy dispenses with the older problem of how the passions are constrained.

It has many times been said that Smith's rather limited references to an invisible hand mean that it is in fact of limited interest and a rather unimportant constituent of his thought, which lacks anything like a coherent 'theory' of the hand.[34] Nonetheless, despite the limited explicit references to the invisible hand in Smith's work, the kind of process that we are expected to imagine with the hand pervades the whole of his work. Essentially, the invisible hand operates across all systems, from astronomy to morality to political economy and, ultimately, the system of natural liberty. Smith's 'system' is unintelligible if one disregards the

32 Smith, *Theory*, 184–85.

33 Catherine Packham, 'System and Subject in Adam Smith's Political Economy: Nature, Vitalism, and Bioeconomic Life', in Barney and Montag (eds.), *Systems of Life*, 102.

34 For example, J. Ronnie Davis, 'Adam Smith on the Providential Reconciliation of Individual and Social Interests: Is Man Led by an Invisible Hand or Misled by a Sleight of Hand?', *History of Political Economy*, vol. 22, no. 2 (1990), 341–52; Emma Rothschild, *Economic Sentiments: Adam Smith, Condorcet, and the Enlightenment* (Cambridge, MA: Harvard University Press, 2001), 116, 136; Gavin Kennedy, 'Adam Smith and the Invisible Hand: From Metaphor to Myth', *Econ Journal Watch*, vol. 6, no. 2 (May 2009), 239–63, (239); William D. Grampp, 'What Did Smith Mean by the Invisible Hand?', *Journal of Political Economy*, vol. 108, no. 3 (2000), 441–65.

invisible hand and the assumptions that lie behind it. Such assumptions would have been clear to his readers, given that the same idea, or at least some version of it, pervades the whole of eighteenth-century liberalism. In *Moll Flanders* (1722), for example, Daniel Defoe describes 'a sudden Blow from an almost invisible Hand [that] blasted all my Happiness', and in *Colonel Jack* (1723) he says that 'it has all been brought to pass by an invisible hand'. For William Leechman, in his 1755 preface to Francis Hutcheson's *A System of Moral Philosophy*, there is a 'silent and unseen hand of an all-wise Providence which over-rules all the events of human life'.[35] For Montesquieu in *The Spirit of the Laws*, some kinds of government are 'like the system of the universe, where there is a force constantly repelling all bodies from the center and a force of gravitation attracting them to it', and hence order is the result, as he puts it in *The Persian Letters*, of 'a secret chain which remains, as it were, invisible'.[36] In his account of the concept of self-interest in intellectual history, Pierre Force gives example after example of references to an invisible hand in eighteenth-century thought: Nicolas Lenglet Dufresnoy comments in 1735 that the 'economy of the universe is overseen by an "invisible hand"'; Charles Bonnet, a friend of Smith's, comments in 1765 on 'the economy of the animal . . . led towards its end by an invisible hand'; Jean-Baptiste Robinet, a translator of Hume's work, refers in 1761 to 'those basins of mineral water, prepared by an invisible hand'. Force himself points out the close connection between the use of 'economy' and 'invisible hand', rooted in the idea that the 'economy' of something is the relationship between the whole and the parts.[37] Understanding the economy of a system means understanding how the parts of that system work together in the system as a whole and how that system works within its wider environment. Even when not mentioned specifically, the general idea of an invisible hand pervades the whole liberal notion of self-sustaining systems, since the idea points to a process that works behind the backs of the System's conscious components. This is why related notions such

35 William Leechman, 'Preface', in Francis Hutcheson, *A System of Moral Philosophy*, vol. I (London, 1755), xii.

36 Montesquieu, *Spirit of the Laws* (1748), trans. Anne M. Cohler et al. (Cambridge: Cambridge University Press, 1989), 27; 'Some Reflections on the Persian Letters' (1754), in Montesquieu, *Persian Letters*, trans. C. J. Betts (Harmondsworth: Penguin, 1993), 283.

37 Pierre Force, *Self-Interest before Adam Smith: A Genealogy of Economic Science* (Cambridge: Cambridge University Press, 2003), 72–73.

as 'spontaneous order' and 'self-organization' also come to the fore in intellectual debate in the first half of the eighteenth century.[38] For Bernard Mandeville, civilized humans and the disciplined state occur 'without reflection, and Men by degrees, and great Length of Time, fall as it were into these Things spontaneously'. For Adam Ferguson, the establishment of social order occurs as 'the result of human action, but not the execution of any human design'. For David Hume, order occurs by 'secret and unknown causes'. For Turgot, 'the ambitious . . . have contributed to the designs of Providence . . . Their passions, even their fits of rage, have led them on their way without their being aware of where they were going.' For Edmund Burke, the system 'is placed in a just correspondence and symmetry with the order of the world . . . by the disposition of a stupendous wisdom' and works best when it works to 'its own operation'. For Burke's arch-rival, Thomas Paine, the system of civilization 'governs itself' according to 'the natural operation of the parts upon each other'.[39]

It is clear that at the heart of eighteenth-century liberalism lay the idea of a self-generating and self-regulating system created and sustained in some unknown and essentially unknowable way, offering what Jonathan Sheehan and Dror Wahrman describe as 'a common language for narrating stories of order in a world whose design could not be taken for granted'.[40] That the social order could be imagined as self-organizing was a crucial ideological trope within the cultural belief that the commercial system functions as a system that needs no control, is most

38 Jacob Viner, *The Role of Providence in the Social Order: An Essay in Intellectual History* (Princeton, NJ: Princeton University Press, 1972), 82; Jonathan Sheehan and Dror Wahrman, *Invisible Hands: Self-Organization and the Eighteenth Century* (Chicago: Chicago University Press, 2015), 99.

39 Bernard Mandeville, *The Fable of the Bees; or, Private Vices, Publick Benefits*, vol. II (1729), ed. F. B. Kaye (Indianapolis: Liberty Fund, 1988), 139; Adam Ferguson, *An Essay on the History of Civil Society* (1767), ed. Duncan Forbes (Edinburgh: Edinburgh University Press, 1966), 122; David Hume, 'Of the Rise and Progress of the Arts and Sciences' (1742) in Eugene F. Miller (ed.), *Essays: Moral, Political and Literary* (Indianapolis: Liberty Fund, 1985), 112; Anne-Robert-Jacques Turgot, 'On Universal History' (1751), in Ronald L. Meek (ed.), *Turgot on Progress, Sociology and Economics* (Cambridge: Cambridge University Press, 1973), 69; Edmund Burke, *Reflections on the Revolution in France* (1790), ed. Conor Cruise O'Brien (Harmondsworth: Penguin, 1968), 120, 282; Thomas Paine, *Rights of Man* (1791–1792), in *Rights of Man, Common Sense and Other Political Writings* (Oxford: Oxford University Press, 1995), 216.

40 Sheehan and Wahrman, *Invisible Hands*, 186.

successful when not controlled, and that operates according to a series of automatic ('invisible' but self-regulating) mechanisms around which we need to adapt ourselves.

Harold Laski once pointed to a 'mysterious alchemy' that permeates so much liberal thinking, and it is easy to see what he meant.[41] Indeed, some writers made this very point at the time, referring, for example, as François Quesnay does, to the 'magic of well-ordered society'.[42] In his essay on astronomy, Smith refers to the invisible hand of Jupiter as a device used by 'savages' to explain irregular events such as comets, eclipses, thunder, and lightning. The terror they experience induces them to ascribe the events to 'some invisible and designing power', and this for Smith is the origin of the 'vulgar superstition' which ascribes irregular events of nature to the pleasure or displeasure of invisible beings such as gods, demons, witches, genies, and fairies.

> It may be observed, that in all Polytheistic religions . . . it is the irregular events of nature only that are ascribed to the agency and power of their gods. Fire burns, and water refreshes; heavy bodies descend, and lighter substances fly upwards, by the necessity of their own nature; nor was the invisible hand of Jupiter ever apprehended to be employed in those matters. But thunder and lightning, storms and sunshine, those more irregular events, were ascribed to his favour, or his anger.[43]

In contrast to the savage, who finds an invisible hand in irregular and apparently supernatural events, the political economist finds the invisible hand in the regular events of the system of natural liberty (the market).

Smith's Calvinist background meant that he was very familiar with the argument that God is more than a momentary Creator who completed his work and then left it to be, but rather that the 'presence of the divine power' is evident in the 'perpetual condition' of the world; that is, in 'God's Providence'. For Calvin, God is not just Creator. God

41 Harold J. Laski, *The Rise of European Liberalism: An Essay in Interpretation* (London: George Allen and Unwin, 1936), 178.

42 François Quesnay, 'Extracts from *Rural Philosophy*' (1764), in Ronald L. Meek, *The Economics of Physiocracy: Essays and Translations* (London: George Allen and Unwin, 1962), 70.

43 Smith, 'Astronomy', 49–50.

engages in 'sustaining, cherishing, superintending, all the things which he has made, to the very minutest'. God constantly governs the universe in such a way that things do not happen fortuitously but are 'directed by the immediate hand of God'. The argument that everything that happens 'is by the ordination and command of God . . . under his hand' applies to the social and economic world, which is itself a manifestation of the 'secret' and 'invisible' workings of the spirit.[44] God is the *Governor*, but governs with an invisible hand. We find in this argument the Calvinist connection to capitalism, in which worldly activity becomes a religious imperative, in such a way as to reconcile the new religion and the desires of the expanding bourgeois class, as Max Weber argued at great length in *The Protestant Ethic and the Spirit of Capitalism* (1904–1905). In that sense it might appear as though we also find some Calvinist roots to Smith's argument. And yet in classical political economy the general motion of the system and its continuous order are explained not by reference to God, keen as those thinkers were to move away from Newton's system which needed the occasional intervention of God to 'wind up his watch from time to time', as Leibniz put it. As Foucault points out, for the economic world to be presented as appearing naturally opaque, economics had to constitute itself as a discipline that can demonstrate both the pointlessness and impossibility of a sovereign vision and control over the totality of the social body. In so doing it had to be constituted as an atheistic discipline.[45] Instead of God, the system is regulated by a 'hand' so invisible as to make it appear as though the System regulates itself and that nothing does or can control it. All of which begins to take us back to the body and its order, since these are ways of saying that the System is *natural*, rooted in a *vital* principle to the extent that it appears to have a *life of its own*. With this in mind, we can begin to slowly make our way back round to the question of immunity.

Smith's notion of a 'vital principle' is adopted from eighteenth-century vitalism and the physiological theories of Edinburgh physicians. Scottish philosophy in the eighteenth century adopted and adapted a range of

44 John Calvin, *The Institutes of the Christian Religion* (1536), trans. Henry Beveridge (Edinburgh: Calvin Translation Society, 1846), 51, 154–56, 166.

45 Michel Foucault, *The Birth of Biopolitics: Lectures at the Collège de France, 1978–1979*, trans. Graham Burchell (Houndmills: Palgrave, 2008), 282.

bodily images from natural philosophy in developing its moral and political claims, finding human physiology and anatomy especially rich as a metaphorical resource. Robert Whytt, for example, an occasional speaker at the Edinburgh Philosophical Society of which Smith was also a member, and about who we will say more in the following chapter, regarded the body as a self-regulating system operating independently of any controlling power such as a mind. Even a process that appears to be a question of *moral sentiment*, such as 'sympathy', turns out to be rooted in *physiology and vitality*. Smith begins *The Theory of Moral Sentiments* with a chapter on sympathy and observes that although several writers have conceived of man as selfish, there must be 'some principles in his nature, which interest him in the fortune of others, and render their happiness necessary to him'. His argument leads us very quickly to the body and pain: we lack the immediate experience of what other people feel, but we can imagine how we would feel in that situation. A person may be suffering 'upon the rack', but our senses themselves cannot tell us what they are suffering. Only in the imagination can we form any conception of their sensations.

> By the imagination we place ourselves in his situation, we conceive ourselves enduring all the same torments, we enter as it were into his body, and become in some measure the same person with him, and thence form some idea of his sensations, and even feel something which, though weaker in degree, is not altogether unlike them. His agonies, when they are thus brought home to ourselves, when we have thus adopted and made them our own, begin at last to affect us, and we then tremble and shudder at the thought of what he feels. For as to be in pain or distress of any kind excites the most excessive sorrow, so to conceive or to imagine that we are in it, excites some degree of the same emotion, in proportion to the vivacity or dulness of the conception.[46]

The fact that we are unable to literally feel the other's pain but can imagine it is also made about our relation to the dead. Our sympathy with the dead person lies in our ability to 'put ourselves in his situation' and

46 Smith, *Theory*, 9.

'*enter, as it were, into his body*'[47] When we experience joy with others, Smith claims that 'their joy *literally* becomes our joy: we are, for the moment, as happy as they are: our heart swells and overflows with real pleasure: joy and complacency sparkle from our eyes, and animate every feature of our countenance, and *every gesture of our body*'.[48]

On a broad level, then, the body is central to Smith's account of the sympathy that is so fundamental to the system of natural liberty, but which turns out to be a *vital* principle in ensuring the life of the body politic through the physical bodies of the subjects. On a more specific level, what is also central to Smith's thought is the body as labour. We noted above that Shaftesbury sought to imagine 'an Animal-Order or Oeconomy'. In his *Dictionary of the English Language* (1755), Samuel Johnson observes that 'OEconomy' concerns the management of a family or household but also the more general 'disposition of things' and their 'regulation', and, as well, a 'system of motions' or the 'distribution of every thing in its proper place'. 'OEconomy' is a question of the distribution and disposition of bodies. Wealth especially is the product of a very particular kind of body: the labouring body. In England, labourers employed in manual or unskilled work were often known as 'hands' (for example 'farmhands'), a usage that had been in place since the sixteenth century. The seemingly immaterial invisible hand on which the wealth of nations depends might therefore turn out to be of less importance than the labour conducted by the very material hands and bodies of the workers. At the same time, in opting to use 'wealth' rather than, say, 'opulence', in the title of his major text, Smith was opting for a word with roots in the Old English term 'weal', which has both physical and material connotations, but which also, with connotations of a common well-being ('commonwealth'), points to something very social. It points, in other words, to the body of the people.

Smith several times makes reference to the 'the whole body of the people' and 'the great body of the people', and I have elsewhere (in *Imagining the State*) shown that these terms are not straightforwardly identical to the older image of the 'body politic'. That older image underwent a huge transformation during the eighteenth century. Most of Smith's uses of the phrase 'the great body of the people' leave its meaning

47 Smith, *Theory*, 71, emphasis added.
48 Smith, *Theory*, 47, emphasis added.

undefined, but it would appear that the great body of the people is the labouring subgroup of the 'whole body of the people'. After outlining the ways in which the division of labour makes men stupid, renders them incapable of taking part in rational conversation, and leaves them lacking in tender sentiment, for example, Smith comments that 'in every improved and civilized society this is the state into which the labouring poor, that is, the great body of the people, must necessarily fall'. The 'whole body of the people', in contrast, would appear to refer to 'society' in general. The point, nonetheless, is that the whole body of the people is a *body*, which is why important features of that body's system, such as money, must be free to circulate like blood, and why monopolies threaten the 'natural balance' of the body's circulation system. This was just one of many such ideas that came to the fore during the period: 'stagnation' moves in the 1730s from connoting something about the physiology of blood and other liquids into the realm of political economy; 'revulsion' moves from the medical action or process of withdrawing blood or other liquids from one part of the body, to connote a sudden violent change of feeling or strong emotional reaction, and then, by the 1770s, also implies a sudden or marked reaction or reversal in trade and commerce; 'convulsion' moves from being an involuntary contraction, spasm, or stiffening of a muscle to being a violent social or political agitation; the old medical term 'fever' starts in the 1760s to be applied to markets ('feverish speculation', 'tulip fever'); 'epidemic' comes to be applied to processes such as industry (an 'epidemic of overproduction'); 'crisis' adds to its medical meaning the idea of a 'critical' and transformative moment in politics or economics. For Smith, the solution to such physio-political problems lies in restoring all the different branches of industry to 'that natural, healthful, and proper proportion which perfect liberty necessarily establishes.'[49]

This notion of a healthy body of the whole people is one reason why the *natural* propensity to truck, barter, and exchange is so important, generating as it does a 'natural balance' and 'healthy proportions' and therefore avoiding becoming an 'unwholesome body' such as Smith thought Britain was becoming.[50] The constant effort of every person to better their condition is

49 Smith, *Wealth*, 606.
50 Smith, *Wealth*, 604.

> enough to maintain the natural progress of things toward improvement ... Like the unknown principle of animal life, it frequently restores health and vigour to the constitution, in spite, not only of the disease, but of the absurd prescriptions of the doctor.[51]

Note, in this passage, the political reading of the 'doctor' or, as he sometimes calls mercantilist thinkers, the 'pretended doctors'. It is their 'absurd prescriptions' which intervene in the natural balance of health. 'There is no commercial country in Europe of which the approaching ruin has not frequently been foretold by the pretended doctors of this system, from an unfavourable balance of trade.'[52]

Smith was acutely aware of the long line of thinkers who argued that the sovereign must 'correct the people's errors in medical fashion', as John of Salisbury put it in *Policraticus*. The trouble with such people is that they always want to 'doctor' things, arranging and rearranging bodies and their parts in different ways and thereby adjusting what are in fact naturally self-regulating systems. One sees this in those systems in which the 'idea of the perfection of policy and law' is employed by those 'political speculators [and] sovereign princes' who believe that they should 'doctor' the social order.[53]

> Quesnai, who was himself a physician, and a very speculative physician, seems to have entertained a notion ... concerning the political body, and to have imagined that it would thrive and prosper only under a certain precise regimen, the exact regimen of perfect liberty and perfect justice. He seems not to have considered that in the political body, the natural effort which every man is continually making to better his own condition, is a principle of preservation capable of preventing and correcting, in many respects, the bad effects of a political oeconomy, in some degree, both partial and oppressive.[54]

What appears to be at stake in the body of people is the idea of a natural *balance* in no need of the intervention of political doctors, an idea which we raised in chapter 1 and about which we can now say a little more.

51 Smith, *Wealth*, 343.
52 Smith, *Wealth*, 496.
53 Smith, *Theory*, 234.
54 Smith, *Wealth*, 674.

For eighteenth-century liberal Enlightenment, 'balance' was a crucial term; we have already noted Leibniz's focus on the idea of the balance of a clock. The word 'balance' was borrowed from the Italian invention of double-entry bookkeeping, in which merchants could assess their 'balance of accounts', but the word also had connotations of the balance scales and hence the Aristotelean sense of harmony. The balance was a mean between two extremes, and hence could be applied in all sorts of ways: 'balance of power', 'balance of interests', 'balance of trade', 'balance of forces', 'balance of payments', and so on. So, balance was something to be *achieved*. The idea of balance could apply to external relations of state and internal relations between social forces, as observed by Jonathan Swift at the beginning of the century: 'It will be an eternal rule in politicks among every free people, that there is a balance of power to be carefully held by every state within itself, as well as among several states with each other.'[55] In other words, the balance that was expected to exist *within the System* was to be matched by a balance *between systems*. But for the liberal political economy underpinning the system of natural liberty, the idea of a balance of trade is very different to the balance of power. Whereas the balance of power is a conscious decision made by those with power and interest – states opt to establish a balance of power, constitutions opt for checks and balances between institutions – at the heart of the idea of a 'balance of trade' is an assumption of adjustment and equilibrium, ideas which together underpinned the belief that a system might be self-regulating and that order could be established as if by some kind of invisible hand. Balance would ideally be achieved through some kind of *self-regulating mechanism*.[56]

Inherent in the image of the self-regulating system is the idea that instability within a system can be turned into order. Imbalance becomes balance. The *system corrects itself*; it *maintains equilibrium* of its own accord. 'Balance shifts, everything gradually gets nearer and nearer to an equilibrium', observes Turgot;[57] people and things *adjust themselves*

55 Jonathan Swift, 'A Discourse of the Contests and Dissensions between the Nobles and the Commons in Athens and Rome' (1701), in Swift, *Major Works*, eds. Angus Ross and David Woolley (Oxford: Oxford University Press, 2003), 25.

56 Andrea Finkelstein, *Harmony and the Balance: An Intellectual History of Seventeenth-Century English Economic Thought* (Ann Arbor: University of Michigan Press, 2000), 90, 184.

57 Anne-Robert-Jacques Turgot, 'A Philosophical Review of the Successive Advances of the Human Mind' (1750), in Meek (ed.), *Turgot on Progress*, 44.

and things 'return to their level', notes Smith.[58] There is a *natural balance*. The market as a system of *natural* liberty will find its own balance. This is the reason why so many discussions of a 'balanced' market order fall back on examples of 'natural' balancing. For example, in his *Dissertation on the Poor Laws* (1786), Joseph Townsend tells the story of two goats placed on a South Seas island by its 'discoverer'. The goats, one male and one female, survive and multiply, and despite various hardships, especially if the number of goats gets too high, they find a degree of balance. When outside intervention occurs, as when a pair of greyhound dogs are placed on the island by the Spanish, the goats are threatened, and the balance disrupted. But the goats retire to the craggy rocks, where the dogs cannot follow them, and '*a new kind of balance was established*'. This is what social policy needs to replicate. Poor laws which seek to help the poorest and weakest members of a system end up interfering with the natural mechanism for maintaining balance. To offer 'a certain and a constant provision for the poor' would 'destroy the harmony and beauty, the symmetry and order of that system, which God and nature have established in the world'.[59] The logic of self-organizing systems in Townsend's *Dissertation*, as with so many other tracts of the period, moves from the natural to the social and back again, in order to treat the social as natural: to take seriously the idea of the *social body*.

Smith's account of the principle of natural balance in the body of the people was entirely consistent with his account of the principle of natural balance in the bodies of people, which are also systems; 'what I call *myself* is an organized system of matter', as Joseph Priestley put it.[60] In June 1776, Smith's close friend David Hume was seriously ill and travelled to Bath to sample the mineral waters there as a way of managing the illness. Smith's advice to his friend runs as follows:

> A mineral water is as much a drug as any that comes out of the Apothecaries Shop. It produces the same violent effects upon the Body. It occasions a real disease . . . over and above that which nature occasions. If the new disease is not so hostile to the old one as to

58 Smith, *Wealth*, 116, 687.

59 Joseph Townsend, *A Dissertation on the Poor Laws* (1786) (London: Ridgways, 1817), 40–45, emphasis added.

60 Joseph Priestley and Richard Price, *A Free Discussion of the Doctrines of Materialism, and Philosophical Necessity* (London: 1778), 75–76.

> contribute to expell it, it necessarily weakens the Power which nature might otherwise have to expell it.

Rather than actively intervening with even something as mild as spa waters, never mind professionally 'doctoring' it, Smith recommends a more 'natural' or 'balanced' path. 'Change of air and moderate exercise' will 'preserve the body in as good order as it is capable of being during the continuance of that morbid state'. Ultimately, one must allow 'the power of Nature to expel the disease'. A few weeks later, Hume died from his illness. Smith's advice reflects the views of his friend and doctor, William Cullen, probably the most important and influential teacher to emerge from Edinburgh Medical School.[61] It reflects too the more general ideas in medical practice and the new developments in 'vitalist' physiological theory of the Enlightenment that were so dominant in Edinburgh in the mid-eighteenth century, in contrast to the main strands of physiology in England which continued to be dominated by mechanical models. But then what of the 'mechanical model'?

In 1763, as Smith was conducting his lectures on jurisprudence at the University of Glasgow, James Watt was working there as an instrument maker, focusing on the inefficiency of Newcomen engines. Watt and Smith were acquainted from at least 1762 and were close enough for Smith to later become a subscriber to the copying machine patented by Watt in 1780.[62] From 1762 to just after the publication of *The Wealth of Nations* in 1776, Watt had more or less completed the development of a feedback system to enable the regulation of steam into an engine's cylinder. The feedback system he employed was called a centrifugal governor and the fact that it appeared just a few years after the publication of Smith's major book is worth noting, because Watt's steam engine was understood to be of revolutionary importance and would soon be seen as marking the start of a new industrial age. People would come from miles to see the engine in operation, and few who saw it would have failed to inquire about the purpose of the rapidly rotating centrifugal weights mounted over the machine. The purpose? Self-regulation. As

61 Smith, letter to David Hume, 16 June 1776, in *Correspondence*, 201; letter to Lord Shelburne, 15 July 1760, in *Correspondence*, 69.

62 Ian Simpson Ross, *The Life of Adam Smith* (Oxford: Oxford University Press, 1995), 146–47, 425.

Otto Mayr puts it, 'The steam engine governor probably did more than any other agent to publicize the concept of self-regulation among engineers and the general population.'[63] From this point onwards, if one wanted to explain the concept of self-regulation, one pointed to the steam engine governor.

Now, we have encountered a similar 'governor' in the passages from Calvin's *Institutes*, and returning briefly to twentieth-century systems theory, we can note that 'cybernetics' has its roots in the ancient term 'gubernator' (governor). But there is a wider dimension to this, about which we can now say a little more. The ancient Greek word κυβερνητικός (*kybernetikos*) means literally 'steering' or 'the work of a ship's helmsman'. The Greek name for this role was *kubernator* and can be found in both Plato and Aristotle, and in Latin it was rendered *gubernator*, as we have noted. The first definition of 'governor' in the *Oxford English Dictionary* is taken from the Latin *gubernator* and listed as 'steersman, pilot, person who directs or controls'. The image of the helmsman in ancient Greece is 'almost entirely confined to individuals in a position of sole direction', notes Roger Brock.[64] As Brock suggests, this might make it look 'suspiciously like an exhortation to tyranny', yet that is not quite the whole story. Although the image of the 'helmsman' is naturally ascribed to those exercising power or authority, thereby linking the term with the kind of guidance or control seen by many as the macrocosm of the universe being 'steered' by the gods, by the fourth century the language of 'steering' also comes to be applied to the *microcosm of the body*. In this regard, the idea of 'steering' or 'piloting' comes to straddle the three dimensions of medicine, politics, and the government of the self. The three dimensions point us towards a certain kind of knowledge and practice which together constitute an art applicable to bodies corporeal and bodies politic. Or as Foucault observes, this is an art applicable to the Prince, insofar as he must govern others, govern himself and cure the ills of the city; applicable to the person who governs himself as one should govern a city, that is by curing his own ills; and applicable to the Doctor, who has to give his views not only on bodies and how

63 Mayr, *Authority, Liberty*, 194–95.

64 Roger Brock, *Greek Political Imagery: From Homer to Aristotle* (London: Bloomsbury, 2013), 55–56.

they might be cured but also souls and how they might be steered.[65] The political, physiological, and personal come together, the ills of the body and the ills of the body politic governed in one and the same process, each subject to the same forces and principles, each 'piloted' in the same way.

Behind every concept of a self-regulating system lies a concept of the Governor, but as a model of leadership and control which connotes a kind of *authoritative cosmic guidance* somewhat akin to an *invisible hand*, keeping the system stable and predictable. Just as Watt's steam engine employed a centrifugal governor which facilitated the machine's self-regulation, so the same image permeated all self-regulating systems. This is precisely what would come back into play in the systems theory of the second half of the twentieth century. As Gregory Bateson would put it, 'In the steam engine with a "governor", the very word "governor" is a misnomer if it be taken to mean that this part of the system has unilateral control.' The Governor maintains stability by processing information from the system but 'has no control over these factors'. Even the Governor of a social system is itself 'controlled' by information from the system.[66] Hence one finds systems theorists making claims such as the one we identified in the previous chapter, drawing a functional equivalence between the Governor of a steam engine and the Governor of a prison.

Enlightenment liberalism and bourgeois political economy, then, encourage us to imagine society as constituted through a system of natural liberty operating as a vast, orderly, and living system in which economic behaviour and vital need go hand in hand. Economic 'life' is biological 'life', forms of 'natural life' which overlap or perhaps even coincide physiologically, medically, and organically. Along with others, Smith helped construct the idea of a system of natural liberty, understood through physiological principles such as 'preservation' and 'animal life', as an alternative to one founded on sovereign regulation. This allowed the idea of system to eventually take centre stage, by the end of the century, as *the System*.

---

65 Michel Foucault, *The Hermeneutics of the Subject: Lectures at the Collège de France, 1981–1982*, trans. Graham Burchell (New York: Picador, 2005), 249.

66 Gregory Bateson, *Steps to an Ecology of Mind* (Chicago: University of Chicago Press, 1972), 315–16.

The *OED* suggests that the first reference to 'the System' to describe the prevailing political, economic, or social order coincides with the emergence of 'system' as an *oppressive* force, adding that this then occurs 'frequently with capital initials': The System. This facilitates the possibility of a critique of the System as a whole. In the first decades of the nineteenth century, radical commentators began to use the idea of the System to grapple with, criticize, and challenge it as a system of *political domination and economic exploitation*. This became the basis of socialist and communist arguments, some of which, most notably Marx's, sought to identify the ways in which the System's totalizing logic also enables it to appear natural and thus independent of the control of human beings. We are encouraged to follow the procedures embedded in the System and behave as though it really is the System that maintains and regulates itself. At the same time, however, such criticism might be understood as part of the process through which the System preserves and sustains itself, in that the System's reform becomes part of the System's autopoietic self-regulation and balance. The System absorbs information and processes that information as part of its self-regulation, or perhaps even self-defence, as we discussed in chapter 2. The system of natural liberty turns out to be a new form of Leviathan, made all the more powerful by what seems to be its naturalness and its invisibility. The invisible compulsion within the System perpetuates a technology of submission to the System. The System requires the submission of subjects to the System, rooted in the subject's ignorance of the System. Control exists by appearing to not be control; control is everywhere and nowhere. This is the reason why the idea of systems *out of control* is so threatening. What happens when the System is out of control? Might it get so out of control that it ignores its own feedback, loses any autopoietic balance, and *turns on itself* and *destroys itself*?

Such questions really take us back to questions asked in previous chapters, but not answered there. What kind of system is it that collapses at the slightest noise? Better still, what kind of system is it that will destroy its own body? What kind of *chaos* would ensue if, for example, a system started attacking the very thing it is meant to secure? It would be no exaggeration to say that the three-hundred-year frenzy of systems exhibited not only a fear of chaos but also, and in particular, a fear of the chaos that would ensue when the System cracks and starts to be overturned from within. And what would *thinking* be if it did not confront this chaos?[67]

'Shipwrecked Passengers on a Doomed Planet'

'Chaos' is a word commonly used to describe states, events, or movements regarded as disorderly. In bourgeois thought, this includes everything from the state of nature to revolution. For Thomas Hobbes, 'chaos' connotes 'the unformed matter of the world' in the state of nature prior to the social contract and the creation of the sovereign power, but chaos always threatens to return in the shape of political rebellion, which reduces the order of sovereignty to the first chaos. For Edmund Burke, revolution dissolves everything into 'unsocial, uncivil, unconnected chaos'. In the twentieth century, liberal thinkers such as Ludwig von Mises (*Planned Chaos*, 1947) and F. A. Hayek (*The Constitution of Liberty*, 1960) derided as chaos everything they sought to resist. Chaos points to the dissolution of an ordered structure. As such, it threatens the idea of predictability, the founding myth of science.[68] But it also threatens the idea of order, the founding myth of ideology. If God created the world by bringing order to the chaos, then the bourgeoisie has taken it upon itself the responsibility of policing that order and protecting it from further chaos. Defeating chaos becomes the ultimate task of the police power, which is precisely why 'disorder' is so powerful a political idea and 'law-and-order' a seemingly irrefutable demand. The law becomes a bulwark against the chaos of disorder. Bourgeois order and chaos are understood to be in true binary opposition. At the same time, disease also feels like a chaotic disorder, and never more so than the autoimmune disease.

For systems theory, and for the wider culture, 'system' is a synonym for 'order' and an antonym of 'chaos'. Chaos theory has therefore been described as a specialized application of systems theory.[69] Key systems

67 Gilles Delueze and Félix Guattari, *What Is Philosophy?* (1991), trans. Hugh Tomlinson and Graham Burchill (London: Verso, 1994), 208.

68 Ilya Prigogine, *From Being to Becoming: Time and Complexity in the Physical Sciences* (San Francisco: W. H. Freeman and Co., 1980), 214.

69 Stephen H. Kellert, *In the Wake of Chaos: Unpredictable Order in Dynamical Systems* (Chicago: University of Chicago Press, 1993), 5, 81–83; N. Katherine Hayles, *Chaos Bound: Orderly Disorder in Contemporary Literature and Science* (Ithaca, NY: Cornell University Press, 1990), 8.

concepts such as 'autopoiesis' are always imagined as mechanisms for warding off chaos. The idea of system thus expresses a belief deeply embedded in our culture: that system is our security against chaos.

Martin Meisel explains that the tendency in the Newtonian system was to banish chaos from the physical world by stressing the uniformity of physical laws. Chaos could be held at bay. Systems were imagined as organized in such a way to enjoy stability so long as no *external* causes troubled them. This belief helped maintain the idea of a universal harmony. As such, chaos was pushed *outside* the system.[70] This vision was transformed with the emergence of modern energetics. In contrast to the Newtonian system, modern energetics proposes a system that contains within it the *energy and propensity of its own destruction*. Where is there room for chaos in the post-Newtonian universe? asks Meisel. The answer turns out to be *everywhere*, so that even the smallest of changes within a system can result in very large differences in that system's behaviour. Can the flap of a butterfly wing in New York set off a tornado in China? Can some unusual practices with live bats in China set off a global pandemic? It was in the heyday of twentieth-century systems theory in the early 1970s that the famous butterfly question was posed (by Edward Lorenz in a paper at the American Association for the Advancement of Science in 1972), but the problem goes back a lot farther. Because once chaos can be imagined as *everywhere*, the system can be imagined as always already under threat from *within*. This is why the *cracks* in the system matter. Much as our major representation of chaos follows the Christian-liberal tradition of thinking about it as the original void at the beginning of creation and then as the void created by the lack of law and order, it is important to recall the parallel history, stemming from the Greek *khaos* as a self-opening abyss, a yawning, a gaping open: chaos stems from the cracking open of the system.

Despite banishing chaos from the physical world, the idea of a chaos within was beginning to become apparent even during the eighteenth century, as it became increasingly clear that human systems were by no means as predictable as systems such as the planetary. What brought it home were the financial bubbles of 1719–1720, especially the South Sea Bubble of 1720. Generally taken to be the first-ever 'crash' on the stock

70 Martin Meisel, *Chaos Imagined: Literature, Art, Science* (New York: Columbia University Press, 2016), 295.

market, the South Sea Bubble seemed to many at the time to reveal how the order of a seemingly self-regulating system could quickly turn to disorder, and in which the 'system' in question was of a decidedly material nature and rooted in the actions of human beings.[71] Newton himself sensed this possibility of disorder but reduced it to human irrationality: the South Sea Bubble is said to be the reason for his famous quip that he could calculate the motions of erratic stars, but not the madness of the multitude.[72] But Montesquieu was to make the broader point, in a comment attributed to Rica in a 1720 entry in *The Persian Letters* (Letter 138), when Rica observes that the economic disorder has meant people who were rich six months ago are now in poverty while others have made fortunes. The nobility seems ruined, the classes are in confusion. 'The State is in chaos!' Rica exclaims. It looks like the system contains within itself threats and contradictions that it might not be able to contain. This, it turns out, is a rather significant problem, as the global pandemic alluded to in the second question brought home: if the chaos, or virus, or disease, or general disorder, begins within the System, with a crack in the System, then maybe it is the System that is the problem.

'In many cases it is difficult to disentangle the meaning of words such as "order" and "chaos"', write Ilya Prigogine and Isabelle Stengers. 'Is a tropical forest an ordered or a chaotic system?'[73] Nonetheless, there is something in chaos theory that gets us to the heart of the contemporary cultural condition. There is a palpable social, political, and cultural impasse between order and chaos, which in turn generates fear and insecurity. To put it bluntly: chaos makes us nervous. It makes us nervous about systems, about our bodies and about the body politic. The butterfly and the bat examples demand that we imagine a system of systems, but also a body of bodies. In that sense, it provokes us to think further about immunity and the idea of an immune system. Chaos theory therefore gets to the heart of the contemporary cultural condition because it makes demands on how we imagine immunity and security as well as system.

71 Sheehan and Wahrman, *Invisible Hands*, 93–110.

72 The source of the quip is usually traced to H. R. Fox Bourne, *The Romance of Trade* (London: Cassell Petter and Galbin, 1871), 292.

73 Ilya Prigogine and Isabelle Stengers, *Order out of Chaos: Man's New Dialogue with Nature* (London: Flamingo, 1984), 169.

The reason that the fear of chaos is writ large in bourgeois ideology is clear: a system cannot regulate itself in an orderly fashion if there is chaos. So long as chaos can be bracketed by insisting that it is outside, then that need not be such a major problem. But another way to bracket it out is to endlessly insist on self-regulation: the commercial order can self-regulate, the steam engine can self-regulate, the system of natural liberty can self-regulate, there are no cracks, so what exactly are you worrying about?

In *What Is Life?* Lynn Margulis and Dorion Sagan make the following observation:

> In steam engines, coal was burned in carbon joined with oxygen, a reaction that, generating heat, made machine parts move. The left-over heat that was generated was unusable. The heat in a cabin on a snow-covered mountain seeks with seeming purpose any available crack or opening to mix with the cold air outside. Heat naturally dissipates. This dissipative behavior of heat illustrates the second law [of thermodynamics]: the universe tends toward an increase in entropy, toward even temperatures everywhere, as all the energy transforms into useless heat spread so evenly that it can do no work.[74]

Their reference to entropy and the second law of thermodynamics points us to the fact that by the late nineteenth century, mechanics had been supplanted by thermodynamics, transforming the way the world was imagined. The *crack* points to a system reaching its limits. But their reference to the dissipation of heat from a steam engine in a book on 'life' points to the fact that life (and its protection) and capital (and its security) are often imagined together. How so?

During the nineteenth century, attempts by engineers such as Nicolas Carnot to improve the efficiency of the steam engine led to the question of energy becoming one of the fundamental questions of physics. The central concepts of Newtonian physics – space, time, mass, and force – were increasingly displaced by space, time, mass, and energy. The range of scientists who were seeking a solution to the problem of energy loss – Thomas Kuhn points to at least twelve different scientists who might

74 Lynn Margulis and Dorian Sagan, *What Is Life?* (London: Weidenfeld and Nicolson, 1995), 22.

be identified in the 'simultaneous discovery' of energy conservation[75] – were interested in the way that the heat which drives an engine does not get lost but *moves* from a hotter into a cooler body; hence 'thermodynamics', a term coined by William Thomson (Lord Kelvin) in 1854. Among those researchers, Rudolf Clausius wanted a word to name a certain quantity that was related to energy but was not energy itself. Clausius adopted a Greek word meaning *transformation-contents*, namely 'entropy', which was intended to be etymologically close to 'energy' (from the Greek *energia*, usually implying *work-contents*). He thereby helped formulate a second law of thermodynamics with the cosmic reach of the first law but pointing in a very different direction to the first law. Clausius:

> One hears it often said that in this world everything is a circuit. While in one place and at one time changes take place in one particular direction, in another place and at another time changes go on in the opposite direction; so that the same conditions constantly recur, and in the long run the state of the world remains unchanged. Consequently, it is said, the world may go on in the same way for ever.
>
> When the first fundamental theorem of the mechanical theory of heat was established, it may probably have been regarded as an important confirmation of this view.[76]

This first theorem was the idea of the conservation of energy. '*One form of Energy can be transformed into another form of Energy, but the quantity of Energy is thereby never diminished; on the contrary, the total amount of Energy existing in the universe remains just as constant as the total amount of Matter in the universe.*'[77] The truth of this theorem was beyond doubt and was widely assumed to expresses the fundamentally unchanging nature of the universe. Yet, according to Clausius, a second fundamental theorem contradicts it.

75 Thomas Kuhn, 'Energy Conservation as an Example of Simultaneous Discovery' (1959), in *The Essential Tension: Selected Studies in Scientific Tradition and Change* (Chicago: University of Chicago Press, 1977).

76 Rudolf Clausius, 'On the Second Fundamental Theorem of the Mechanical Theory of Heat: Lecture, 23 Sept. 1867', *London, Edinburgh, and Dublin Philosophical Magazine and Journal of Science*, vol. 35, no. 239 (1868), 405–19 (417).

77 Clausius, 'On the Second Fundamental Theorem', 418.

> I have endeavoured to express the whole of this process by means of one simple theorem, whereby the condition towards which the universe is gradually approaching is distinctly characterized. I have formed a magnitude which expresses the same thing in relation to transformations that energy does in relation to heat and ergon – that is, a magnitude which represents the sum of all the transformations which must have taken place in order to bring any body or system of bodies into its present condition. I have called this magnitude *Entropy*.[78]

In contrast to the principle that the quantity of energy is never diminished, Clausius sought to show that when the positive transformations exceed the negative, an increase of entropy occurs.

> Hence we must conclude that in all the phenomena of nature the total entropy must be ever on the increase and can never decrease; and we thus get as a short expression for the process of transformation which is everywhere unceasingly going on the following theorem:-
>
> *The entropy of the universe tends towards a maximum.*[79]

Whereas the first law of thermodynamics points to the changing forms of energy in a universe in which the total energy is conserved, the second law points to a *universal tendency to the dissipation of energy*; prior to Clausius's work on entropy, the second law was known as the 'dissipation law'.

The implications of this were stark. In contrast to the mechanics of something like a waterwheel, which simply uses the motion of the water to engage the motion of the wheel's blades, a steam engine transforms heat into motion, and the transformation is irreversible. Imagined as a steam engine, the Earth itself appears to be losing heat and dissipating energy. This utterly transforms the way the Earth is imagined. As stated by Clausius: 'The more the universe approaches this limiting condition in which the entropy is a maximum, the more do the occasions of further changes diminish', from which it would seem that 'no further change could evermore take place'. In effect, 'the

78 Clausius, 'On the Second Fundamental Theorem', 419.

79 Clausius, 'On the Second Fundamental Theorem', 419.

universe would be in a state of unchanging death'. Or, as Thomson put it in an early essay on the subject: 'Within a finite period of time to come the earth must again be unfit for the habitation of man as at present constituted.'[80]

Such stark formulations appeared to many at the time to have 'tossed the universe into the ash-heap', as Henry Adams put it. Time and again, we find formulations similar to those of Thomson and Adams, as just a few examples show: the universe is 'condemned to a state of eternal rest' and the 'life of men, animals, and plants could not of course continue' (Hermann von Helmholtz, 1854); 'the terroristic nimbus of the second law . . . has made it appear to be an annihilating principle for all living beings of the universe' (Josef Loschmidt, 1876); 'all attempts at saving the universe from this thermal death have been unsuccessful' (Ludwig Boltzmann, 1886); 'if the Solar System is slowly dissipating its energies . . . are we not manifestly progressing towards omnipresent death?' (Herbert Spencer, 1900).[81] The *fin de siècle* looked rather like the *fin du monde*. The new century that followed could not shake off the looming sense of an end, as a few more examples show: 'the sun must ultimately grow cold and this earth must become a dead planet moving through the intense cold of empty space' (Ernest Rutherford, 1905); 'time's arrow' points towards the cataclysmic death of the universe (Arthur Eddington, 1928); the law suggests that we are 'a stage in the decay of the solar system' and that 'human life and life in general on this planet will die out in due course' (Bertrand Russell, 1927). As Erwin Schrödinger put it in *What Is Life?*

80 Prof. W. Thomson, 'On a Universal Tendency in Nature to the Dissipation of Mechanical Energy', *London, Edinburgh, and Dublin Philosophical Magazine and Journal of Science*, vol. 4, no. 25 (1852), 304–06; also Sir William Thomson (Lord Kelvin), 'On the Age of the Sun's Heat', *Macmillan's Magazine*, March 1862, in *Popular Lectures and Addresses*, vol. I (London: Macmillan, 1889), 349–68.

81 Henry Adams, *A Letter to American Teachers of History* (Washington: J. H. Furst Press, 1910), 4; H. Helmholtz, 'On the Interaction of Natural Forces: A Lecture Delivered February 7, 1854, at Königsberg in Prussia', in *Popular Lectures on Scientific Subjects*, trans. E. Atkinson (New York: D. Appleton and Co., 1885), 172–73; Loschmidt in Stephen G. Brush, *The Kinetic Theory of Gases: An Anthology of Classic Papers with Historical Commentary* (London: Imperial College Press, 2003), 190; Ludwig Boltzmann, 'The Second Law of Thermodynamics' (1866), in Ludwig Boltzmann, *Theoretical Physics and Philosophical Problems*, ed. Brian McGuiness (Dordrecht: D. Reidel Publishing, 1974), 19; Herbert Spencer, *First Principles*, 6th edn. (London: Williams and Norgate, 1900), 471.

(1944): entropy tells us that 'the whole system fades away into a dead, inert lump of matter'.[82]

It is perhaps important to note that the meaning of 'entropy' was far from agreed upon in the nineteenth century. In 1868, for example, in the first book with 'thermodynamics' in the title, Peter Guthrie Tait was already suggesting that

> we shall . . . use the excellent term Entropy in the opposite sense to that in which Clausius has employed it, *viz* – so that the *Entropy of the Universe tends to zero* . . . rather than the unmodified nomenclature of Clausius, according to which the *Entropy tends to a maximum*.[83]

The multiplicity of meanings has added to the sense that entropy is excruciatingly difficult to grasp; even physicists acknowledge that it is 'not easily understood even by physicists'.[84] This situation became even more difficult when twentieth-century systems thinking appropriated the concept. When Claude Shannon's *Mathematical Theory of Communication* was published in 1949, one of the earliest reviews noted that the book 'develops a concept of *information* which, surprisingly enough, turns out to be an extension of the thermodynamic concept of *entropy*'.[85] Indeed, it is apparently the case that Shannon did not initially intend to use such a complicated and highly charged term for his information measure, preferring instead the simpler word 'uncertainty', but changed his mind after a discussion with John von Neumann, who

82 Ernest Rutherford, 'Radium – The Cause of the Earth's Heat' (1905), in *The Collected Papers of Lord Rutherford of Nelson*, vol. I (London: George Allen and Unwin, 1962), 585; Sir Arthur Eddington, *The Nature of the Physical World* (London: Macmillan, 1928), 68–69; Bertrand Russell, *Why I Am Not a Christian* (1927), in *Why I Am Not a Christian and Other Essays on Religion* (London: Routledge, 2004), 8; Erwin Schrödinger, *What Is Life?* (1944), in *What Is Life? with 'Mind and Matter' and 'Autobiographical Sketches'* (Cambridge: Cambridge University Press, 1992), 69.

83 P. G. Tait, *Sketch of Thermodynamics* (Edinburgh: Edmonston and Douglas, 1868), 29.

84 D. ter Haar, 'The Quantum Nature of Matter and Radiation', in R. J. Bling-Stoyle, et al. (eds.), *Turning Points in Physics* (Amsterdam: North-Holland Publishing Co., 1959), 37. Likewise, Nicholas Georgescu-Roegen, *The Entropy Law and the Economic Process* (Cambridge, MA: Harvard University Press, 1971), 9; James Gleick, *The Information: A History, A Theory, A Flood* (London: Fourth Estate, 2011), 269.

85 Arthur W. Burks, 'Review of *The Mathematical Theory of Communication*', *The Philosophical Review*, vol. 60, no. 3, 1951, 398–40 (399).

encouraged Shannon to use 'entropy' precisely because no one knew what it was.[86] To add to the confusion, where Shannon and many information theorists treat information and entropy as identical, Wiener opposes the two terms, arguing that whereas entropy measures a system's degree of randomness and disorganization, information gives order ('in-forms') and is thus a measure of organization; 'information is information, not matter or energy'.[87] In one sense, these differences and complications are not of particular relevance to the discussion here, which concerns the general sense of foreboding and nothingness entailed by 'entropy' and the fact that the tendency towards nothingness is understood as *irreversible*. The laws of thermodynamics and the concept of entropy point to the disorder in any system and the fact that all systems, including the System, come to an end. 'In a very real sense we are shipwrecked passengers on a doomed planet', Wiener observed.[88] Hope must be abandoned. No future.

On the one hand, this idea of irreversible loss and thus cosmic heat death is precisely why the second law is the most *metaphysical* of the laws of physics, as Henri Bergson put it. It is why a poetics of energetics emerged and why entropy appeared to have religious as well as metaphysical overtones.[89] And yet at the same time, and on the other hand, there is also something palpably *human* about entropy. Entropy 'smells of its human origin', as one scientist put it in 1941.[90] What smell is that? Might it be the smell of death?

Given our discussions of suicide and the death drive in chapter 2, it is probably worth noting that Freud will develop his own account of

86 Shannon in conversation with Myron Tribus, cited in Myron Tribus and Edward McIrvine, 'Energy and Information', *Scientific American*, vol. 225, no. 3 (1971), 179–88 (180).

87 Norbert Wiener, *Cybernetics, or, Control and Communication in the Animal and the Machine*, 2nd edn. (Cambridge, MA: MIT Press, 1961), 132.

88 Norbert Wiener, *The Human Use of Human Beings: Cybernetics and Society* (1950) (Boston: Da Capo Press, 1954), 40.

89 Henri Bergson, *Creative Evolution* (1907), trans. Arthur Mitchell (London: Macmillan, 1960), 256; Stewart and P. G. Tait, *The Unseen Universe, or Physical Speculations on a Future State* (London: Macmillan, 1875); Yehuda Elkana, *The Discovery of the Conservation of Energy* (Cambridge, MA: Harvard University Press, 1974), 54; Barri J. Gold, *ThermoPoetics: Energy in Victorian Literature and Science* (Cambridge, MA: MIT Press, 2010).

90 P. W. Bridgman, *The Nature of Thermodynamics* (Cambridge, MA: Harvard University Press, 1941), 3.

energy in his early *Studies on Hysteria* written with Josef Breuer, and his later concept of the death drive could be understood as an entropic tendency directing the organism towards nothing, to its own death. This would explain why *Beyond the Pleasure Principle* alludes at various points to Helmholtz and Clausius on the degradation of energy. When writing about neurotic behaviour in 'From the History of an Infantile Neurosis' (1918), Freud comments that we have to think of people who exhibit such behaviour in terms of 'the conversion of psychical energy no less than of physical', and in this light he suggests that 'we must make use of the concept of an *entropy*, which opposes the undoing of what has already occurred'. In a later essay, 'Analysis Terminable and Interminable' (1937), he returns to the point and uses a very distinct phrase: 'psychical entropy'. This adoption of a key term from energetics was no doubt due to the fact that 'energy' is a crucial term for Freud. The pleasure principle is a matter of psychical *economy* and *energy*: 'energy mechanism', 'energy investment', 'energy transformation', 'energy work', and 'energy reserve' are all terms he uses to situate the Self within what Teresa Brennan calls an 'intersubjective economy of energy'.[91] But it is also no doubt due to the fact that Freud was sensitive to a fundamental tension between conservation and death, and this tension would appear to replicate the fundamental tension between the two laws of thermodynamics. As Lacan puts it, Freud starts with a conception of the nervous system according to which it tends towards some kind of homeostatic equilibrium, but once he comes up against the dream, he begins to treat the brain as a dream machine, and this requires thinking through the energy of the machine in question. It took him another twenty years to realize what this meant in terms of energy, which is what required him to elaborate about what is beyond the pleasure principle: the death drive.[92] In other words, and as psychoanalysts other than Lacan have pointed out, the death drive can be understood through the lens of entropy.[93]

91 Teresa Brennan, *Exhausting Modernity: Grounds for a New Economy* (London: Routledge, 2000), 65–67.

92 Jacques Lacan, *The Seminar of Jacques Lacan, Book II: The Ego in Freud's Theory and in the Technique of Psychoanalysis 1954–1955* (1978), trans. Sylvana Tomaselli (New York: W. W. Norton and Co., 1988), 75–76.

93 Siegfried Bernfeld and Sergei Feitelberg, 'The Principle of Entropy and the Death Instinct', *International Journal of Psycho-Analysis*', vol. 12 (1931), 61–81; Leon J. Saul, 'Freud's Death Instinct and the Second Law of Thermodynamics', *International Journal of Psycho-Analysis*, vol. 39 (1958), 323–25.

The picture is as clear as it is daunting, and captured in the mnemonic device for remembering the laws of thermodynamics in the passage from Thomas Pynchon which provides the epigraph to this chapter: you can't win, things are going to get worse before they get better, who says they're going to get better? The tendency to destruction is in systems and the System as well as the Cells and the Self. The implications of the new science 'weighed principally with me', observed Thomson, and the physicist James Clerk Maxwell wrote about the personal state of 'disgregation' which he experienced as he became conscious of the implications of the entropic process. In similar fashion, Herbert Spencer wrote to John Tyndall, a prominent physicist who had worked on various aspects of thermodynamics, that the discovery that equilibration is death had so 'staggered' him that he was 'out of spirits for some days afterwards' and that he remained 'unsettled about the matter'.[94] In this regard, we might note that a statistically significant number of the thinkers who worked in the field of thermodynamics either committed suicide or attempted to: German physicist Julius Robert von Mayer jumped from a window; Austrian physicist Ludwig Boltzmann hanged himself; Boltzman's student Paul Ehrenfest shot himself; American chemist Gilbert Lewis took cyanide; more examples could be given.

As the nineteenth-century physicists recognized, the apocalyptic message of the entropy law directs us well beyond the steam engines, the factories that house them, and the laboratories that study them. The message directs us to a homologous pattern affecting all systems: disorder, chaos, death.

The moment at which entropy enters the intellectual vocabulary was a time of deep crisis in Western cultural and intellectual life. This enabled the second law to be quickly applied to all systems. If 'the tendency of heat is towards equalization', as Balfour Stewart and Peter Guthrie Tait put it in *The Unseen Universe* (1875), then this 'will no doubt ultimately

94 Thomson, Comment of 1 March, 1851, in the 'Preliminary Draft from the Dynamical Theory of Heat', reprinted in Crosbie W. Smith, 'William Thomson and the Creation of Thermodynamics: 1840–1855', *Archive for History of Exact Sciences*, vol. 16, no. 3 (1977), 231–88 (281); Maxwell, letter to Peter Guthrie Tait, 12 Feb. 1872, in *The Scientific Letters and Papers of James Clerk Maxwell*, vol. II: *1862–1873* (Cambridge: Cambridge University Press, 1995), 710. Spencer's letter is from late 1858 or early 1859, in David Duncan, *The Life and Letters of Herbert Spencer* (London: Methuen, 1908), 104.

bring the system to an end'.[95] A vision of the future as the heat death of the universe had obvious social, political, and cultural implications. (In that sense, it can easily be read as a precursor of the idea of the Anthropocene; an 'Entropocene', in Bernard Stiegler's formulation.) The physicists Balfour Stewart and Norman Lockyer were already writing in 1868 that the problem of energy in the social world 'is well understood', and a few years later Maxwell noted that 'Moral and Intellectual Entropy are noble subjects'.[96] Such comments are a reflection of the extent to which the cosmic despair inherent in the language of thermodynamics coincided with, or perhaps even offered a scientific foundation for, a culture of pessimism associated with modernity's dissipation and degradation. We noted this in chapter 2, in the context of Metchnikoff's observations about the marked pessimism of the period, which he thought was prompted by the dread of disease and death. Here we can also point to a flurry of books on the subject, including Agnes Taubert, *Pessimism and its Opponents* (1873); Edmund Pfleiderer, *Modern Pessimism* (1875); James Sully, *Pessimism: A History and a Criticism* (1877); Elme-Marie Caro, *Pessimism in the 19th Century* (1880); Eduard von Hartmann, *Towards a History and Foundation of Pessimism* (1880); Olga Plümacher, *Pessimism in the Past and Present* (1884); Edgar Saltus, *The Philosophy of Disenchantment* (1885); and R. M. Wenley, *Aspects of Pessimism* (1894). Entropy appeared to be confirmation that such pessimism was not misplaced.

A large part of this debate filtered into the question of the 'degeneration' that was said to occur with mass society. Heat death was regarded as degeneration on a universal scale, while degeneration was understood as the social and cultural counterpart of the second law. The 'degeneration' thesis that came to the fore in the late nineteenth century built on the idea that 'things as they are totter and plunge', as Max Nordau put it.

> Over the earth the shadows creep with deepening gloom, wrapping all objects in a mysterious dimness, in which all certainty is destroyed

95 Stewart and Tait, *Unseen Universe*, 84.

96 Balfour Stewart and Norman Lockyer, 'The Sun as a Type of the Material Universe', *Macmillan's Magazine*, August 1868, 319–27 (319); Maxwell, letter to Peter Guthrie Tait, 10 March 1873, in *Scientific Letters and Papers*, vol. II, 833–34.

> and any guess seems plausible. Forms lose their outlines, and are dissolved in a floating mist. The day is over, the night draws on.[97]

Similarly-minded thinkers would later make the same point: in *The Decline of the West*, for example, Oswald Spengler treats entropy as the most conspicuous symbol of decline and evidence that the 'system is manifestly approaching to some final state, whatever this may be'.[98] The politics of cultural despair revelled in the idea of an entropic social system. The bourgeois order, on which it had long been thought that the sun never set, was now clearly an order for which one day the sun would not rise (to paraphrase John Ruskin on entropy's threat to the British Empire).[99] Which is why the steam engine was so central to the debate.

The 'philosopher's stone' of the modern world was, according to Helmholtz, the endeavour to construct machines of 'perpetual motion'.

> Under this term ['perpetual motion'] was understood a machine, which, without being wound up, without consuming in the working of it falling water, wind, or any other natural force, should still continue in motion, the motive power being perpetually supplied by the machine itself. Beasts and human beings seemed to correspond to the idea of such an apparatus, for they moved themselves energetically and incessantly as long as they lived, and were never wound up; nobody set them in motion. A connexion between the supply of nourishment and the development of force did not make itself apparent. The nourishment seemed only necessary to grease, as it were, the wheelwork of the animal machine, to replace what was used up, and to renew the old. The development of force out of itself seemed to be the essential peculiarity, the real quintessence of organic life. If, therefore, men were to be constructed, a perpetual motion must first be found.[100]

97 Max Nordau, *Degeneration* (1892), ed. George L. Mosse (Lincoln: University of Nebraska Press, 1993), 5–6.

98 Oswald Spengler, *The Decline of the West* (1926–1928), trans. Charles Francis Atkinson (New York: Modern Library, 1965), 216–20.

99 John Ruskin, 'The Storm-Cloud of the Nineteenth Century', Two Lectures delivered at the London Institution, 4 and 11 Feb. 1884.

100 Helmholtz, 'Interaction', 155–56.

Ultimately, of course, it is the machine of the universe that was of interest to the physicists ('We may regard the Universe in the light of a vast physical machine', states the opening line of Balfour Stewart's *Conservation of Energy* in 1875), but one source of the motion in question lay in the steam engine. Yet the steam engine was not just any engine. It was an engine at the heart of the industrialization of capital, the basis of capitalist expansion and the exploitation of workers. It was also therefore the focus of class struggle. The reason the general strike organized by the Chartists in 1842 is known as the 'Plug Plot Riots' is because of the way in which the striking workers pulled the plugs out of the steam engines. So integral was the steam engine to British capitalism that any assembly organized for the purpose of damaging a steam engine was punishable by death. 'The history of the proletariat in England begins with . . . the invention of the steam-engine', observes Engels in the opening lines of *The Condition of the Working Class in England in 1844*. At the heart of thermodynamics, then, lies a *physics of economic value*, as Nicholas Georgescu-Roegen has shown.[101] At the heart of thermodynamics lies a political problem: capitalism is losing steam.

Yet the problem of capital required not just protecting the steam engine from workers. A well-organized police power can do that well enough. It also required addressing the far more problematic issue of the energy of machines and the energy of workers as part of the wider energy of the universe. 'The idea of work for machines, or natural process, is taken from comparison with the working power of man', Helmholtz puts it.

> A raised weight can produce work, but in doing so it must necessarily sink from its height, and, when it has fallen as deep as it can fall, its gravity remains as before, but it can no longer do work.
>
> A stretched spring can do work, but in so doing it becomes loose. The velocity of a moving mass can do work, but in doing so it comes to rest. Heat can perform work; it is destroyed in the operation. Chemical forces can perform work, but they exhaust themselves in the effort.

101 Georgescu-Roegen, *Entropy Law*, 3, 277; John Tresch, *The Romantic Machine: Utopian Science and Technology after Napoleon* (Chicago: University of Chicago Press, 2012), 104; Amy Wendling, *Karl Marx on Technology and Alienation* (Houndmills: Palgrave, 2009), 66.

> Electrical currents can perform work, but to keep them up we must consume either chemical or mechanical forces, or heat.
>
> We may express this generally. *It is a universal character of all known natural forces that their capacity for work is exhausted in the degree in which they actually perform work.*[102]

In another of his essays on physics, Helmholtz writes,

> The perpetual motion [of the machine] was to produce work inexhaustibly without corresponding consumption, that is to say, out of nothing. Work, however, is money. Here, therefore, the great practical problem which the cunning heads of all centuries have followed in the most diverse ways, namely, to fabricate money out of nothing, invited solution. The similarity with the philosopher's stone sought by the ancient chemists was complete. That also was thought to contain the quintessence of organic life, and to be capable of producing gold.[103]

The closest one could get to the philosopher's stone and fabricating money out of nothing is not a *machine* of perpetual motion but a machine which consumes the energy of workers without exhausting that energy. We are back to the body: to the individual bodies of the workers and to the body of the people. The closest capital can get to the philosopher's stone is the body of workers. But like all bodies, workers suffer exhaustion, illness, disease.

As already noted, *energy* is derived from the Greek word *energia*, implying 'work', and in the nineteenth century energy came to be defined as the power of doing work; undergraduate physics students are still taught that this is how to understand energy.[104] The energy concept, as such, is one which demands we think of the working capacities of bodies. In the early nineteenth century, energy started to be used explicitly to

---

102 Hermann von Helmholtz, 'On the Conservation of Force: Introduction to a Series of Lectures Delivered at Carlsruhe in the Winter of 1862–1863', in *Popular Lectures*, 359.

103 Helmholtz, 'Interaction', 155–56.

104 Peter Guthrie Tait, *Lectures on Some Recent Advances in Physical Science* (London: Macmillan, 1876), 18; Philip Mirowski, *More Heat Than Light: Economics as Social Physics, Physics as Nature's Economics* (Cambridge: Cambridge University Press, 1989), 13, 62, 108, 125.

refer to the power of bodies to work. The *OED* cites Thomas Young's *Lectures on Natural Philosophy and the Mechanical Arts* (1807) as the first book to use it in this way, but other leading thinkers in thermodynamics also treated expenditure of energy as identical to work.[105] As Ernst Mach put it in *History and Root of the Principle of the Conservation of Energy* (1872), the 'vanishing of heat' is intimately connected to 'the performance of work'.[106] This is one reason why so many physicists would make their point in economic terms pertaining to labour, as we have already seen. When the physicists Stewart and Lockyer comment that 'energy in the social world is well understood', they explicate the point by suggesting that the analogy between the social and physical worlds can be illustrated through the idea of energy. 'When a man pursues his course undaunted by opposition, unappalled by obstacles, he is said to be a very energetic man. By his energy, we mean the power he possesses of overcoming obstacles.' But this energy and ability to overcome obstacles is measured 'by the amount of work which he can do'. That said, such a man may find himself defeated by another who does not possess anywhere near the same amount of energy. Why? Because this second person may possess a 'higher position', though less energy. Similarly, if two men of the same energy oppose each other, the one with the higher standing will be successful. The reason for this success is because the 'high position means energy in another form':

> It means that at some remote period a vast amount of personal energy was expended in raising the family into this high position. The founder of the family had doubtless greater energy than his fellow-men, and spent it in raising himself and his family into a position of advantage.

The original energy 'has been transmuted into something else'.[107] Not only can the physics of energy be explained through social analogy, then, but it would appear that the class structure of industrial capitalism is part of this same physics.

105 Helmholtz, 'Interaction', 157; Boltzmann, 'Second Law', 19.

106 Ernst Mach, *History and Root of the Principle of the Conservation of Energy* (1872), trans. Philip E. B. Jourdain (Chicago: Open Court Publishing, 1911), 37–41.

107 Stewart and Lockyer, 'Sun as a Type', 319.

The implications for capital of the energy question were therefore huge, given capital's fundamental requirement for human bodies to expend energy in labour. Hence the degradation of energy and the decline of work or the capacity to work – that is, the decline of the working class to labour and thus for capital to survive – would necessarily go hand in hand once the second law and the idea of entropy emerged. How does the System keep an exhausted working class going so that the System can itself keep going?

Anson Rabinbach writes that the concept of entropy and its intimate yet paradoxical relationship with energy is at the very heart of the nineteenth-century revolution in modernity because it threatens the idea of evolution and progress and undermines the vision of a stable and productivist universe.[108] Instead of stability and ever-expanding production, one gets decline and degradation. By pointing to the dissipation of energy from the industrial system, entropy posits 'the rise or fall of political systems, the freedom or bondage of nations, the movements of commerce and industry, the origin of wealth and poverty, and the general physical welfare of the race'.[109] (And as energy which cannot be reabsorbed back into the machine it points to that other bourgeois obsession, *waste*, because when heat is allowed to pass from one body to another 'there is an absolute waste of energy available to man'.[110]) No amount of mythologizing about social progress, self-regulation, or autopoiesis can hide the decline or prevent the decay of the social body and the death of the System. Time's arrow points to the death of capital.

The fact that energetics has this class dimension is why Rabinbach describes Helmholtz as 'the first great bourgeois philosopher of labor power'.[111] It is also why Marx and Engels built, albeit tentatively, concepts from thermodynamics into the critique of political economy, as they grappled with the ways in which capitalism treats working human bodies as energy-machines. On the one hand, this concerned labour power as energy transferred to the human organism through the

108 Anson Rabinbach, *The Human Motor: Energy, Fatigue, and the Origins of Modernity* (Berkeley: University of California Press, 1992), 3.

109 Frederick Snoddy, *Matter and Energy* (New York: Henry Holt, 1912), 9, 11.

110 Thomson, 'Universal Tendency', 304. On waste as bourgeois obsession see chapter 2 of my *War Power, Police Power* (Edinburgh: Edinburgh University Press, 2014).

111 Rabinbach, *Human Motor*, 61.

material of nature. In that sense, the question of energy was crucial to the whole issue of 'metabolism', one of Marx's key concepts, since in the course of labour human energy is expended and has to be replaced. On the other hand, Marx and Engels were aware of the wider questions about nature and machinery raised by thermodynamics. Marx was working on his critique of political economy at the very same time as the debates about the thermodynamics of heat engines were feeding into the work on industry of thinkers such as William Stanley Jevons, whose book *The Coal Question* (1865) pointed to the probable *exhaustion* of the coal mines as the start of the reduction of vital processes and, as a consequence of this, the danger of the end of civilization. 'The working individual is not only a stabiliser of *present* but also, and to a far greater extent, a squanderer of *past*, solar heat', as Engels put it, on one of the many occasions he and Marx addressed the topic.[112] This is the basis of Marx's connection between the energy catastrophe and the crisis of capitalist production in the final section of chapter 15 of *Capital*, on 'large-scale industry and agriculture'. Capitalist production

> disturbs the metabolic interaction between man and the earth, i.e. it prevents the return to the soil of its constituent elements consumed by man in the form of food and clothing; hence it hinders the operation of the eternal natural condition for the lasting fertility of the soil.

Capitalist production, Marx continues, 'only develops the techniques and the degree of combination of the social process of production by simultaneously undermining the original sources of all wealth – the soil and the worker'. In his correspondence with Engels, Marx refers the former to a book by Karl Nikolaus Fraas published in 1847, which Marx shows that the climate changes 'in *historical* times' and that if cultivation is not consciously controlled it will 'leave deserts behind it'.[113] Such conscious control was in part what Marx and Engels understood by

112 Engels, Letter to Marx, 21 March 1869, Marx and Engels, *Collected Works*, vol. 43 (London: Lawrence and Wishart, 1988), 245–46. Engels, Letter to Marx, 19 December 1882, Marx and Engels, *Collected Works, vol. 46* (London: Lawrence and Wishart, 1962), 411.

113 Karl Marx, *Capital: A Critique of Political Economy, vol. 1* (1867), trans. B. Fowkes (Harmondsworth: Penguin, 1976), 637–38; Marx, Letter to Engels, 25 March 1868, *Collected Works, vol. 42*, 558–59.

communism and one reason why they thought that the problem of 'universal exhaustion' created 'the conditions for the ultimate victory of the working class'.[114] We have already seen the second law being described as terroristic, but now the terror has a specific name: 'Heat is *par excellence* the communist of our universe.'[115]

Demons and Parasites: Towards a Politics of the Nervous State

Entropy is the reason the energy state struggles in its attempt to manage capitalist modernity. It turns out that even if we believe the System is autopoietically self-regulating, it is still regulating itself to death. Policing itself to death, we might say. This is the reason why capital works with what Thomas Richards calls *an economy of controlled information*.[116] The more capital develops, so the more it generates information about itself and then more information systems *for* itself. This is what lies behind the twentieth-century expansion of informatics, as we saw in the previous chapter. By the end of the nineteenth century, with information coming in more and more quantities and at faster and faster speeds, many of the problems with communication and administration came to be defined as examples of entropy, as did the whole notion of the state as a domain of comprehensive knowledge. Herein lies one way to think about the creature known as Maxwell's Demon.

First imagined in a letter from James Clerk Maxwell to Peter Tait, in December 1867, and then presented in his *Theory of Heat* in 1871, the Demon captured the imagination of scientists at the time and continues to do so. Attempting to hypothetically violate the second law, Maxwell imagined a being able to control the movement of heat. As he explains it in *Theory of Heat*:

> If we conceive of a being whose faculties are so sharpened that he can follow every molecule in its course, such a being, whose attributes are as essentially finite as our own, would be able to do what is impossible

---

114 Frederick Engels, 'Introduction [to Borkheim's Pamphlet, *In Memory of the German Blood-and-Thunder Patriots*]', 1887, *CW26*, 451.

115 Stewart and Tait, *Unseen Universe*, 84.

116 Thomas Richards, *The Imperial Archive: Knowledge and the Fantasy of Empire* (London: Verso, 1993), 74–76.

> to us. For we have seen that molecules in a vessel full of air at uniform temperature are moving with velocities by no means uniform, though the mean velocity of any great number of them, arbitrarily selected, is almost exactly uniform. Now let us suppose that such a vessel is divided into two portions, A and B, by a division in which there is a small hole, and that a being, who can see the individual molecules, opens and closes this hole, so as to allow only the swifter molecules to pass from A to B, and only the slower molecules to pass from B to A. He will thus, without expenditure of work, raise the temperature of B and lower that of A, in contradiction to the second law of thermodynamics.[117]

Through such action, this being appears able to violate the second law. Ostensibly engaged in a simple valve operation, the being is operating an *information-gathering process* through which it develops an intelligence and knowledge of the system, carries out an act of judgement, and thereby somehow controls or offsets the entropic process.[118] The being appears to be able to *govern the system*. As Maxwell put it in a lecture in 1868, this being is the part of a machine by means of which the machine is kept nearly uniform:

> If, by altering the adjustments of the machine, its governing power is continually increased, there is generally a limit at which the disturbance, instead of subsiding more rapidly, becomes an oscillating and jerking motion, increasing in violence till it reaches the limit of action of the governor. This takes place when the possible part of one of the impossible roots becomes positive. The mathematical investigation of the motion may be rendered practically useful by pointing out the remedy for these disturbances.[119]

When it was first imagined, then, this creature was a 'neat-fingered' and 'finite' being, and then known as a Governor; Maxwell's lecture of 1868

117 James Clerk Maxwell, *Theory of Heat* (1871) (London: Longman, 1902), 338–39.

118 Letter from Maxwell to Tait, 11 Dec. 1867, cited in Cargill Gilston Knott, *The Life and Scientific Work of Peter Guthrie Tait* (Cambridge: Cambridge University Press, 1911), 213–14.

119 J. Clerk Maxwell, 'On Governors' (1868), in *The Scientific Papers of James Clerk Maxwell*, vol. II (Cambridge: Cambridge University Press, 1890), 106.

has the title 'On Governors'. Only later did the being become known as Maxwell's Demon.[120]

The career of this Demon-Governor is quite telling, as is the fact that intellectual history has retained the idea of a Demon. Even though Maxwell and others insisted time and again that the 'Demon' is not intended to imply malignancy but is simply a mediating figure, differing 'from real living animals only in extreme smallness and agility', as Thomson claimed at the time, the fact that it is known as a Demon needs comment. After all, aside from the fact that even Thomson was to concede that this simple creature could in fact push or pull atoms in any direction in such a way as to make it somewhat distinct from 'real living animals', and aside from the fact that this is why accounts of the Demon more often than not refer to its duplicitous nature and magical powers,[121] there is the far from insignificant history of demons in the Christian West:[122] the demonic forces of disorder and chaos that run amok in the world, against which universal wars are fought, along with the demonic forces of madness and unreason that run amok in our psyche, against which personal wars are fought (the outcome of which is more often submission rather than victory, which is no doubt one reason why Freud, when writing of the energetics of the death drive in *Beyond the Pleasure Principle*, discusses it as a *demonic* force driving the psyche). Here, we can simply observe that just as it is hard to say 'Governor' without thinking of Governors (and thus the Governed), so it is hard to say 'Demon' without at least considering that the figure referred to might be a little, well, demonic.

The power of the Demon-Governor is indicated by the fact that it manages the dissipation of energy through an exercise of sheer will, thereby saving the world from the spectre of entropy, exhaustion, communism, and death. The Demon thus appears to conjure order out of chaos. Imagined as a *Demon*, the figure operates at a level of fantasy.

120 J. Clerk Maxwell, 'Note to Tait "Concerning Demons"' (circa 1875), in *Scientific Letters and Papers*, vol. II, 185–86.

121 Sir William Thomson, 'The Sorting Demon of Maxwell' (1879), in *Mathematical and Physical Papers*, vol. V (Cambridge: Cambridge University Press, 1911); Brian Clegg, *Professor Maxwell's Duplicitous Demon: The Life and Science of James Clerk Maxwell* (London: Icon Books, 2019); Paul Davies, *The Demon in the Machine* (London: Penguin, 2019), 27, 49.

122 See Mark Neocleous, *The Universal Adversary: Security, Capital and 'The Enemies of All Mankind'* (Abingdon: Routledge, 2016).

But imagined as a *Governor*, the Demon acts as the police power. This explains why the creature is variously described as *guarding the allotment* (Thomson) or acting as the *customs officer* of the universe (Henri Poincaré). The key question concerning the Demon is the question of police: how to *guide and control the motions of bodies*.[123] The Demon-Governor manages information, controls movement, and determines which molecules move across the border. Hence Victorian accounts of the entropic process frequently fall back on seeing the economy of controlled information as but one part of a *controlled economy*, a version of capitalism in which the seemingly 'spontaneous' social order is fabricated and in which 'self-regulation' is, in fact, administered politically, a police process designed to render capital immune from its own decline.[124] The economy of controlled information is a mechanism designed to secure an impossible immunity for a system that contains within itself the seeds of its own death.

'Organism is opposed to chaos, to disintegration, to death, as message is to noise', says Wiener in *The Human Use of Human Beings*. Why noise? Noise is the name we give to the disruption of messages in an informational circuit. We encountered such disruption in the previous chapter as 'disturbance' (Wiener's telephone engineer) and then as 'protest' (Luhmann's futile gesture). In French, the 'noise' of information theory is translated as *parasite*. The Parasite distorts information and dissolves signals. But the French *parasite* also retains our understanding of the Parasite as a creature that takes without giving. Now, one could discuss the Parasite in the terms of warfare discussed earlier – as we noted, officials within the security state have considered the Parasite's strategic powers – and one could equally discuss the Parasite in relation to the image of the body that takes us through the process (or failure) of

123 Maxwell, *Theory of Heat*, 153; William Thomson, 'The Kinetic Theory of the Dissipation of Energy', *Proceedings of the Royal Society of Edinburgh*, vol. 8 (1874), 325–34; Henri Poincaré, *Leçons sur les hypothèses cosmogoniques professées á la Sorbonne* (Paris: Librarie Scientifique, 1911), 253.

124 It is also the reason why the Demon remained such a powerful presence in systems theory, information science, and communication studies, such as in Wiener's *Cybernetics* and *The Human Use of Human Beings*, and Monod's *Chance and Necessity*. Heinz von Foerster would eventually anoint Maxwell's Demon as the patron saint of cybernetics – Heinz von Foerster, 'Responsibilities of Competence' (1971), in Heinz von Foerster, *Understanding Understanding: Essays on Cybernetics and Cognition* (New York: Springer, 2003), 193.

ingestion, mastication, digestion, absorption and elimination. Yet the Parasite is a far more complex figure than fighting-killing or eating-shitting allows. Indeed, in its complex relationship to the host, the Parasite appears to undermine both those processes, jamming the machine. It does so by posing a fundamental problem for immunity.

Although the word 'parasite' has ancient origins, derived from *parasitus* (in Latin) and *parasitos* (in Greek), meaning 'a person who eats at the table of another', from *para* (beside) and *sitos* (wheat, flour, bread, or food), it is really in early modernity that the Parasite starts to appear in European thought and culture, and it does so as a derogatory name for a political figure. 'Live loathed and long, / Most smiling, smooth, detested parasites, / Courteous destroyers', Shakespeare has Timon of Athens say. In this guise, the Parasite comes to shape the whole image of the body politic, as threats to sovereignty from within come to be interpellated as parasitical (such as Hobbes's depiction of corporations as 'like the little Wormes, which physicians call *Ascarides*', in chapter 29 of *Leviathan*). Following its political use, the parasite would come to be used in scientific discourse to refer to an organism living on, in, or at the expense of another, and a science of parasitology would eventually emerge to add biological weight to the concept. By conjuring images of microbiological creatures that penetrate our borders, and social groups or individuals deemed to be living off the health and wealth of others, sucking the life and energy from the body, the Parasite figures as an enemy but also as a source of disorder and disintegration from within. This image plays across several registers, as Michel Serres points out. The Parasite is a biological presence that will take without giving, weakening the host's systems though often without killing the host; the Parasite is noise and thus an 'interference' or 'disturbance' in the running of a system; and the Parasite is the visitor or guest who hangs around, taking without giving, living off social systems of others.[125] If we were to try to think these ideas together, we conjure up an image of the Parasite as a figure that interrupts the System.

The Parasite thereby draws our attention to the weight attached to systems and threats to bodies. The Parasite plugs into the body and by

---

125 Michel Serres, *The Parasite* (1980), trans. Lawrence R. Schehr (Baltimore: Johns Hopkins University Press, 1982). Also see Elling Ulvestad, *Defending Life: The Nature of Host-Parasite Relations* (Dordrecht: Springer, 2007).

plugging into the body it either brings into question the viability of the body's systems or seeks to make itself a permanent presence within them. We might say that, by plugging into the body, the parasite reveals the gaps and weaknesses in the system's security-immunity. The *biological Parasite* enters the host and disrupts the system by living off the material intended to sustain the body, *noise* enters into an informational circuit and disrupts the system of communication, and the *social Parasite* disrupts the body politic by taking without giving and thereby disrupting the system of exchange. All of which could imply the end of the system or the need to eradicate the Parasite. But eradication seems impossible. 'There is no system without parasites', notes Serres. 'This constant is a law.' Just as 'noise gives rise to a new system', so the Parasite always 'invents anew'.[126] The Parasite moves through the system, helping it fluctuate and sustain equilibrium, even to the extent of giving rise to a new system. The system adapts or dies.

The Parasite is thus a figure of disorder within the order. Yet the Parasite can also establish new orders. Herein lies the whole problem: Do we seek to destroy the Parasite over and over, securing the borders of the body to keep it out and seeking out the Parasite within in order to expel it? But this task seems impossible. Do we therefore incorporate it and learn to live with it? And how do we live with the fear that it remains within and might eventually destroy us?

The Parasite is therefore an expansive figure, seeking power by threatening to expand and fill the environment. As such, it becomes associated with similarly expansive dangers and threats. In that, it is rather like the Devil. 'Here is the Devil then', Serres writes in the final sentences of his book *The Parasite*, which turns out, he says, to have been 'a book of Evil'. The Parasite's meal, taken from the host, is the feast of the Devil.[127] The political Parasite is thus much more than simply an application of the biological trope of an organic attack on the body politic, for it plays on nothing less than a form of power that seeks to somehow master and control the System. We start thinking of Parasites and we end up thinking of Demons.

'Behold, this is the enemy of the world, the destroyer of civilizations, the parasite among the nations, the son of chaos, the incarnation of evil,

126 Serres, *Parasite*, 12, 14, 33, 35.

127 Serres, *Parasite*, 93, 253.

the materialised demon of mankind's decay.' So said Joseph Goebbels in a speech at the 1937 Nazi Party rally in Nuremberg, making a connection between the Parasite and the Demon that was far from uncommon.[128] As is well known, Nazism is replete with descriptions of the body politic containing disease, poison, cancer, sickness, toxins, plague, viruses, and, of course, parasites. 'By nature, the Jew is a parasite', Hitler announced, time and again.[129] This idea of the Parasite is explained by Alfred Rosenberg in perhaps the most sustained statement of Nazi ideology, *The Myth of the Twentieth Century* (1930):

> The concept [of parasite] will not be grasped as a moral evaluation but as the characterization of a biological fact, in exactly the same way as we speak of parasitical phenomena in the plant and animal world. The sack crab bores through the posterior of the pocket crab, gradually growing into the latter, sucking out its last life-strength. This is an identical process to that in which the Jew penetrates into society through the open wounds in the body of the people, feeding off their racial and creative strength until their decline.[130]

As 'biological fact', the enemy as Parasite works through the blood and the cells. This image grounds the enmity in the most basic of biological tropes. It also lends itself to a form of dehumanization well beyond the tropes of cockroaches or rats, for the Parasite is *within* the body. Yet the Parasite in these texts spirals out beyond the 'biological fact' to become the very moral evaluation that Rosenberg denies. In so doing, the Parasite merges with the Demonic.[131] 'Half-demonic . . . he [the Jew] operates under the Satanic name and remains always the same, always

128 Cited in Alex Bein, 'The Jewish Parasite: Notes on the Semantics of the Jewish Problem, with special Reference to Germany', *Leo Baeck Institute Year Book*, vol. 9, no. 1 (1964), 3–40 (24).

129 Hitler, 13 Feb. 1945, in *The Testament of Adolf Hitler: The Hitler-Boorman Documents*, trans. Colonel R. H. Stevens (London: Icon Books, 1960), 60; *Mein Kampf*, trans. Ralph Manheim (Boston: Houghton Mifflin Company, 1943), 150, 231–33, 243, 303–35, 337, 623.

130 Alfred Rosenberg, *The Myth of the Twentieth Century* (1930), trans. Vivian Bird (Newport Beach, CA: Noontide Press, 1982), 299.

131 The argument here picks up on themes in my books *The Monstrous and the Dead: Burke, Marx, Fascism* (Cardiff: University of Wales Press, 2005) and *The Universal Adversary*.

fervently believing in his "mission" and yet forever a barren and condemned parasite'.[132] The biological fact of the Parasite is restated on a metaphysical or theological register. 'Where any kind of wound is torn open in the body of a nation, the Jewish demon always eats itself into the infected part, and as a parasite, it exploits the weak hours of the great nations of this world.' The mission is rooted in the fact that the Parasite also has its *mythos* and is thus fighting a war of souls as well as a war of bodies.[133] Each race has its soul – a 'race-soul' – and, for Rosenberg, the race-soul is being infected and consumed by the Jew's Demonical as well as Parasitical power. The Demonic forces align themselves against the Godlike in a battle to win control of Europe. Examine the Jewish Demon and one finds the Parasite; examine the Parasitical Jew and one finds a Demon.[134] At the same time, just as the Parasite can be found in the financial system as well as the physiological system – in the body of capital and the body of the people – so the Demon appears in the body as well as the soul. The Demon is always a devouring force. In such a politics, the Parasite moves from being a biological agent living within the host to a metaphysical or even theological power standing over and above the same host yet using the host in the same way, while the Demon becomes flesh in the form of a Parasite. The Parasite has a demonical power; the Demon acts as a parasitical force. The Demon is materialized as the Parasite; the Parasite takes on the immaterial power of the Demon. Either way, the figure of our own demise exists *inside the body*.

That Nazism articulated its fears in this way should not surprise, given the centrality of security to Nazi politics, about which I will say more in the following chapter. The Parasite is a security issue as well as a problem for immunity. This is the reason why the Parasite appears time and again in security politics, hand-in-hand with the Demon, as illustrated during the intensification of security politics in the early twenty-first century, when the force behind 9/11 was interpellated as both a 'terrorist parasite' and 'the darker demon of our new world'.[135] In the world of security, the enemy is the Demon-Parasite, inside the System.

132 Rosenberg, *Myth*, 167.

133 Rosenberg, *Myth*, 298–99.

134 Rosenberg, *Myth*, 186, 298–99, 311, 460.

135 The first is from President Bush, Jan. 2002, the second from his national security adviser, Condoleezza Rice, Sept. 2005, both discussed in Neocleous, *Universal Adversary*.

In his survey history *Europe's Inner Demons* (1976), Norman Cohn points out that the fear of the Demon is a fear of the Demon's mastery. It is a fear that the Demon resides among us, within us, and is thus able and willing to take what it wants from us, from within, and thus take us over, from within, becoming our master. Like a Parasite. We fear the Demon-Parasite, then, because we fear its mastery. We fear that the Demon-Parasite and Demon-Governor will become one.[136] Where do we find security against a figure with such demonic power? How might we develop immunity against a figure with such parasitical techniques? The fear of mastery rests heavily on an image of movement through the system. This draws our attention time and again to parasite-immunology or immuno-parasitology, but also draws attention to the ways in which the Parasite might stimulate the immune system into acting as the mediator between itself and the host organism, working *for* the Parasite rather than defeating or destroying it. No sovereign *decision* (in Schmittian terms) to eliminate or expel the Parasite will work, and neither will the *ban* (in Giorgio Agamben's terms). In the immunological terms laid out in chapters 1 and 2 above, we might reformulate the standard claim about the Parasite (that it moves from outside to inside) by suggesting instead that it is Other that seeks to become One: in crude immunological terms, the Parasite enters the host as non-Self, and then appears to function as Self.

As an expansive power, a force within the body, and a threat of mastery against which no security-immunity measures appear to work, this figure makes us nervous, to say the least. We know the immune system cannot cope, cannot expel it, cannot provide the security we need, and we are too attached to the idea of an immune Self to accept the non-Self. The Parasite's demonic power makes us nervous; the Demon's parasitical power makes us nervous. The System's disruption, noise,

136 At the end of one of his books of poetry, Christopher Dewdney describes 'the complex battle between *the Parasite* and *the Governor*', an undecidable power differential seemingly without end, as the Governor acts as barrier – inside/outside, A/B – which the Parasite seeks to cross or even obliterate. Dewdney suggests that whereas the Governor is a mechanistic force that seeks to programme conceptual limits and operate classificatory schema, controlling movement, language and imagination, the Parasite is 'a special condition of intelligence outside the realm of . . . the "Governor"'. But this sets the Parasite and the Governor too far apart. It is the figure of the Demon that holds them together. Christopher Dewdney, 'Parasite Maintenance', in *Alter Sublime* (Toronto: Coach House Press, 1980), 75–78.

imbalance, disorder, dissolution, chaos, entropy, and suicidal death drive make us nervous. The System's openness – open like an immune system, open like a security system; open like a body, open like a body politic – means cracks will appear, and they too make us nervous. More than just *Unruhe*, the System, like the Self, is disturbed and disturbing, nervous, and maybe even a little mad.

5

Nervous States: Politics of Systems III

Since a man can go mad, I do not see why a system cannot do so.
Georg Christoph Lichtenberg, *The Waste Books* (1765–1799)

This system has something disturbing about it.
Jacques Lacan, *The Ego in Freud's Theory and in the Technique of Psychoanalysis* (1954–1955)

The system is demented, yet works very well at the same time.
Gilles Deleuze, 'Capitalism: A Very Special Delirium' (1973)

In his book *Understanding Media: The Extensions of Man* (1964), Marshall McLuhan observes that, in the same year of the publication of Søren Kierkegaard's *Begrebet Angest* (1844), people were playing chess and lotteries on the first American telegraph. This technology seemed to initiate a significant 'outering' or 'extension' of man, particularly man's central nervous system, and McLuhan suggests that 'to put one's nerves outside, and one's physical organs inside the nervous system . . . is to initiate a situation – if not a concept – of dread'.[1] *Begrebet Angest* can be translated as *The Concept of Dread*, but *The Concept of Anxiety* is the more standard translation. The 'age of anxiety' was upon us, McLuhan

1 Marshall McLuhan, *Understanding Media: The Extensions of Man* (New York: New American Library, 1964), 222.

suggests, because something was happening with the nerves and the nervous system.

Like the immune system, with which it is frequently connected, a connection to which we will shortly turn, the nervous system is one of the systems through which we imagine bodies. However, the concept 'nervous system' has a much longer history than 'immune system'. In the seventeenth and eighteenth centuries, nerves were considered 'the substratum of life in physiology and the principle of classification in natural history', writes Karl Figlio. As such, the idea of a nervous system offered a powerful metaphor for thinking about organization.[2] As the substratum of life and a way of thinking about organization, the nervous system was also integral to how the body politic was imagined. Here is Hobbes's definition of the mechanical body, that we have already cited in the introduction:

> Seeing life is but a motion of Limbs, the begining whereof is in some principall part within; why may we not say, that all *Automata* (Engines that move themselves by springs and wheeles as doth a watch) have an artificiall life? For what is the *Heart*, but a *Spring*; and the *Nerves*, but so many *Strings*; and the *Joynts*, but so many *Wheeles*, giving motion to the whole Body.

This definition applies to all bodies, including the body of the Commonwealth, which 'is but an Artificiall Man; though of greater stature and strength than the Naturall'. In the Leviathan:

> The *Soveraignty* is an Artificiall *Soul*, as giving life and motion to the whole body; The *Magistrates*, and other *Officers* of Judicature and Execution, artificiall *Joynts*; *Reward* and *Punishment* . . . are the *Nerves*, that do the same in the Body Naturall.

Hobbes was far from alone in thinking politically with the physiology of nerves. James Harrington in the *Art of Lawgiving* (1659), to give just one further example from the period, suggests a 'political anatomy' which

---

2 Karl M. Figlio, 'The Metaphor of Organization: An Historiographical Perspective on the Bio-medical Sciences of the Early Nineteenth Century', *History of Science*, vol. 14, no. 1 (1976), 17–53 (30).

must 'imbrace all those muscles, nerves, arterys and bones, which are necessary to any function of a well-order'd commonwealth'.[3] Hobbes and Harrington were finessing the more general notion of 'the very nerves of state', as Lucio calls it in Shakespeare's *Measure for Measure* (act 1, scene 4). What emerges from such comments and claims is an image that has been part of our political imagination for centuries: the nerves of state. This begs a question first raised in the introduction: What happens to our image of the body when it shifts from nerves to immunity? Better still, can one imagine a body through the lens of *both* nerves *and* immunity? If we can, then what does this mean politically, for example in thinking about the body politic?

When immunology was reaching its maturity as a science during the heyday of immune 'system' research, comparison was often made with the nervous system. Niels Jerne's foundational 1973 essay 'The Immune System' ends by comparing the relatively new idea of the immune system with the much older idea of the nervous system:

> The immune system and the nervous system are unique among the organs of the body in their ability to respond adequately to an enormous variety of signals. Both systems display dichotomies: their cells can both receive and transmit signals, and the signals can be either excitatory or inhibitory. The two systems penetrate most other tissues of the body, but they seem to avoid each other . . .
>
> . . . [Their] elements can recognize as well as be recognized, and in so doing they form a network. As in the case of the nervous system, the modulation of the network by foreign signals represents its adaptation to the outside world. Both systems thereby learn from experience and build up a memory, a memory that is sustained by reinforcement but cannot be transmitted to the next generation.[4]

A year later, Jerne concluded another essay by observing that when viewed as a functional network open to stimuli from the outside, the immune system bears a striking resemblance to the nervous system.

3 James Harrington, *Art of Lawgiving* (1659), in *The Oceana and Other Works*, ed. John Toland (London: A. Millar, 1737), 429.

4 Niels Kaj Jerne, 'The Immune System', *Scientific American*, vol. 229, no. 1 (1973), 52–60 (60).

Both systems display dichotomies and dualisms, the cells of both systems receive as well as transmit signals, the signals in both systems can either excite or inhibit, and both systems penetrate most other tissues of our body. More than anything, the two systems in the body that are capable of learning are the nervous system and the immune system.[5] It has since become commonplace to find statements to the effect that 'the complexity of the immune system has frequently been compared to that of the nervous system',[6] but the only thing surprising about such statements is how long it took for this to become a commonplace. After all, the central nervous system includes the brain, and the brain has long been treated as playing a role in one's identity.

> Our brain is the great coordinator of our actions, our behaviour, our consciousness, our thinking, our hopes, our fears and our dreams. Our immune system protects us from germs and infectious diseases. At first glance, the two organs do not have much in common. However, the brain and the immune system both have, to varying degrees and in different ways, the same property, one that is both essential and mysterious: to ensure our unique identity.[7]

Or, in the language of chapter 2, to ensure a unity of the Self.

Since then, many people have suggested that rather than simply comparing the nervous and immune systems, we should be considering them conjointly. To do so opens our field of analysis considerably. First, because the physiology of nerves raises the question of 'nervousness' and associated terms such as 'nervous exhaustion' and 'nervous breakdown', just as the physiology of immunity raises the question of autoimmune disease. But also, second, because as we have seen, the image of the body politic can never escape the image of the body. We find, then, that any account of the nerves of state cannot escape becoming an account of the nervous state. And if nerves and immunity are not only inherently connected but actually form a unity, a development within

5 Niels Jerne, 'Towards a Network Theory of the Immune System', *Annales d'immunologie*, 125C (1974), 373–89 (387).

6 Peter Melander, 'How Not to Explain the Errors of the Immune System', *Philosophy of Science*, vol. 60, no. 2 (1993), 223–41.

7 Jean Claude Ameisen, *La sculpture du vivant: le suicide cellulaire ou la mort créatrice* (Paris: Seuil, 1999), 70.

biological sciences known as psychoneuroimmunology, then what might this tell us about the nervous state over-responding immunologically to the things it is nervous about? In other words, what does the combining of nerves with the immune process do for our understanding of immunity and/as security?

Nervous System, Nervous Self

If the nerve was considered the substratum of life in the seventeenth century, then by the eighteenth century it had become a pervasive cultural and political as well as physiological sign.[8] It was, for example, fundamental to the Scottish Enlightenment's idea of self-regulation and vital balance discussed in the previous chapter. One reason for this was because the nerve had come to dominate Edinburgh physiology and social theory. Illnesses that had been previously ascribed to foul vapours or bad humours were increasingly connected to the nervous system, which was seen as the location of the 'vital principle'. For Robert Whytt, writing in 1765, 'all diseases may, in some sense, be called affections of the nervous system because, in almost every case, the nerves are more or less hurt'. Likewise for William Cullen, writing in 1769, more or less all the disturbances of the body 'depend in such a manner upon the motions of the nervous system, that almost all diseases may, in a sense, be called *nervous*'.[9] (The extent to which this notion was pursued is evident from a letter written by Cullen to the head of the Scottish police, advising them on how to 'recover' drowned persons. Cullen advised that some of these persons only *seem* to be dead, and the police must understand the nerves as the body's vital principle and thus maintain the seemingly dead person's heat, lay them in the warm sun, breathe air into them and

8 George S. Rousseau, *Nervous Acts: Essays on Literature, Culture and Sensibility* (Houndmills: Palgrave, 2004), 21, 257.

9 William Cullen, *A Methodical System of Nosology* (1769) (Stockbridge: Cornelius Sturtevant, 1807), 103; Robert Whytt, *Observations on the Nature, Causes and Cure of those Disorders which are Commonly Called Nervous, Hypochondriac, or Hysteric* (London: T. Becket and P. Du Hondt, 1765), 93; Christopher Lawrence, 'The Nervous System and Society in the Scottish Enlightenment', in Barry Barnes and Steven Shapin (eds.), *Natural Order: Historical Studies of Scientific Culture* (Beverly Hills, CA: Sage, 1979), 24.

inject them with water.[10]) To imagine the vital principle of nerves as operating through the body as a whole rather than in one location, responding to stimuli, avoiding pain and pursuing its own preservation, was to imagine the nerves as a *system*. In the previous chapter we noted that one of Smith's criticisms of Descartes concerned the idea that the latter's 'system' relied on a rational mind controlling the body. In contrast, to imagine the nervous system as the body's vital principle was to imagine it as coextensive with the body and unified through the sensitivity of each part of the body to the body's other parts.

This sensitivity was a way of imagining the nerves as the material basis of 'sensibility' and hence a certain kind of 'sympathy', the category that was at the heart of the social and political philosophy of the Scottish Enlightenment. The full title of Whytt's 1765 book on the nerves is *Observations on the Nature, Causes and Cure of Those Disorders Which Are Commonly Called Nervous, Hypochondriac, or Hysteric, to Which Are Prefixed Some Remarks on the Sympathy of the Nerves*. The 'prefix' in question constitutes the whole of the book's first chapter, 'On the Sympathy of the Nerves', and explains how the various parts of the body connect and communicate with each other *sympathetically*. Later in the book he connects this to sympathy between persons. 'There is a remarkable sympathy, by means of the nerves, between the various parts of the body; and now it appears that there is a still more wonderful sympathy between the nervous systems of different persons.'[11] The physiological account of the nervous system as the foundation of sensitivity, sensibility, and sympathy meant that the nervous system could be imagined as equally the foundation of the kind of sensitivity, sensibility, and sympathy required in the system of natural liberty as envisaged by the Scottish Enlightenment. The nervous system was imagined as the bridge uniting mind and body, but also as mediator between man and environment. Hence the idea of sensibility allowed man's higher attributes to be connected to the body: the physiological and social conditions of existence coincide; body and body politic have the same grounding. Sensibility became the foundation of the body's balance, integrating it

10 William Cullen, *A Letter to Lord Cathcart, President of the Board of Police in Scotland, Concerning the Recovery of Persons Drowned and Seemingly Dead* (London: J. Murray, 1776), 2–3.

11 Whytt, *Observations*, 219.

into a harmonized whole, with the nervous system acting as a kind of corporeal pathway for the communication of sympathy throughout and between bodies.

In the section 'Of Liberty and Necessity', in part III of book II of his *Treatise of Human Nature* (1739–1740), David Hume spells out the importance of government and property for maintaining a regular social order. He points to the way that the 'different stations of life' and the fundamental differences between 'ranks of men' influence the whole fabric of life and affect both body and character. Just as 'the skin, pores, muscles, and nerves of day-labourers are different from those of a man of quality', so too are 'his sentiments, actions and manners'. Rude and uncivilized peoples in other nations and the less civilized labouring poor in Britain are 'insensible'. They lack *physical* feeling and therefore *emotional* feeling, and because of this they also lack *social* feeling. In his *Theory of Moral Sentiments*, Smith develops Hume's point by suggesting that the hard life experienced by rude and barbarous nations enables them to endure labour, hunger, and pain in a way that affects their sensitivity to others and leaves them lacking the sympathy needed in civilized states.

> Every savage undergoes a sort of Spartan discipline, and by the necessity of his situation is inured to every sort of hardship . . . His circumstances not only habituate him to every sort of distress, but teach him to give way to none of the passions which that distress is apt to excite. He can expect from his countryman no sympathy or indulgence for such weakness . . . All savages are too much occupied with their own wants and necessities, to give much attention to those of another person. A savage, therefore whatever be the nature of his distress, expects no sympathy from those about him.[12]

Smith gives an example of a savage taken as prisoner of war and subjected to 'dreadful torments' such as being 'hung by the shoulders over a slow fire'. Smith writes that the savage can bear such torments to the extent that he even 'derides his tormentors' and can seem almost 'indifferent' to his situation. But Smith's main point is that the savage

12 Adam Smith, *The Theory of Moral Sentiments* (1759), ed. D. D. Raphael and A. L. Macfie (Indianapolis: Liberty Fund, 1982), 205.

spectators of such scenes '*express the same insensibility*'. The sight of the torture 'seems to make no impression upon them; they scarce look at the prisoner, except when they lend a hand to torment him. At times they smoke tobacco, and amuse themselves . . . as if no such matter was going on'.[13] In contrast, civilized peoples possess a social sensibility and therefore have sympathy for such a person. They do so because they are 'persons of delicate fibres'. So delicate are their fibres that they respond even to the sight of a beggar with ulcers or sores on their body. The *fibres* here point to the fact that their sympathetic response is physiological as well as moral:

> Persons of delicate fibres and a weak constitution of body complain, that in looking on the sores and ulcers which are exposed by beggars in the streets, they are apt to feel an itching or uneasy sensation in the correspondent part of their own bodies. [14]

Thus, when Smith suggests that 'the amicable virtue of humanity requires . . . a sensibility, much beyond what is possessed by the rude vulgar of mankind',[15] the sensibility in question is again simultaneously physiological, emotional, and social. 'Sympathy' is, in that sense, a moral and psychological expression of 'sentiment'. Such claims resonate with similar claims in medical texts of the Scottish Enlightenment. Cullen, for example, claimed to have observed differences between the nervous systems of the inhabitants of rude and less 'sensible' nations and those of civilized and polished nations that paralleled their different cultural habits.

The system of natural liberty is thus *rooted* in the nervous system and is itself a *form* of nervous system. Just as the version of the mechanically controlled body was being overturned by physiological conceptions of a balanced body regulating itself through the nervous system, so the whole body of the people could be imagined as a balanced and self-regulating system that needs no 'control' by external forces because it has – or even *is* – its own nervous system. This vision of the nerves meant that the nervous system could be imagined as a kind of corporeal

13 Smith, *Theory*, 206.
14 Smith, *Theory*, 10.
15 Smith, *Theory*, 25.

version of the social body's organization. The nerves of state constitute the life of the social body.

The idea of nerves imagined as the substratum of vital life and vital politics was then reinforced by the gradual development of ideas about 'electricity'. Electricity had been understood as one of the substances said to be the body's animating power ever since William Gilbert, physician to Elizabeth I and author of *De Magnete* (1600), coined the terms 'electrics' and 'electricity' to describe the 'animistic forces'.[16] This was reinforced with a series of technological developments and discoveries. Luigi Galvani in the late eighteenth century showed that when a frog's legs are touched by a copper probe and piece of iron at the same time they twitch as though an electric current were running through them, which he eventually thought of as biologically generated electricity. In 1800 Alessandro Volta constructed a chemical battery enabling electricity to be produced by chemical reactions. This was followed by the finding that electric current can modify chemical affinities and produce chemical reactions. In 1820, Hans Christian Oersted discovered the magnetic effects produced by electrical currents, and two years later Thomas Johann Seebeck showed that heat could produce electricity. In 1831, Michael Faraday introduced an electric current by means of magnetic effects, and in 1847, James Prescott Joule identified the qualitative transformation of electricity as 'conversion', finding a way of measuring such conversions in terms of the 'energy' which we discussed at length in the previous chapter. Electricity was also increasingly welcomed into houses as a purer, odourless, and invisible form of energy. Indeed, it was seen almost as a sort of vitamin, almost synonymous with energy and life. Such radical developments around the idea of electricity and the 'electrical' nature of the nervous impulse had a cultural impact. It underpinned, for example, the 'spark of life' that animates the creature in Mary Shelley's *Frankenstein* and Walt Whitman's desire to 'sing the body electric', and it remains with us when we describe ourselves as 'wired up', 'electrified', 'charged up', 'recharging our batteries', 'plugging into new ideas', and acting like 'human dynamos'; 'dead' is how we describe an electrical machine that has stopped working.

16 William Gilbert, *De Magnete* (1600), trans. P. Fleury Mottelay (New York: John Wiley, 1893), 82.

Just as electricity became a key concept for understanding nerves, so nerves became a key concept for understanding the 'electrical' nature of life itself, so that 'life' could be understood as a dynamic, changeable, and elusive 'vital' force, the body animated by nervous electrical impulses. During the nineteenth century, this connection had multiple dimensions, but one of the key ones concerned the connection between the nervous system as a kind of *biological wiring* and new technologies such as the telegraph as electrical *nerve systems*. Scientists studying organic forms and engineers working on communication systems inspired one another. 'From the first time that investigators first began studying the nervous system, they have described it in terms of contemporary technologies', Laura Otis writes, and, conversely, the design of communication systems has often been inspired by the structures of living bodies.[17] The ways in which the telegraph system enabled information to move more quickly and independently of physical entities suggested to many observers an 'analogy . . . with the functions of the motor nerves'. That was William F. Channing, an American physician who was also prominent in the development of telegraph and other forms of electric communication, writing in 1852 about how to improve the police system. For Channing, 'the Electric Telegraph is to constitute the nervous system of organized societies'. Its functions 'are analogous to the sensitive nerves of the animal system'.[18]

Comments along the same lines as Channing's can be found time and again: German physiologist Emil Du Bois-Reymond commented in 1851 that the similarity between the nervous system and the electric telegraph 'is more than similarity; it is kinship'. Samuel Morse, the figure behind the Morse code, wrote that 'it [will] not be long ere the whole surface of this country [is] channeled for those nerves which are to diffuse, with the speed of thought, a knowledge of all that is occurring throughout the land'. In an address by John Tyndall to the British Association for the Advancement of Science delivered in Belfast in 1874, an address renowned for its materialist attack on religion, Tyndall

17 Laura Otis, *Networking: Communicating with Bodies and Machines in the Nineteenth Century* (Ann Arbor: University of Michigan Press, 2001), 3–4, 12.

18 William F. Channing, 'On the Municipal Electric Telegraph', *American Journal of Science and Arts*, vol. 13 (1852), 58–83 (58–59).

elaborates on materialism in the following way: 'A telegraph-operator has his instruments, by means of which he converses with the world; our bodies possess a nervous system, which plays a similar part between the perceiving power and external things.' When the first transatlantic cable was completed, its financier Henry Field described it as 'the nerve of international life', and at the end of the century C. R. Henderson could write that 'telegraphs and telephones . . . are really the nerve system of the social body'.[19] In the *Handbook for the Instruction of Attendants on the Insane*, published in London in 1885, there runs the following description:

> The grey skin of the brain may be compared to a great city, the head-quarters of the telegraph system, and the grey clusters through the white substance of the brain are the suburbs of the city, the grey clusters of the spinal cord are the towns, and the points of skin, muscle, organs, &c, where nerve fibres end, are the villages. The nerve fibres connect villages, towns, suburbs, and the great city with one another; for no station can exist without a connection . . .
>
> The internal nerves and the nerve cells of the mind connect with each other so as to form a network, which, while we are awake or dreaming, is in a state of busy activity, telegraphing ideas from cell to cell.[20]

Commenting in 1881 on the Chicago police's use of a combined telegraph and telephone system, which at that point was one of the most advanced of its kind, the journal *Scientific American* suggested,

> When the entire area of the city shall have been covered by the system the analogy between the civic organization and the nervous organization of an individual animal will be curiously complete. The civic organization will become sensitive, so to speak, at every point, and the transmission of intelligence therefrom to the brain and subordinate

19 John Tyndall, 'Address Delivered before the British Association Assembled at Belfast, with Additions', 1874; C. R. Henderson, *The Social Spirit of America* (New York: Chautauqua Century Press, 1897), 96. The others cited in Otis, *Networking*, 11, 120.

20 Medico-Psychological Association, *Handbook for the Instruction of Attendants on the Insane* (London: Ballière, Tindall and Cox, 1885), 13, 15.

> nervous ganglia – that is, the central and district police stations – will be practically instantaneous.[21]

This is the nerves of state in the form of an information-processing, message-transmitting, and computing police machine. Such an image enables the policing of the system to be imagined as operating throughout the whole system, albeit with a central organ, with the nerves of state a control mechanism for the social body.[22]

All told, then, the idea that society could be imagined as an immense nervous system forged new ways of thinking about the social body: the new communication systems appeared to make possible a coordination of society similar to the kind of coordination that the nervous system brought to the body, while the body's nervous system appeared to be an apparatus of total communication. With those thoughts in mind, I want to first segue back to the question of systems and then, via a discussion of nervousness, and not simply nerves, reintegrate the question of immunity.

The main thrust of these ideas about nervous systems was adopted wholesale by systems theory in the twentieth century and given fresh impetus by the development of neurological research. The idea that the nervous system transmitted information and, later, digital code, along with the idea that the same system facilitated communication between elements of the various other systems and subsystems, meant that it was ripe for adoption by systems theorists. In a letter written to a colleague in January 1945, Norbert Wiener commented that 'the subject embracing both the engineering and neurology aspects is essentially one'. Such an idea underpins systems theory, as witnessed by key contributions such as Warren S. McCulloch and Walter Pitts's article 'A Logical Calculus of the Ideas Immanent in Nervous Activity', published in 1943 in the *Bulletin of Mathematical Biophysics*, and the article 'Behavior, Purpose and Teleology' by Wiener, Arturo Rosenblueth, and Julian Bigelow, published in *Philosophy of Science* the same year. The one major shift was the replacement of the telegraph with the computer as the

21 [No author], 'The Chicago Police Telephone and Patrol System', *Scientific American*, vol. 44, no. 17 (1881), 255–58 (255).

22 Kai Eriksson, *Communication in Modern Social Ordering: History and Philosophy* (New York: Continuum, 2011), 54.

closest mechanical analogue to the nervous system. 'The computing machine furnishes us the greatest promise for an adequate central nervous system in future automatic-control machines', Wiener wrote in a 1948 article published in the *Annals of the New York Academy of Sciences*. From then onwards, there was a growing interest in the idea that, on the one hand, the nervous system functions as a machine operating through information storage and communication, which is precisely what makes it a system, and, on the other hand, the idea that machines can be understood in terms of the running of the nervous system. 'The nervous system and the automatic machine are fundamentally alike', Wiener wrote in *The Human Use of Human Beings*, later adding that he, along with many others, were compelled to regard the nervous system and computer in the same light. Eventually, the conjunctive form of 'computing machines and the nervous system' was dropped and replaced with a singular: 'nervous computing machines'.[23] (Such ideas were forerunners of the kind of 'digital nervous system' articulated much later by people such as Bill Gates in *Business @ the Speed of Thought*.)

Computing machines: nervous systems: systems: societies. Complex systems, one and the same. In one sense, the image before them was the brain, as articulated in William Ross Ashby's *Design for a Brain* (1952), William Grey Walter's *The Living Brain* (1953), and John von Neumann's *The Computer and the Brain* (1958). But the brain soon came to be treated as one adaptive system among all the others, and as systems theory developed, these ideas were applied across more and more fields, in texts such as Heinz von Foerster's *Understanding Understanding* and

23 Wiener to Rosenblueth, 24 January 1945, cited in Steve J. Heims, *John Von Neumann and Norbert Wiener: From Mathematics to the Technologies of Life and Death* (Cambridge, MA: MIT Press, 1980), 185; Warren S. McCulloch and Walter Pitts, 'A Logical Calculus of the Ideas Immanent in Nervous Activity', *Bulletin of Mathematical Biophysics*, vol. 5, no. 4 (1943), 115–33 (117); Arturo Rosenblueth, Norbert Wiener, and Julian Bigelow, 'Behavior, Purpose and Teleology', *Philosophy of Science*, vol. 10, no. 1 (1943), 18–24; W. R. Ashby, 'The Nervous System as Physical Machine: With Special Reference to the Origin of Adaptive Behavior', *Mind*, vol. 56, no. 221 (1947), 44–59; Norbert Wiener, *The Human Use of Human Beings: Cybernetics and Society* (1950) (Boston: Da Capo Press, 1954), 33; Norbert Wiener, *Cybernetics; or, Control and Communication in the Animal and the Machine*, 2nd edn. (Cambridge, MA: MIT Press, 1961), 4, 8, 14–15, 18, 96, 116, 130, 145; Norbert Wiener, *I Am a Mathematician* (New York: Doubleday, 1956), 269, 323–25; Norbert Wiener, *God and Golem, Inc. a Comment on Certain Points Where Cybernetics Impinges on Religion* (Cambridge, MA: MIT Press, 1964), 75–76.

The Beginning of Heaven and Earth Has No Name; in the development of the concept of autopoiesis (Maturana and Varela's *Autopoiesis and Cognition* and *The Tree of Knowledge: The Biological Roots of Human Understanding*); in biology (Jacques Monod's *Chance and Necessity*); management science (Stafford Beer's *Brain of the Firm*, *Cybernetics and Management*, and *The Heart of Enterprise*); in the application of autopoiesis to social systems (Luhmann's many works); and of course, to politics too, which brings us to Karl Deutsch's *Nerves of Government: Models of Political Communication and Control* (1963).

Deutsch's book was the period's most directly political account of systems of command and communication, information and learning, and hence, ultimately, mechanisms of control.[24] Deutsch takes seriously the etymology of 'cybernetics' that we have discussed earlier, suggesting that we view government less as a problem of power and more as a problem of steering, and that steering is first and foremost a matter of communication. Although the subtitle of Deutsch's book points to the influence of Wiener's *Cybernetics* in stressing control and communication, there is a sense in which the book was more generally a culmination of three centuries of thinking about the 'nerves of state' going back to at least Hobbes. But instead of an account of fear, diffidence, instinct, violence, force, and darkness (to pick just a few of Hobbes's favourite and most interesting topics), what we get with Deutsch is information, communication, feedback, learning, growth, and stability. The political system, like all systems, functions by gathering information, storing information, learning from information, and acting on information. It does this all while maintaining system homeostasis, the system exhibiting a plasticity in continuity (that is, systems order). A myriad range of publications were spawned by Deutsch's book, the phrase 'nerves of government' joining 'nerves of state' in becoming a popular political phrase. One can easily see why. For a start, the longevity of 'nerves' in thinking about bodies and the body politic meant it had an appeal as a classical political trope. A second appeal was that the nervous system was already conceptualized *as a system*, reinforcing the whole systems analytic. And third, the systems theorists wanted to think about nervous systems simply in terms of *neural networks* of information communication.

24 Karl W. Deutsch, *The Nerves of Government: Models of Political Communication and Control* (New York: Free Press, 1963).

Yet there is something very odd about this work. In accounts such as Deutsch's, and the parallel arguments in other work, the systems in question are 'nervous' in the sense that they transmit impulses from one part of an organism to another part, such that the organism can adjust and adapt accordingly, with sensors processing information and ensuring that as little information as possible is lost during the process. The nerves of government, like the nervous system, is a pattern of information flow, communication, feedback, equilibrium, and efficiency: a political system running like an electrical or digital machine in a way that stabilizes the social body. It is a vision of the social body imagined 'from the mathematician's point of view'.[25]

One can see this in some of the famous experiments conducted by some of key figures in systems analysis, from shock therapy on patients who were thought to have a mental illness – a 're-programming' – to the design of machines with brains. Take the experiment conducted on a live frog by McCulloch, Pitts, Jerome Lettvin, and Maturana. Treating the frog's body as a system, the researchers cut open the frog's eye to try to detect nervous impulses concerning information and its processing. Finding that when a volley of disks are twitched a series of electrical impulses runs through the nerve fibres, they concluded that the frog hunts by vision and escapes enemies by seeing them, and they use this to show that the frog's nervous system acts on algorithmic principles, identifying objects as 'enemy' or 'prey'. The whole issue of the nerves is one of transmission of information through the system.[26] What is odd about the experiment is precisely what is odd about the whole logic of nerves within systems theory as a whole: the experiment does not allow for the idea that an organism might not see an enemy but might *imagine* it and, as a consequence, might become rather *nervous* in the other way that we use this word. The idea seems to be that we should treat 'the activity of the nervous system as determined by the nervous system itself', as Maturana and Varela put it.[27]

25 John von Neumann, *The Computer and the Brain* (New Haven, CT: Yale University Press, 1958), 1, 39.

26 J. Y. Lettvin, H. R. Maturana, W. S. McCulloch, and W. H. Pitts, 'What the Frog's Eye Tells the Frog's Brain' (1959), in Warren S. McCulloch, *Embodiments of Mind* (Cambridge, MA: MIT Press, 1965), 230–55.

27 Humberto R. Maturana and Francisco J. Varela, *Autopoiesis and Cognition: The Realization of the Living* (Dordrecht: D. Reidel Co., 1980), xv.

The 'nerves' are understood simply in terms of the information passed through neurons, so the idea that an organism might *construct the idea of an enemy* in a fashion that we might call *nervous* – might nervously imagine enemies, or indeed might nervously imagine anything else, in such a way as to provoke an aggressive reaction or even overreaction – is not relevant. Indeed, it is not thinkable. One response to this might be to say that frogs do not experience 'nervousness' in the way we use that term. But what about the following example from Deutsch's *Nerves of Government*? Deutsch offers as an example of 'recognition' a piece of electronic equipment then being developed called the 'Identification of Friend or Foe (IFF)', without ever considering that the friend-enemy distinction is always already overdetermined by processes, practices, assumptions, and beliefs that are way beyond simple 'recognition'.[28] To be sure, Deutsch does hint at the idea that the system could in theory become overloaded with information and so respond accordingly, but it remains merely a hint.

The nervous system in systems theory is imagined as a machine communicating information, calculating what is needed for the system to self-regulate and hence to maintain organizational homeostasis, which it does by transmitting information. The System steers itself through its nervous system. 'The nervous system as a whole is organized in such a way (organizes itself in such a way) that *it computes a stable reality*.'[29] The system is never 'nervous' in the sense that it *feels nervous*; it has nerve impulses, but not nervous impulses. This is the reason the stability of the system is always highlighted. Unstable realities – nervously unstable realities – must not and perhaps even cannot be imagined. The possibility that the organism might *nervously* and even *imaginatively* interpret, and perhaps even *over-interpret*, its environment and the incoming information, does not arise. And so, the possibility of *over-protecting* itself, of instituting a series of measures to secure itself from that which does not exist, of possibly even creating an enemy by its *nervous construction of protection*, perhaps even of *over-protection* and *excessive immunity-security measures*, does not compute.

28 Deutsch, *Nerves*, 87.

29 Heinz von Foerster, 'Cybernetics of Epistemology' (1973), in Heinz von Foerster, *Understanding Understanding: Essays on Cybernetics and Cognition* (New York: Springer, 2003), 244, emphasis added.

The singular meaning of 'nerves' as connoting the central nervous system explains why Freud and psychoanalysis are such important targets in much of the literature in systems theory. In a forthright attack on Freud called 'The Past of a Delusion', McCulloch suggests that the only thing to understand in human psychology is 'how brains work'. All the rest, especially any ideas about *unconscious* fear, *unstable* realities, and *anxious* tendencies, are dismissed as nonsensical, false, specious, delusional, oversimplified, invented, unverified, unverifiable, and even bordering on communist nonsense.[30] This might also explain why the one early systems thinker who did see some value in psychoanalytical ways of thinking, Lawrence S. Kubie, is said to have been the group's whipping boy and the butt of their jokes. Kubie's papers to Macy and other conferences were usually met with fierce resistance, no doubt due to his desire to introduce ideas about reflexivity and ambiguity into the discussions. The discussion following his paper at the 1949 Macy Conference gives an indication of how hostile the systems approach is to such ideas. In the paper, called 'Neurotic Potential and Human Adaptation', Kubie discusses the irrational, the dichotomy between the conscious and unconscious, the dissociation of unconscious energy, the symbolic representation of that energy, and the repetition of neurotic psychological experiences. The essentially Freudian argument sought to press home the importance of what he was trying to bring to the understanding of systems and to introduce some notion of 'neurosis' into the account of nerves. Kubie opens the discussion following the paper by suggesting that without taking account of such issues 'we are constantly in danger of oversimplifying the problem so as to scale it down for mathematical treatment'. Yet within barely a minute or two, Wiener, Bateson, and Pitts are scaling the discussion down in that very way, treating the physiological system as an 'analogic calculating machine'.[31] In the systems view, the nervous system has nothing to do with *nervousness*. Even memory is treated as an 'activity in a circular chain of neurons which keeps up indefinitely, once started', rather like 'the function of the

30 McCulloch, *Embodiments of Mind*, 276–302.

31 Kubie's paper and ensuing discussion are in Heinz von Foerster (ed.), *Cybernetics: Circular, Causal, and Feedback Mechanisms in Biological and Social Systems. Transactions of the Sixth Conference March 24–25, 1949* (New York: Josiah Macy Jr. Foundation, 1950).

telegraph repeater'.[32] As anyone who has a memory knows, memory is a little more complicated than that. The position was clear: if scientific claims were tinged with unconscious and subjective thoughts, feelings, and concerns, then science, and thus any science of systems, would be severely limited. Hence Kubie's position, and essentially any other such psychoanalytic approaches or assumptions, had to be discounted for promising everything and delivering nothing, and Freud dismissed as a charlatan.

The outcome of treating nerves solely as a technical system communicating information and nothing else means that any of the emotional, subjective, psychological, or irrational connotations associated with the term 'nerves' are ignored. It is an outcome that follows naturally from a conception of systems without subjects. This was the intuition of Hegel when he criticizes the brain-gazers in *The Phenomenology of Mind* (1807): if you want to understand mind, you will not get far by simply dissecting brains. Or, to put it in terms of our discussion in previous chapters, to reduce the nervous system to electrical impulses is to ignore the problem of the Self, its divisions, and its ability to turn aggressively against itself. Is Self-destruction really just an information error that can be corrected through feedback? To put it bluntly: what is imagined in these accounts is the *nerves of state*, but never a *nervous state*. Yet is it not impossible to ignore the nervous state, once we bring into the picture the other meaning of 'nerves' and its connotations of anxiety, stress, fear, insecurity, or dread?

This 'nervous-less' conception of the nervous system is even more remarkable given two wider features of the political culture of the post-war period in which systems analysis develops. The first is that 'stress', a concept which emerged in an engineering context in the nineteenth century to denote the effects of a force acting against a resistance and which was used to reinforce the idea of a system, became a commonplace term within intellectual circles, not only to account for the body's response to shock but also to the 'wear and tear' of life, following the development of the concept by Hans Selye.[33] 'Before the war, no one

32 Norbert Wiener, 'Time, Communication and the Nervous System', *Annals of the New York Academy of Sciences*, vol. 50 (1948), 197–220 (211).

33 Selye's publications on the topic were large in number and spanned decades, but a good late overview is the revised edition of *The Stress of Life* (New York: McGraw-Hill, 1976).

spoke of stress; after it, increasingly, everyone did', notes Robert Kugelmann.[34] Picking up some of the meanings attached to nervousness and related terms, 'stress' also had connotations of energy, control, balance, and stability and slotted nicely into an intellectual world dominated by systems. Yet, for systems theory, 'stress' would again only ever be an engineering concept connoting the force or combination of forces exerted to maintain a state of equilibrium. The second wider feature of the political culture that is pertinent is that this was also the period described by many as an 'age of anxiety'. The poet W. H. Auden called his post-war book-length poem *The Age of Anxiety* (1947), the liberal intellectual Arthur Schlesinger would treat anxiety as the official emotion of the age in his bestselling book *The Vital Center* (1949), and existential psychotherapist Rollo May would call his book on the post-war world *The Meaning of Anxiety* (1950).

The idea that the age in which they were living was being described as an age of anxiety, an age of stress, and hence a *nervous age*, mattered not one bit to systems theory. Systems theory says a lot about nerves and yet is a million miles from being able to use this for any kind of meditation on the human condition and its social and material foundations. For systems theory, people have a 'nervous system' which helps them maintain homeostasis and the stability of the organism as a whole, like a fish (or rather, like a 'fish-system'), and the idea of 'nerves' can be used only to describe the work of the central nervous system conceived of as an information machine and homeostatic device. The idea of a 'nervous complaint' or the idea that one might suffer from 'nerves' makes no sense. The comparison between the nervous system and the brain with the computer or a technology such as the telegraph fails to register that no computing machine or telegraphic communication system ever suffered the fear of losing their home or job. Nor does a computer or telegraph ever contemplate their own death. No computer ever experienced or imagined the kind of insecurity which might lead it to act in certain ways rather than others. Put bluntly: if someone asks you, 'Are you nervous?', they are expecting you to discuss your feelings, not your neurons.

To understand the notion of 'nerves' solely in terms of the nervous system imagined as an electrical system is to reduce 'nerves' to nerve

34 Robert Kugelmann, *Stress: The Nature of Engineered Grief* (Westport, CT: Praeger, 1992), 54.

fibres as the prime conductive mechanism of the system and to imply that 'nerves' has no meaning other than the purely neurological. This is to obscure completely the philosophical, psychological, and political issues attached to the idea of 'nerves' and 'nervousness'. What we need, then, is an alternative account of the nervous system which brings into the picture the idea of nervousness. Such an account will allow us to reconnect to the politics of immunity.

'If a patient had gone to his doctor in Hippocratic times and complained of being "nervous" his doctor would have expected to see a sinewy fellow, strong and vigorous', observes W. F. Bynum.[35] This was because the earliest meaning of 'nervous' related to fibres, sinews, or tendons, and their distinguishing feature was strength. By the middle of the eighteenth century, however, a shift had begun to occur. Thinkers and physicians treated the nervous system as an essential constituent of health, as we have seen, and so 'nervous disease' and 'nervousness' came to be regarded as sickness. In this context, the term 'neurosis', coined by Cullen in the eighteenth century to capture large branches of medicine, including what would later be called neurological and psychiatric disorders, increasingly came to refer to a plethora of disorders ranging from melancholy to insanity and rooted in a general weakness of the nervous system or lack of nervous energy. For Cullen, the idea of 'nervous' problems played on the older idea of the physiology of fibres and sinews but broadened the class of diseases and thus enabled him to speak of 'neuroses' that occur when there is an imbalance of 'sensibility', excessive 'irritability', and a shift in the 'tone' of the mind. Cullen's category of 'neuroses' was meant to capture emotional states and 'disorders of judgment' that were symptoms of 'nervousness'. The idea of 'neurosis' would eventually lose its reference to physical nerves as such and become a 'disease of the mind', connected to the 'nervous soul' or 'nervous temperament', terms which became equally common.[36] As the eighteenth century ended and the nineteenth century progressed, describing people as 'nervous' became increasingly common as a way of describing some kind of emotional disorder.

35 W. F. Bynum, 'The Nervous Patient in Eighteenth- and Nineteenth-Century Britain: The Psychiatric Origins of British Neurology', in W. F. Bynum, Roy Porter, and Michael Shepherd (eds.), *The Anatomy of Madness*, vol. I (London: Tavistock, 1985), 91.

36 Klaus Doerner, *Madmen and the Bourgeoisie: A Social History of Insanity* (1969), trans. Joachim Neugroschel and Jean Steinberg (Oxford: Basil Blackwell, 1981), 48–49, 307.

'At the beginning of the nineteenth century, we do not hesitate to affirm, that *nervous disorders* have now taken the place of fevers, and may be justly reckoned two thirds of the whole, with which civilized society is afflicted.' Thus began Thomas Trotter's *A View of the Nervous Temperament*, published in 1807.[37] It was an indication of the extent to which the language would become part of medical terminology, as for example in Moritz Heinrich Romberg's *Manual of the Nervous Diseases of Man* (1840, translated into English in 1853) and the eventual emergence of publications such as the *Journal of Nervous Exhaustion*, which ran for a period in the 1880s. It was also an indication of the extent to which terms such as 'nervous disorder', 'nervousness', and 'nervous state' (états nerveux, *Nervosität*) were becoming a common feature of the language and culture of bourgeois modernity. The term 'nervous breakdown', for example, appeared in the 1870s, and the first book with that title was published in 1901. 'True! – nervous – very, very dreadfully nervous I had been and am', begins Edgar Allan Poe's 'The Tell-Tale Heart' (1843); 'we all say it's on the nerves' is Wilkie Collins's observation about Mr Fairlie's problems in *The Woman in White* (1860); 'a frail young man of thirty, nervous and anaemic' is the main protagonist in Joris-Karl Huysmans's *Against Nature* (1884); it was a 'nerve-racked age', observed Robert Musil of *fin de siècle* Vienna in his novel *The Man Without Qualities*. As George Steiner observes about the image of the nineteenth century as a period of progressive and ambitious change, 'For every text of Benthamite confidence, of proud meliorism, we can find a counter-statement of nervous fatigue.'[38]

A key figure in pinpointing this nervous fatigue was George M. Beard. In a series of publications beginning with an article on 'nervous exhaustion' in the *Boston Medical and Surgical Journal* in 1869, followed by *A Practical Treatise on Nervous Exhaustion (Neurasthenia)* (1880) and then *American Nervousness* (1881), Beard developed the idea of nervousness as a certain kind of electrical force or energy coursing through

37 Thomas Trotter, *A View of the Nervous Temperament* (London: Longman, 1807), xvii.

38 George Steiner, *In Bluebeard's Castle: Some Notes Towards the Re-Definition of Culture* (London: Faber and Faber, 1974), 18; Peter Gay, *The Bourgeois Experience, Victoria to Freud*, vol. II, *The Tender Passion* (Oxford: Oxford University Press, 1986), 331–35.

the body carrying messages. But an excess of expenditure or environmental pressure could diminish such energy, leading to a 'lack of nerve-force'. On this basis, Beard encouraged the use of what was by then a prevalent medical practice, namely the use of electrotherapy to replenish a depleted nervous system; he had previously published a book on *The Medical and Surgical Uses of Electricity* (1871), and the ideas there fed into his account of nervous energy. For good measure, he also incorporated into his thinking some of the prevalent ideas about energy from within thermodynamics. In other words, he developed a *thermodynamic account of mental energy*: work and other uses of energy resulted in heat loss, and entropy was the result of an exhausted body. He was far from alone in this, as there was a general recognition that the second law of thermodynamics pointed to the decline of 'nervous energy' in the body. The term Beard used for this was 'neurasthenia'. Although this term had been used in medical dictionaries to describe insanity, forms of mental disorder, and nervous weakness, Beard extended and expanded the term to include people suffering from depression, fatigue, melancholy, bereavement, sexual dysfunction, the overuse of narcotics, the loneliness of retirement, and a whole host of other such ailments; the list of symptoms provided by Beard in *American Nervousness* covers two pages. In effect, 'neurasthenia' became synonymous with 'nervous exhaustion', lack of 'nerve force', or just 'nerves' and 'nervousness'. By the end of the century, neurasthenia was being defined as the 'disease of the century', with American journals alone carrying over three hundred articles on the subject between 1870 and 1900. 'The name of neurasthenia is on everybody's lips; it is the fashionable disease', wrote French psychiatrist Paul Dubois in 1905.[39]

For Beard, the problem was quintessentially one of 'American civilization', but he eventually conceded that it was a problem of modern civilization in general. 'All this is modern', he writes at one point, and we have already seen others point this out. 'The Greeks were certainly

39 Léon Bouveret, *La Neurasthénie: Épuisement Nerveux* (Paris: Publisher Baillière, 1891), 8–9, 16; Guthrie Rankin, 'Neurasthenia: The Wear and Tear of Life', *British Medical Journal*, vol. 1 (2 May 1903), 1017–20; Paul Dubois, *The Psychic Treatment of Nervous Disorders: The Psychoneuroses and the Moral Treatment* (1905), trans. Smith Ely Jelliffe and William A. White (New York: Funk and Wagnalls, 1909), 18. Figures for journals in Megan Barke, Rebecca Fribush, and Peter N. Stearns, 'Nervous Breakdown in 20th-Century American Culture', *Journal of Social History*, vol. 33, no. 3 (2000), 565–84.

civilized, but they were not nervous, and in the Greek language there is no word for that term', he notes, adding that 'no age, no country, and no form of civilization' in history 'possessed such maladies'.[40] The author of the first book with *Nervous Breakdown* as its title, Albert Abrams, commented there that the disease identified by Beard was 'a disease of the whole civilized world'.[41]

One reason it was a modern rather than American problem lay in the fact that it was a question of *energy* and the *transformation of work*. For Beard, the problem predominantly affected the 'brain-workers' of the professional and white-collar classes of the larger cities, illustrated by the metaphors used by Beard and others.[42] Beard compares the 'millionaires of nerve-force', those 'who never know what it is to be tired out or feel that their energies are expended', with those whose 'inheritance is small' and who 'have been able to increase it but slightly, if at all'. The latter are in a precarious position, because through 'overtoil, or sorrow, or injury, they overdraw their little surplus' and find that it can take months or even years to make up the deficiency.

> The man with a small income is really rich, as long as there is no overdraft on the account; so the nervous man may be really well and in fair working order as long as he does not draw on his limited store of nerve-force. But a slight mental disturbance, unwonted toil or exposure, anything out of and beyond his usual routine, even a sleepless night, may sweep away that narrow margin, and leave him in nervous bankruptcy, from which he finds it as hard to rise as from financial bankruptcy.[43]

This theme of 'nervous bankruptcy' resonated widely in late nineteenth-century intellectual culture. Théodule-Armand Ribot in *Les maladies de la volonté* ([*The Diseases of the Will*] 1883) calls this the loss of 'nervous

40 George M. Beard, *American Nervousness: Its Causes and Consequences. A Supplement to Nervous Exhaustion (Neurasthenia)* (New York: G. P. Putnam's Sons, 1881), vi–viii, 96, 176.

41 Albert Abrams, *Nervous Breakdown* (San Francisco: Hicks-Judd Company, 1901), 4.

42 Beard, *American Nervousness*, ix, xiv, 193–291; George M. Beard, *A Practical Treatise on Nervous Exhaustion (Neurasthenia), Its Symptoms, Nature, Sequences, Treatment* (New York: William Wood and Co., 1880), 1, 44.

43 Beard, *American Nervousness*, 9–10.

and psychic capital', while Abrams begins *Nervous Breakdown* by observing that 'every individual is endowed at birth with a definite amount of nerve capital' which can then be depleted.[44]

As well as 'nervous capital', however, fatigue had become a key concept in the medical policing of workers: the possibility of their exhaustion, discussed in the previous chapter, was now the possibility of their *nervous* exhaustion. The 'human motor' appeared unable to simply motor on and on. The nervous exhaustion found among the working class was sometimes treated as a kind of *inverted work ethic* – an 'ethic of resistance to work' as Rabinbach puts it, or a 'rebellious nervous body', in Michael Cowan's phrase – as the state of nervous exhaustion or weakness/absence of will was seen to threaten the whole structure of capital; it would eventually give rise to the 'scientific management' of the early twentieth century.[45] On the other hand, fatigue and exhaustion were considered universal problems, in that they were imagined as capable of consuming the energies of society in its entirety. 'The nerves and their disorders were inextricably interwoven with . . . assumptions about more than health and illness', observes Janet Oppenheim. They 'were interlaced with attitudes toward success and failure, civilization and barbarism, order and chaos, masculinity and femininity'. More generally, 'just as nerves pervaded the physical body, so did they permeate the images with which the Victorians evoked their society'.[46] The interest in nervous exhaustion was a reminder of, but also perhaps a symptom of, the *exhaustion of natural resources*, the *physical exhaustion of the universe*, and, as a consequence, the *exhaustion of bourgeois modernity*.

For some, such as Ribot, nervousness could be overcome through an exertion of willpower. The nerves need to be 'fully charged' with will,

---

44 Théodule-Armand Ribot, *The Diseases of the Will* (1883), trans. Merwin-Marie Snell (Chicago: Open Court, 1903), 102; Abrams, *Nervous Breakdown*, 7. For discussion see the chapter 'On the Verge of Bankruptcy' in F. G. Gosling, *Before Freud: Neurasthenia and the American Medical Community, 1870–1910* (Chicago: University of Illinois Press, 1987).

45 Anson Rabinbach, *The Human Motor: Energy, Fatigue, and the Origins of Modernity* (Berkeley: University of California Press, 1992), 167; Michael Cowan, *Cult of the Will: Nervousness and German Modernity* (University Park: Pennsylvania State University Press, 2008), 11; Anson Rabinbach, *The Eclipse of the Utopias of Labor* (New York: Fordham University Press, 2018), 2–19, 85.

46 Janet Oppenheim, *'Shattered Nerves': Doctors, Patients, and Depression in Victorian England* (Oxford: Oxford University Press, 1991), 3.

and energy would be restored.[47] But the general feeling was that nervousness was a phenomenon of modern life stretching from the individual Self to the structure and system of capital. Nervous exhaustion and all the other 'nervous diseases' were agreed to be a product of the shock and speed of capitalist industry, of the demands that liberal freedoms imposed on the subjects of capital, of the insecurities associated with such freedoms, and of the constant agitation of social relations brought about by the permanent and ever-increasing changes in the social world. By the end of the nineteenth century, 'nervousness' had become a disease of modernity.[48] We can point to three brief and well-known examples from the period to spell out the common themes.

First, Max Nordau in *Degeneration* (1892) sought to describe the 'feverish restlessness' produced by the telegraph, the train, electricity, the industrialized press (18,000 new publications and 6,800 newspapers), increased travel (the number of travellers in Germany, France, and England increasing from 2.5 million in 1840 to 614 million in 1891), and the postal system. 'All these activities', he writes, 'involve an effort of the nervous system and a wearing of tissue. Every line we read or write, every human face we see, every conversation we carry on, every scene we see through the window of the flying express, sets in activity our sensory nerves and our brain centres.' The result is exhausted nervous systems, nervous complaints, and nervous diseases as features of a degenerating civilization.[49] Second, in *The Division of Labour in Society* (1893), Émile Durkheim adopts some of Beard's arguments, suggesting that 'our nervous system . . . has become more delicate' than our forefathers, who 'were of a coarser grain', and the 'hyperactivity' of general life makes us 'wearisome, tensing up our nervous system'. In particular, a more extensive market and expanding large-scale industry brings

47 Ribot, *Diseases of the Will*, 40–42, 46, 71, 123.

48 For the subtle differences according to national context, see George Frederick Drinka, *The Birth of Neurosis: Myth, Malady and the Victorians* (New York: Simon and Schuster, 1984); Tom Lutz, *American Nervousness, 1903* (Ithaca: Cornell University Press, 1991); Joachim Radkau, *Das Zeitalter der Nervosität: Deutschland zwischen Bismarck und Hitler* (Munich: Carl Hanser, 1998); Andrea Killen, *Berlin Electropolis: Shock, Nerves, and German Modernity* (Berkeley: University of California Press, 2006); Petteri Pietikainen, *Neurosis and Modernity: The Age of Nervousness in Sweden* (Leiden: Brill, 2007).

49 Max Nordau, *Degeneration* (1892), ed. George L. Mosse (Lincoln: University of Nebraska Press, 1993), 37–41.

'greater fatigue . . . to the nervous system'. Four years later, in *Suicide*, Durkheim suggests that suicide can often result from 'a deep affection of the nervous system', a 'violent shock to the nervous system' or 'extreme sensitivity of the nervous system'.[50] Third, in an essay published in 1903 called 'The Metropolis and Mental Life' Georg Simmel observes that the increasingly complex sensory mental imagery that comes with the urbanization of the world has resulted in 'the intensification of nervous stimulation'. The rapidity and contradictory nature of the impressions on our minds 'tear the nerves so brutally hither and thither that their last reserves of strength are spent'. The modern metropolis 'stimulates the nervous system of the individual to its highest achievement.[51]

'Nerves' had thus become central to the way the modern political condition was imagined; people were nervous about the nerves. Nervousness appeared to threaten both subject and state, Self and System, body and body politic. This is perhaps the point alluded to by McLuhan in his suggestion that technology initiates an 'extension of man' through the nervous system, connecting the human subject and their anxious dread to the world as object and its material and technological transformations. It was also an argument made by Walter Benjamin in his account of the shock of modernity and its impact on the nerves. Benjamin points to the ways in which Charles Baudelaire placed the 'shock experience' of modern life 'at the very center of his artistic work', but, for Benjamin, the shock is experienced nowhere more powerfully than in the factory, where workers become subordinate to the speed and unceasing motion of the machine.[52] As Susan Buck-Morss puts it, commenting on Benjamin's account of the shock of modernity and Hegel's critique of the brain-gazers:

> The nervous system is not contained within the body's limits. The circuit from sense-perception to motor response begins and ends in the world. The brain is thus not an isolable anatomical body, but part

---

50 Émile Durkheim, *The Division of Labour in Society* (1893), trans. W. D. Halls (Houndmills: Macmillan, 1984), 187, 276, 282, 305–06; *Suicide: A Study in Sociology* (1897), trans. John A. Spalding and George Simpson (London: Routledge, 1970), 67–69, 299.

51 Georg Simmel, 'The Metropolis and Mental Life' (1903), trans. Kurt H. Wolff, in *The Sociology of Georg Simmel* (Glencoe, IL: Free Press, 1950), 410, 414–15.

52 Walter Benjamin, 'Some Motifs in Baudelaire' (1939), trans. Harry Zohn, in *Illuminations* (London: Fontana, 1970), 163, 165, 177–78.

> of a system that passes through the person and her or his (culturally specific, historically transient) environment. As the source of stimuli and the arena for motor response, the external world must be included to complete the sensory circuit. (Sensory deprivation causes the system's internal components to degenerate.) The field of the sensory circuit thus corresponds to that of 'experience', in the classical philosophical sense of a mediation of subject and object, and yet its very composition makes the so-called split between subject and object (which was the constant plague of classical philosophy) simply irrelevant.[53]

Might the same not also be said of the immune system? Especially, that is, if it is increasingly difficult to separate the two systems? Let us work towards that argument with a little more nervous history.

The term 'neurasthenia' would eventually disappear from the discussion, gradually considered to be overly laden with assumptions of organicity and eventually going out of fashion as a diagnosis. Freud stresses, in his early work, the importance of the concept of neurasthenia and its existence as a pathological condition of the nervous system, taking on board the connections made by Beard and others concerning neurasthenia's roots in the 'social conditions' of contemporary society, but soon came to treat neurasthenia as a sexual neurosis and instead sought to develop a more generalized account of 'anxiety'. Much later, the American Psychiatric Association would remove 'neurasthenia' from its *Diagnostic and Statistical Manual*, and the World Health Organization's *Classification of Mental and Behavioural Disorders* (*ICD-10*) would list it only under 'other neurotic disorders' and with reference to a vague 'mental fatigue'. In contrast to the demise of neurasthenia, however, the general idea of 'nervous problems' and 'nervousness' remained. 'Nervous breakdown' became detached from the neurasthenia concept and, gaining traction with the idea of 'shell shock' following World War I, became an increasingly common term to capture forms of emotional collapse ranging in severity from 'stress' to 'insanity', and capturing part of neurasthenia's core meaning as regards the emotional distresses and anxieties of modern life.

53 Susan Buck-Morss, 'Aesthetics and Anaesthetics: Walter Benjamin's Artwork Essay Reconsidered', *October*, 62 (1992), 3–41 (12–13).

The key point is that the concept of 'nerves' was bifurcating, a double meaning emerging between, crudely speaking, physicalist and psychodynamic interpretations. On the one side, 'nerves' connoted the somatic nervous system and underpinned the advance of neurology; on the other side, 'nerves' connoted anxiety, fatigue, melancholy, ennui, listlessness, stress, and exhaustion. On the one side, *feeling* in the manner of sensory experience through the convoluted mass of nervous substance; on the other, *feeling* as emotional disturbance and a convoluted mess of nervous emotion. On the one side, biochemical explanation and psychopharmacology; on the other, psychoanalytic explanation and psychotherapy. On the one side, the brain; on the other, the mind. On the one side, medically trained physicians dealing with the organic, anatomical, and chemical body; on the other, lay analysis and its training in mental processes. With this bifurcation, *nervousness* became increasingly detached from neurology and handed over to the psychotherapeutic field.

'Nerves' thus became a concept with a remarkable duality, but 'nervousness', 'nervous exhaustion', and 'nervous breakdown' veered in one direction, towards the kinds of disintegration we discussed in chapter 2. These terms served to underpin the psychological paradigm which in turn became the predominant explanation of a range of what Edward Shorter calls 'small-p psychiatric problems', from depression to fatigue, lethargy to irritability, and emotional exhaustion to mental weakness. Increasingly divorced from the idea of the neurological system itself, 'nervousness' was essentially *dismembered*, as Simon Wessely puts it. Or as Beard himself suggested at the beginning of *American Nervousness* and hinting at the older meaning of nerves as strength: *nervousness* is in fact a kind of *nervelessness*.[54] All told, 'nervous states' came to be considered problems of *psyche* rather than *soma*.[55]

This was captured in colloquial English, in that to say that someone has 'problems with their nerves' became a way of suggesting not that

---

54 Edward Shorter, *From Paralysis to Fatigue: A History of Psychosomatic Illness in the Modern Era* (New York: Free Press, 1992), 222–26, 232–33, 253; Simon Wessely, 'Neurasthenia and Fatigue Syndromes: Clinical Section', in German Berrios and Roy Porter (eds.), *A History of Clinical Psychiatry* (London: Athlone Press, 1995), 514.

55 Jacques Lacan, *The Seminar of Jacques Lacan*, bk. II, *The Ego in Freud's Theory and in the Technique of Psychoanalysis 1954–1955* (1978), trans. Sylvana Tomaselli (New York: W. W. Norton and Co., 1988), 76.

they were seeing a neurologist but that they were suffering from some sort of psychological or emotional disorder. Witness the writer Alan Bennett, describing the depression experienced by his mother, a northern British working-class woman struggling to cope with life after moving to a small English village in the middle of the twentieth century:

> Always there was the shame at the nature of the illness, something Mam was able to overcome or at least ignore when she was well but which became a burden whenever her spirits began to fall . . . Self-consciousness, if not shame, was in such a small community understandable, but the longer my parents lived in the village the more I became aware that my mother was not alone in her condition. Several middle-aged women were similarly afflicted in different degrees, one stumping silently round the village every afternoon, another flitting anxiously into a friendly neighbour's, sometimes in tears, and both suffering from what was still called 'nerves'.[56]

This is language familiar to us all. Many of us would have a story or two to tell about our own 'trouble with nerves', an account of our own 'nervous exhaustion', a little memoir, perhaps, of our own 'nervous illness'.

Memoirs of My Nervous Illness; or, Something Rotten in the State of Denmark

In 1903, a book was published written by Daniel Paul Schreber called *Denkwürdigkeiten eines Nervenkranken, nebst Nachträgen und einem Anhang über die Frage: 'Unter welchen Voraussetzungen darf eine für geisteskrank erachtete Person gegen ihren erklärten Willen in einer Heilanstalt festgehalten werden?'*. The book, translated into English as *Memoirs of My Nervous Illness*, was written during one of Schreber's periods of confinement in an asylum (the 'Devil's Kitchen', as he calls it). Schreber had been a presiding judge in Leipzig, stood for election to the Reichstag in the 1884 election, had some kind of nervous breakdown, recovered, and then had a further breakdown in 1893, which appears to have lasted until 1902. Believing that 'the human soul is contained in the

56 Alan Bennett, *Untold Stories* (London: Faber and Faber, 2005), 37–38.

nerves of the body', Schreber's *Memoirs* report that those nerves start out white but can become 'blackened' through sin, corrupting through 'rays' the relation between the nerves of the subject and divine nerves. The result is soul murder (*Seelenmord*), about which he had many forebodings. Schreber believes that his enlightenment about his own strange forebodings is an indication that God has engaged him in 'nerve-contact', that his body is the centre of divine miracles, and that his mission is the redemption of the world. The last of these was to occur via his transformation into a woman. 'When the rays approach', he informs the reader, 'my breast gives the impression of a pretty well-developed female bosom.' 'I could see beyond doubt that the Order of the World imperiously demanded my unmanning, whether I personally liked it or not, and that therefore it was common sense that nothing was left to me but reconcile myself to the thought of . . . fertilization by divine rays for the purpose of creating new human beings.' This process must happen despite his senior political status: 'Fancy a person who was a *Senatspräsident* allowing himself to be fucked?' ['*Das will ein Senatspräsident gewesen sein, der sich f . . . läßt?*']. He also describes physiological transformations such as losing his stomach, his diaphragm moving to just below his larynx, seeing heads change, foreign beings controlling his conscience, and wasps being provoked into existence by his own appearance in the garden. At the same time, however, Schreber believed he was being persecuted by a range of people: his psychiatrist, the renowned professor Paul Emil Flechsig; Flechsig's accessory, God (at one point they become 'God Flechsig'); and the attendants at the asylum.[57] Schreber's *Memoirs* thus certainly contain enough material to lead some commentators to suggest that 'it is uncertain if he was ever fully sane, in the ordinary social sense'.[58] But whether he was ever fully sane in either the 'ordinary social sense' or in any other sense, he was certainly sane enough to write one of the most famous, compelling, and *political* memoirs of whatever it is that goes by the name 'nervous illness'.

Schreber's book is generally treated as one of the most important documents in psychiatric literature yet is somewhat overshadowed by

57 Daniel Paul Schreber, *Memoirs of My Nervous Illness* (1903), trans. Ida Macalpine and Richard A. Hunter (New York: New York Review of Books, 2000), 9, 19, 141–45, 164, 248, 277, 282.

58 Morton Schatzman, *Soul Murder: Persecution in the Family* (London: Allen Lane, 1973), xi.

the text that helped forge its canonical status, namely Freud's 1911 essay 'Psycho-Analytic Notes on an Autobiographical Account of a Case of Paranoia (Dementia Paranoides)'. Freud's essay became famous because of the connection drawn there between paranoia and repressed feminine and passive homosexual tendencies, a connection which then became central to the history of Freudian psychoanalysis (such as in Melanie Klein's 'Notes on Some Schizoid Mechanisms' and Jacques Lacan's *Seminar* on psychoses) and which made Schreber's case a paradigmatic study in paranoia. Yet, as numerous scholars have pointed out, the radical disjuncture between Schreber's and Freud's texts is so great that it sometimes seems as though Freud was, at times, simply not thinking about what Schreber had actually written; Elias Canetti goes so far as to suggest that we would be best off ignoring Freud's text entirely. On the question of persecution, for example, Freud suggests that Schreber's most significant fear was of his father, famous for his unusual and highly disciplinary child-rearing philosophy and contraptions, rather than the legal and institutional authorities Schreber himself discusses. Schreber's *Memoirs* can be read on many levels: as an autobiographical account of mental illness; as an account of incarceration in an asylum; as a story of a legal battle to regain civil liberty against that incarceration; as a document about the history of psychiatry in late nineteenth-century Germany; and as a document about modernity and contemporary culture.[59] Yet, for Freud, Schreber seemingly lived his life in a vacuum. Even if one were to concede that what is exhibited in Schreber's text is what Freud considers to be the author's paranoia, delusion, megalomania, or father complex, these things only emerge out of what appears to be a desire to pass on some kind of message to the world. More to the point, that message is deeply political: the nerves are the pathways of power, and so any struggle over them is a political one. At the very least, Schreber points out that his periods of nervous illness were directly related to standing as a candidate for the Reichstag and working as a senior judge. In a manner that Beard or Benjamin would have

59 Elias Canetti, *The Conscience of Words* (New York: Seabury, 1979), 25; Schatzman, *Soul Murder*, 93–115; Zvi Lothane, *In Defense of Schreber: Soul Murder and Psychiatry* (Hillsdale, NJ: Analytic Press, 1992); Eric Santner, *My Own Private Germany: Daniel Paul Schreber's Secret History of Modernity* (Princeton, NJ: Princeton University Press, 1996); Peter Goodrich, *Schreber's Law: Jurisprudence and Judgment in Transition* (Edinburgh: Edinburgh University Press, 2018).

appreciated, Schreber's difficulties with nerves stem from his political struggles on the one hand and sheer overwork on the other (being denied the natural right to give one's nerves a rest, as Schreber puts it at one point).

More tellingly, those struggles lead Schreber to a general problem: '*There is something rotten in the state of Denmark*'.[60] Perhaps one message, then, is that there is something rotten at the core of the state. Schreber is naming something: rotten. He is pointing out what is rotten: the System. And he is explaining what makes the System rotten: the fact that it tries to make its victims believe that it is they who are rotten. Such is the game of the System: to lodge rottenness in the Self when it is the System itself that is rotten.[61] Perhaps the author's nervous illness is a result of being *trapped* in a system of power and its trappings – the titles and offices, the roles and procedures, the rights and regulations – and feeling *overawed* or even *overpowered* by such trappings? Might it not be the case that nervous disorders are caused by power and its trappings? Might it not be the case that the subject experiences nervous illness because of a culture in which the *subject* can be treated like an *object*, not least an object of scientific experimentation, as Schreber puts it the letter to Professor Flechsig that is used to open the *Memoirs*?

If we take this point seriously, that is, politically – though not in such a way that reads Schreber's *Memoirs* as some kind of proto-fascist text along the lines laid down by Canetti, Deleuze, and Guattari, in which soul murder is understood as replicating older conceptions of racial struggles and God's chosen people find themselves threatened (we 'find in Schreber a political system of a disturbingly familiar kind', Canetti writes, alluding to the later rise of Nazism)[62] – then there is surely some mileage in treating Schreber's case in terms of the historical form and political condition of nerves and nervousness at the time he was writing. The key part of the *Memoirs* could be taken to be the essay which

60 Schreber, *Memoirs*, 186.

61 I am playing here with some ideas of Michel de Certeau, 'The Institution of Rot', in David B. Allison et al. (eds.), *Psychosis and Sexual Identity* (Albany: State University of New York, 1988).

62 Elias Canetti, *Crowds and Power* (1960), trans. Carol Stewart (London: Victor Gollanz, 1962), 441–43; Gilles Deleuze and Félix Guattari, *Anti-Oedipus: Capitalism and Schizophrenia* (1972), trans. Robert Hurley, Mark Seem. and Helen Lane (London: Athlone Press, 1984), 89.

appears as an addendum to the main book, called 'In What Circumstances Can a Person Considered Insane Be Detained in an Asylum against His Declared Will?', which is, in fact, not *merely* an addendum, since it was important enough to appear as the subtitle to the original. (Its omission from the title of the English translation encourages the reader to read the text through a psychiatric lens.) The essay is a demand for 'release from the Asylum . . . in order to live among civilized people'. In other words, the essay is a demand for freedom. Despite the 'oddities of [my] behaviour', as the book's opening lines have it, the text can be read as a political argument that easily ranks among the best of the anti-psychiatry and decarceration literature that appeared much later in the century; Eric Santner observes that the anti-psychiatric movement in Wilhelmine Germany crystallized around a series of cases, one of which was Schreber's, and Peter Goodrich has argued for reading it as a text of legal argument. Like much of the anti-psychiatry literature that appeared much later in the century, Schreber recognizes that the issue of 'paranoia' is a political question concerning power, raising as it does the police function of the state and the seemingly executive powers of the physician and psychiatrist. Comparing public asylums with other 'welfare institutions', he observes that as a rule such institutions are not forced upon a person unless their work is for a specific general purpose such as compulsory education. 'The same would apply to Public Asylums, if apart from the furtherance of public welfare their work were not in many cases *simultaneously a matter for the security police.*'[63] Schreber holds to the view that the state has no authority over the health of its citizens, for such authority would blend medical power and police power organized around a logic of security and therefore detrimental to human freedom. But it is abundantly clear that the state *does* claim and exercise such a power: it takes the form of a medical police, and fundamental to that police is psychiatry and law. Schreber's 'paranoia', then, or what we might more sympathetically call his generalized nervousness, is based on the nature of the System and not just the Self. It is, after all, hard to overstate the centrality of *system* to the whole logic of paranoia in general and, in particular, a sense that the System is somehow unstable or unhinged. But also, and more importantly, what if Schreber thought

63 Schreber, *Memoirs*, 315, 317, emphasis added.

he was being persecuted not because he was suffering from paranoia but, rather, because he really was being persecuted?

Yet there is something else at stake here. Zvi Lothane suggests that the appointment of Flechsig to the prestigious chair of psychiatry at Leipzig University, and then to the role of director of the psychiatric clinic in which Schreber was being held, signalled a shift that was in many ways paradigmatic, because Flechsig actually had no psychiatric experience and had never worked in a psychiatric setting. He was a brain anatomist. But he was also the author of an influential book in which nerves are treated as a series of circuits and cables. As we know, when imagined as a series of circuits and cables the nervous system is treated as though it is simply a communication system, whether of the 'telegraphic' or 'computing' type. Schreber suggests in his letter to Flechsig that what were described as his 'hallucinations' were in fact '*influences on my nervous system emanating from your nervous system*'. His nervous filaments were 'like telephone wires', and his cries for help were 'presumably a phenomenon like telephoning'. This mode of thinking about nerves follows Flechsig's logic of nerves as circuits and cables: Schreber writes that 'the weak sound of the cries of help coming from an apparently vast distance is received *only by me* in the same way as telephonic communications can only be heard by a person who is on the telephone'. The 'nerve-language', which is 'the real language of human beings', is one that is expressed 'not aloud'.[64] The image of the nervous system in the *Memoirs* is thus an image of lines of communication and the transmission of information. In other words, as Friedrich A. Kittler suggests, the neuroanatomical paradigm at work here dissolves the Self into a System of information transfer. Or as Lothane puts it, 'In one fell swoop . . . the tradition of the soul ended and the reign of the brain began.'[65] Schreber himself was far less circumspect about what he thought had happened to him in this telephonic process: nothing less than the murder of his soul by the state's security forces.

'Soul murder' is a term that had long been used in European languages to describe the desire of the Devil and his associates, and Schreber's

64 Schreber, *Memoirs*, 8, 118, 235, 277.

65 Friedrich A. Kittler, *Discourse Networks 1800/1900* (1985), trans. Michael Metteer (Stanford, CA: Stanford University Press, 1990), 294–95; Lothane, *Defense*, 204–05, 298.

book is in that sense an argument about the Demonic. But also, and more generally, 'soul murderer' is a term used to describe one who so 'dispirits' people that they want to give up on life. Recall from chapter 2 Freud's observation that *psyche* is a Greek word which may be translated as soul (*Seele*), but also that *psyche* is extended. For Schreber, 'the human soul is contained in the nerves'.[66] From our point of view, we might say that such a murder is the killing of the subject who is nervous in the other sense of the term: the anxious, insecure, and fundamentally nervous soul. All this better places Schreber in the wider context but also allows us to return to the point about the social and cultural condition of nervousness.

Discussing Schreber's case as an example of the need to understand 'madness' in its social and historical context, Roy Porter suggests that we need to register the extent to which 'Schreber was essentially inserting his own life-story into the deepest anxieties of late-nineteenth-century culture', most notably in terms of what we have discussed in the previous chapter as the *fin de siècle* as *fin du monde*. The evidence of the death of previous civilizations was reinforced with the idea of the heat death of the universe, and these threats of cosmic disaster imprinted themselves on Schreber's mind: he felt as though 'the earth had lost so much heat that general glaciation had either occurred already or was imminent'. As well as making astute political observations, page after page of the *Memoirs* shows a familiarity with the scientific chiliasm of nineteenth-century research into cosmic catastrophe, and page after page connects this cosmic catastrophe with social crisis.[67] Schreber essentially fuses contemporary scientific theory about the universe and the nervous system and situates both within what were by then the widespread assumptions about the social and cultural effect of the decrease in the warmth of the sun and general glaciation. And Schreber faces up to the key question, the same one faced by those who registered the significance of the second law: What is a soul to do with such knowledge? What is a soul to do when it hears of 'the running out of the clocks of the world',[68] when it thinks it understands what that means, but when it also

66 Schreber, *Memoirs*, 19.

67 Roy Porter, *A Social History of Madness* (London: Weidenfeld and Nicolson, 1987), 156–57.

68 Schreber, *Memoirs*, 88–93.

cannot say what is meant by the expression? Might not such a soul rightly become a little nervous? Might this be what some psychologists would later call 'psychological entropy'? How might the organism as a whole respond to such powers which constitute their environment? Might they not end up engaging in acts of self-destruction? If so, how does this connect to the autoimmune disease, which in chapter 2 we already started to imagine in terms of self-destruction?

At the end of a century which had continued a long tradition of interest in nerves, nervous systems, and nervous disorders, the final decade of which saw people such as Schreber being incarcerated because of a 'nervous illness', Eli Metchnikoff was developing his concept of biological immunity, as we have seen. It often goes unremarked that Metchnikoff was not shy of making links between the immune process and the nervous system. In his published lectures of 1891, he connected the work of phagocytes to the nervous system in general but also to what he termed the 'psychical nervous apparatus'. By the time of his much longer books *The Nature of Man* (1903) and *The Prolongation of Life* (1907), this connection between the nervous system and the immune process had been further developed, and he was also more than willing to push the possible connection between immunity and the *psychical* nervous apparatus as regarding emotions and stress.[69] Metchnikoff's insight was largely ignored within immunology. So too were other insights, such as the observation by William Osler, one of the late nineteenth century's leading physicians, in his major text *The Principles and Practice of Medicine* (1892), that rheumatoid arthritis, later defined as one of the major autoimmune diseases, 'has, in all probability, a nervous origin'. By 'nervous' he was pointing to what he believed was the 'association of the disease with shock, worry, and grief'.[70] Such suggestions went largely undeveloped, as prior to the 1970s immunology tended to focus on the work of cells in a controlled

---

69 Elias Metchnikoff, *Lectures on the Comparative Pathology of Inflammation, Delivered at the Pasteur Institute in 1891*, trans. F. A. Starling and E. H. Starling (London: Kegan Paul, 1893), 150–56, 189, 195; *The Nature of Man: Studies in Optimistic Philosophy* (1903), trans. P. Chalmers Mitchell (New York: G. P. Putnam's Sons, 1903), 241–44; *The Prolongation of Life: Optimistic Studies* (1907), trans. P. Chalmers Mitchell (New York: G. P. Putnam's Sons, 1908), 19, 194, 269.

70 William Osler, *The Principles and Practice of Medicine* (New York: Appleton, 1892) 282–83.

environment, revealing a great deal about how the immune system maintains bodily integrity and deals with threats to the organism, but treating the immune process as largely autonomous. Soviet scientists had in the 1920s conducted research on brain-immune process interactions, although this was not widely reported beyond the Soviet Union and was a matter of dispute even there. In general, the assumption was that the immune process was a self-regulating and autonomous defensive and protective mechanism but otherwise independent of the nervous system. In particular, the idea of a 'blood-brain barrier' which treated the brain as 'immune privileged' (in the sense that the cells of the immune system cannot get at it) was firmly entrenched in the medical world and biological discourse. In relation to the other meaning of 'nerves', connoting emotional stress and anxiety, some research did explore these as possible grounds for immune disorders, and one can find a few speculative research projects in the late 1950s and 1960s on emotions, stress, nerves, and immunity (and also profoundly different to the general tenor and arguments of the search for the 'good Self' of the Cold War that we discussed in chapter 2).[71] But immunology, in general, was still treating immunity as a largely autonomous process; or, if you like, the immune system as an autonomous system. George Solomon, who was responsible for some of that speculative work, commented many years later that in 1963 he jokingly put a sign on his door calling it the 'Psychoimmunology Laboratory'. Hans Selye, who did more than most to put the concept of 'stress' centre stage, had toyed with that same word 'psychoimmunology' in developing the stress concept.[72] But the fact that their provocations went largely unremarked is telling. Psychoimmunology? They must be joking.

71 C. E. G. Robinson, 'Emotional Factors and Rheumatoid Arthritis', *Canadian Medical Association Journal*, vol. 77, no. 4 (1957), 344–45; G. S. Philippopoulos, E. D. Wittkower and A. Cousineau, 'The Etiologic Significance of Emotional Factors in Onset and Exacerbations of Multiple Sclerosis: A Preliminary Report', *Psychosomatic Medicine*, 20 (1958), 458–74.

72 G. F. Solomon and R. H. Moos, 'Emotions, Immunity, and Disease: A Speculative Theoretical Integration', *Archives of General Psychiatry*, vol. 11, no. 6 (1964), 657–74; George F. Solomon, 'Emotions, Stress, the Central Nervous System, and Immunity', *Annals of the New York Academy of Sciences*, vol. 164 (1969), 335–43; G. Solomon, 'Emotions, Immunity and Disease: An Historical and Philosophical Perspective', in R. E. Ballieux, J. F. Fielding, and A. L'Abbate (eds.), *Breakdown in Human Adaptation to 'Stress'*, vol. II (The Hague: Martinus Nijhoff, 1984), 671.

In the late 1970s and into the 1980s, however, the categorical separation between immunity and nerves began to be questioned. 'You may feel that there is no conceivable relationship between the behaviour of our cells (for instance, in inflammation) and our conduct in everyday life. I do not agree.' That was how Selye explained what he was doing with the stress concept, which he wanted people to accept as more than simply 'nervous tension'.[73] The backdrop was an approach to nervous disorders which sought to treat them in terms of a unity of mind and body. In his presidential address to the Royal Society of Medicine in 1973, Richard Hunter, one of the translators of Schreber's book into English, pointed his audience to the dangers in reifying mental symptoms into mental diseases. The earliest physicians simply regarded such symptoms as 'valuable adjuncts to physical diagnosis', in that 'the abnormal mental state is not the illness, nor even its essence or determinant, but an epiphenomenon'. 'Many diseases are ushered in by a lowering of vitality which patients appreciate as irritability and depression. The mind is the most sensitive indicator of the state of the body. An abnormal mental state is equivalent to a physical sign of something going wrong in the brain'.[74] It was a plea for an approach to mind, brain, and body that would eventually have an impact on the way in which immunity was imagined.

At the very least, there was the fact that, as we have noted, the complexity of the immune system has frequently been compared to that of the nervous system. But what was also at stake was the *actual convergence* or *integration* of the two systems. Related research also revealed that communication takes place through hormones that link both systems to the endocrine system. It was found, for example, that lymphocytes in the immune system produce peptides that had once been thought to reside solely in the brain; that interleukins, a deficiency of which is associated with autoimmune diseases, help the immune system communicate with the brain; that immune cells can produce the adrenocorticotropic hormone (ACTH); that the release of a class of steroid hormones called glucocorticoids from the adrenal glands during an initial immune challenge play a key role in triggering resistance to

73 Selye, *Stress of Life*, 414.

74 Richard Hunter, 'President's Address. Psychiatry and Neurology: Psychosyndrome or Brain Disease', *Proceedings of the Royal Society of Medicine*, vol. 66, no. 4 (1973), 359–64.

further inflammatory disease; that components of the immune system such as the thymus gland, spleen, and bone marrow all somehow communicate or exchange information with the nervous system; that signals from the immune system must be able to turn on the brain's stress response; that immune cells communicate with and receive commands from the choroid plexus; that the 'blood-brain barrier' was a myth which merely sustained the Cartesian dualism that ran through medical biology. It was also found that the immune system affects our ability to think, reason, and cope with damage to the central nervous system. Put simply, it quickly became clear that one can no longer imagine immunity distinct from other processes taking place in the body in general and the nervous system in particular. This was reinforced by research on 'immune cognition', generating a more nuanced reading of how the immune system 'recognizes' and 'knows' what needs to be recognized and known. This cognitive paradigm suggests that we need to understand *how we think with our whole body*, including, perhaps even especially, through the immune process, generating a vision of a 'cognitive immune self' and 'immunoknowledge'. The research in question initially picked up on the earlier idea of a 'psychoimmunology', and this was developed under different labels, including immuno-psychiatry, neuroimmunology, cognitive immunology, gestalt immunology and psychoneuroendocrinology. The label that has proved most lasting is 'psychoneuroimmunology' (PNI), with its own journal, *Brain, Behavior, and Immunity*, founded in 1987 as the official journal of the Psychoneuroimmunology Research Society.[75]

Robert Ader was one of the driving forces behind the early research in this area. Reflecting in the year 2000 on the research in the field, Ader was able to suggest that a series of independent lines of research had converged around these links, many of which were becoming increasingly popularized.[76] As Susan Greenfield puts it in one such popular book, PNI practitioners now thrive on a wealth of evidence that 'the

75 Robert Ader, 'Presidential Address – 1980: Psychosomatic and Psychoimmunologic Research', *Psychosomatic Medicine*, vol. 42, no. 3 (1980), 307–22 (312–13); Robert Ader, David C. Felten, and Nicholas Cohen (eds.), *Psychoneuroimmunology* (San Diego: Academic Press, 1991); Vita Vedhara and Michael Irwin (eds.), *Human Psychoneuroimmunology* (Oxford: Oxford University Press, 2005).

76 Robert Ader, 'On the Development of Psychoneuroimmunology', *European Journal of Pharmacology*, vol. 405 (2000), 167–76.

three great control networks of the body – the immune, the nervous, and the endocrine (hormonal) systems – are all interlinked'. This arrangement 'makes intuitive sense'; without it there would be 'biological anarchy'.[77] Or as Candace Pert, one of the leading researchers in the field, put it in the conclusion of one of her jointly authored articles, 'The conceptual division between the sciences of immunology, endocrinology, and psychology/neuroscience is a historical artifact.'[78] This understanding is reinforced by the idea of the essential *plasticity* of the brain and central nervous system, following the evidence that the brain exhibits a remarkable ability to adapt, change, and reorganize, and the same idea now also appears in the literature on psychoneuroimmunology.

Yet what about 'nerves' in the emotional and psychological rather than neurological and physiological sense? Pert's claim was intended to not only reject the conceptual division but also to argue that the various fields of the division also need to be reconnected to an 'older psychologic literature'. This would help us grasp the fact that 'emotional states can significantly alter the course and outcome of biologic illnesses previously considered to be strictly in the somatic realm'.[79] On this view, 'immunity' can be understood not solely in terms of the nervous system but also in terms of nervousness, for the approach highlights the importance of stress, anxiety, and other forms of 'nervous illness'. This is the *psyche* in 'psychoneuroimmunology', which points us back to the unity of *psyche* and *soma* discussed in chapter 2. The titles of some of the research articles in *Brain, Behavior, and Immunity* in the last decade indicate the general thrust of this approach: 'Depression as an Evolutionary Strategy for Defense against Infection' (2013); 'The Role of Immune Genes in the Association between Depression and Inflammation' (2013); 'Depressive Symptoms Enhance Stress-Induced Inflammatory Responses' (2013); 'The Inflammasome: Pathways Linking Psychological Stress, Depression, and Systemic Illnesses' (2013); 'Malaise, Melancholia and Madness: The Evolutionary Legacy of an Inflammatory Bias' (2013). The same connections are made time and again in other major publications: 'Inflammation: Depression Fans the Flames and Feasts on the Heat' (*American Journal of*

77 Susan Greenfield, *Tomorrow's People: How 21st-Century Technology Is Changing the Way We Think and Feel* (London: Penguin, 2004), 201–02.

78 Candace B. Pert et al., 'Neuropeptides and their Receptors: A Psychosomatic Network', *Journal of Immunology*, vol. 135, no. 2 (1985), 820–26 (824).

79 Pert et al., 'Neuropeptides', 824.

Psychiatry, 2015); 'The Role of Inflammation in Depression: From Evolutionary Imperative to Modern Treatment Target' (*Nature Reviews Immunology*, 2016); 'Alterations in Immunocompetence During Stress, Bereavement, and Depression' (*American Journal of Psychiatry*, 1987); 'Stress and Immunity: An Integrated View of Relationships between the Brain and the Immune System' (*Life Sciences*, 1989); 'The Macrophage Theory of Depression' (*Medical Hypotheses*, 1991); 'Evidence for an Immune Response in Major Depression' (*Progress in Neuropsychopharmacology*, 1995). An increasingly large number of books also reinforce the connections between stress, emotions, and immune function, such as Paul Martin, *The Sickening Mind: Brain, Behaviour, Immunity and Disease* (1998), Bruce Rabin, *Stress, Immune Function, and Health* (1999), Esther M. Sternberg, *The Balance Within: The Science Connecting Health and Emotions* (2000), Darian Leader and David Corfield, *Why Do People Get Ill?* (2007), Edward Bullmore, *The Inflamed Mind: A Radical New Approach to Depression* (2018), Antonio L. Teixeira and Moises E. Bauer (eds.), *Immunopsychiatry: A Clinician's Introduction to the Immune Basis of Mental Disorders* (2019), and Gabor Maté, *When the Body Says No: The Cost of Hidden Stress* (2019). Such connections have all been reinforced by, and used in turn to reinforce, research by thinkers such as Antonio Damasio and Joseph LeDoux, on emotions as biological functions of the nervous system and part and parcel of the homeostatic regulation of organic systems.

In other words, the conjunction of immunity and nerves facilitates an enormous shift in thinking: it is easily extended into an account of nervousness; it underpins the idea of a deep and complex relationship between immunity, inflammation, and depression; and it brings 'stress' into the heart of the discussion of immune defence. This sheds further light on the autoimmune disease as we discussed it in chapter 2 and at the same time conceptualizes 'nervousness' and its gamut of associated terms – nervous exhaustion, nervous breakdown, nervous tension – in a way that spans the neurological, immunological, and emotional. Nervous *feelings* constantly interact with and effect changes in the immune process, and their *plasticity* connotes explosiveness as well as malleability, a *destructive* capacity to annihilate and not just adapt.[80]

80 Catherine Malabou, *What Should We Do with Our Brain?* (2004), trans. Sebastian Rand (New York: Fordham University Press, 2008), 5.

Lisa Feldman Barrett comments that understanding inflammatory disorders such as autoimmune disease has 'been a game-changer for our understanding of mental illness'.[81] But why only one way? Returning to our discussion of suicide in chapter 2, we can tweak the question by Sarah Manguso cited there: What comes first, the depression or the autoimmune disease? What comes first, the nervous disorder or the autoimmune disorder? Asking such questions enables us to make sense of the terms which emerge to try to capture anew the kind of experiences that have long dominated these debates, such as the idea that all nervous illness, like madness or depression, is an expression of some kind of anger at the world, and hence a form of rebellion or political desire, but also, perhaps, a form of defeat, knowledge of which simply compounds the 'illness' in question. For example, we hear a lot these days about burnout and the burnout society, and the term appears to capture much of what many of us feel about a general level of nervous exhaustion and the demands of modern life. But research by Jean Philippe Blankert and colleagues in the Stichting Burnout (Burnout Foundation) in the Netherlands, working for several years on biomarkers of burnout, have amassed a fair amount of evidence that 'burnout consists of an *immunological* reaction of body and mind without external, physical pathogen'. Their developing 'grand theory' of burnout covers the role of the immune system in identifying threats and dangers that are not consciously recognized or perceived. Burnout, in other words, can be understood as a form of neuroinflammation and in that sense needs to be understood through the lens of psychoneuroimmunology. This is one reason why books on burnout often rely on immunological models of explanation, such as Byung-Chul Han's *The Burnout Society* (2010). What comes first, the burnout or the autoimmune disease? All of this might go some way to explaining the parallel and exponential rise in the number of people diagnosed with autoimmune diseases, detailed in the introduction, and the number of people diagnosed with nervous exhaustion, burnout, depression or stress, or self-identifying as feeling on the verge of a nervous breakdown. Might these not be connected?

At the same time, might they not also demand a political analysis? The experience of nervous exhaustion or burnout is socially grounded

81 Lisa Feldman Barrett, *How Emotions Are Made: The Secret Life of the Brain* (London: Pan, 2017), 202.

and culturally reproduced, as we know from our earlier discussion. As well as being extended and hence *somatic*, *psyche* is also *social*. The issue is a burnout society – and hence a society that burns us out – not just burnt-out individuals. What is presented to us as a problem of the Self must be understood as a problem presented *for* the Self by the world. All illness is a social condition, and as a social condition, all illness is a statement about the world and not just a physiological or biochemical state. A protest, perhaps. For Blankert and colleagues at the Stichting Burnout, what undergirds their research is not simply an immunological or even psychoneuroimmunological reaction of body and mind, but the *need for justice* and for the *satisfaction of other needs* that one might expect to be *met through work*. Or, as Schreber put it: there is something rotten in the state of Denmark. Might the evidence for this rottenness be the way the System itself exhibits a nervousness and over-responds to its own fears and anxieties, a nervous overreaction that produces nothing else but that security issue par excellence: a terror within. Might such terror lead to a nervous breakdown?

Nervous State, Nervous Terror

Can a political system have a nervous breakdown? Here is a sample of headlines from just a few months in 2018 and 2019 during the writing of this book: 'White House is on the verge of a nervous breakdown'; 'Brexit Britain: Country on the edge of a nervous breakdown'; 'Spain on the verge of a nervous breakdown'; 'China is an economy on the verge of a nervous breakdown'; 'Russia on the verge of a nervous breakdown'; 'Davos arrives as world on verge of nervous breakdown'; 'Is politics on the verge of a breakdown?'; 'Australia's nervous breakdown'; 'Ireland on the verge of a nervous breakdown'; 'Democracies on the verge of a nervous breakdown'. It should be noted that these were all prior to the outbreak of Covid-19. It should also be noted that the journalistic trope holds to the idea that these systems have not *had* a nervous breakdown, and neither is it ever definitively stated that they *going* to have one. That would push the idea too far. For the trope to hold, the system in question is imagined as on the *verge* or the *edge* of a nervous breakdown. It is, after all, the function of the state to manage its social body so that it does not actually break down, to ward off the chaos, to fight the entropy.

Nonetheless, the *fear of breakdown* now appears to manifest itself in the sphere of political life, allowing us to tweak a point gleaned from Winnicott in chapter 2: the fear of breakdown is a fear of disintegration, an unthinkable state of affairs connected to the failure of the mechanisms of security. Fuelled by a nervousness over its own security, immunity and identity, and by its fears of enemies real and imagined, we are confronted by the *nervous state*.

Why the nervous state? In the opening essay of his book *The Nervous System*, Michael Taussig connects the physiological nervous system with the chaos and unpredictability of an unstable social system.[82] The expansion captures some of the ways in which we imagine the double meaning in the phrase 'nervous system' that we have already discussed. On the one hand, the nervous system as a powerful and systemic physical force in our lives: energy. On the other hand, a system that is nervous to the extent that it appears on the verge of breaking down: fear. Considered across the body as a whole, we encounter a highly organized system in which lines of control and communication maintain good order, but we also encounter a system held together by a nervousness and a fear of breaking down. We hear much about 'failed states', 'rogue states', and 'weak states', terms which liberal democracies like to use about other states. But *nervous state*? Even liberal states cannot avoid having that label used to describe them.[83]

In a nervous state, the body exists in a condition of such heightened alertness that its systems run amok. Hyper-intensive levels of fear, compounded by the demands of the subsystems of capital and security, make homeostasis impossible. The security system responds to the ever-greater nervousness by searching for enemies, by finding enemies and by fabricating new enemies. Part of this impact is on the outside (the non-Self, Other, Foreign), and part of it is turned inwards. The *hyper-vigilant security system* turns towards an enemy *within*; the nervous defence of the body gets easily overextended. To ward off the fear of breakdown, more and more security operations ensue but themselves

82 Michael Taussig, *The Nervous System* (London: Routledge, 1992), 1–2.

83 William Davies, *Nervous States: How Feeling Took Over the World* (London: Jonathan Cape, 2018); Ulrike Lindner, Maren Möhring, Mark Stein, and Silke Stroh (eds.), *Hybrid Cultures – Nervous States: Britain and Germany in a (Post)Colonial World* (Amsterdam: Rodopi, 2010); Nancy Rose Hunt, *A Nervous State: Violence, Remedies, and Reverie in Colonial Congo* (Durham, NC: Duke University Press, 2016).

lead to disintegration, increasing the likelihood of breakdown. The system *turns against its own body*.

Much of the description in the previous paragraph captures the autoimmune disease, but it also captures what is happening through the security practices of the contemporary body politic, in which a *hypervigilant and intensified security operation* searches for the enemy *within* the body politic, turning self-defence into self-destruction. A nervous extension of more and more security operations leads to breakdown and disintegration. The state appears to be on the verge of a nervous breakdown, suicidal even, as though it too possesses a death drive.

'Without the concept of a death drive, some events of recent history – German history in particular – are incomprehensible', Bruno Bettelheim once commented.[84] Bettelheim's comment has been repeated in less allusive terms by those historians who have observed that fascism's destructive tendency means that it is driven towards its own destruction as much as the destruction of its enemies. Albert Speer in *Inside the Third Reich* (1969), J. P. Stern in *Hitler: The Führer and the People* (1975) and Sebastian Haffner in *The Meaning of Hitler* (1979) all explored the ways in which Nazism condemned the German body politic to death. Evidence for such a claim tends to involve the flurry of decrees issued by the crumbling Nazi regime from mid-March 1945 ordering the destruction of Germany's infrastructure, most notably the 'Nero Decree' issued on 19 March (the 'Order for Destructive Measures on Reich Territory'); the 'suicide epidemic' of February to May 1945; Hitler's final decision to blow his own brains out, the suicidal death of the leader as emblematic of the suicide state; and ideological slogans such as 'Long Live Death!' More than anything, of course, is the way that the Nazi state presided over the destruction of its own citizens in an attempt to eliminate the diseases, bacteria, parasites, and germs within the national body. In this light, the idea of the 'suicidal state' has been applied many times to fascism by other thinkers as well as historians. For Foucault, for example, 'the objective of the Nazi regime was . . . not really the destruction of other races. The destruction of other races was one aspect of the project, the other being to expose its own race to the absolute and universal threat of death.' The regime was thus murderous and racist,

84 Bruno Bettelheim, *Freud and Man's Soul* (London: Chatto and Windus, 1983), 107.

but also *completely suicidal.*[85] Paul Virilio, Gilles Deleuze, and Félix Guattari have made similar points. Building on an essay by Virilio called 'The Suicidal State', Deleuze and Guattari comment,

> In fascism, the State is far less totalitarian than it is *suicidal* . . . Fascism is constructed on an intense line of flight, which it transforms into a line of pure destruction and abolition. It is curious that from the very beginning the Nazis announced to Germany what they were bringing: at once wedding bells and death, including their own death, and the death of the Germans. They thought that they would perish but that their undertaking would be resumed, all across Europe, all over the world, throughout the solar system. And the people cheered, not because they did not understand, but because they wanted that death through the death of others . . . Suicide is presented not as a punishment but as the crowning glory of the death of others . . . Virilio's analysis strikes us as entirely correct in defining fascism not by the notion of the totalitarian State but by the notion of the suicidal State . . . A war machine that appropriates the State and channels it into a flow of absolute war whose only possible outcome is the suicide of the State itself.[86]

In *The Monstrous and the Dead* (2005), I discussed the limitations of this notion of fascism as suicidal. Here I want to turn the argument around a little and suggest that the notion has some mileage, but only if one is willing to *treat fascism as an expression of the modern security state.*[87] Fascism was, after all, and beyond all else, a politics of *security*, which is the very reason it took the *immunity* of the body politic so seriously.

Everything the suicidal state does to bring about its suicide is done through the measures it enacts in the name of security: containment

---

85 Michel Foucault, *Society Must Be Defended: Lectures at the Collège de France, 1975–76*, trans. David Macey (London: Allen Lane, 2003), 260.

86 Gilles Deleuze and Félix Guattari, *A Thousand Plateaus: Capitalism and Schizophrenia* (1980), trans. Brian Massumi (London: Athlone Press, 1987), 230–31; Paul Virilio, 'The Suicidal State' (1993), in James Der Derian (ed.), *The Virilio Reader* (Oxford: Blackwell, 1998).

87 Mark Neocleous, 'Inhuman Security', in David Chandler and Nik Hynek (eds.), *Critical Explorations of Human Security* (London: Routledge, 2011).

and control, police power and legal terror, war and war again. Hyper-security operations seep into every institution and practice, destroying the life of the social body itself. The state is a security machine that is willing and able to engage in the destruction of its own body in the name of that very security. The security machine becomes a suicide machine. The point, however, is that such a suicidal tendency haunts all security projects, liberal as well as fascist. It is the fate of a body organized around security *to turn against itself*. If there was ever one word capable of summing up a society trying to live under the ever-increasing hyper-intensification of security, it was always going to be *suicidal*. And this, it turns out, can help us unravel the politics of immunity and the nervous state.

Addressing what he calls the 'homicidal temptation of biopolitics' that took place in the Third Reich, Roberto Esposito asks, 'Why did Nazism (and *only Nazism*) reverse the proportion between life and death in favor of the latter to the point of hypothesizing its own self-destruction?' The answer he offers lies in his notion of immunization:

> It is only immunization that lays bare the lethal paradox that pushes the protection of life over into its potential negation. Not only, but it also represents in the figure of the autoimmune illness the ultimate condition in which the protective apparatus becomes so aggressive that it turns against its own body (which is what it should protect).[88]

Esposito's tendency to jumble things is once again apparent, here slipping rather too easily from a Nazi desire to immunize the German body politic from the enemy to the idea that immunization somehow turns against the German body politic as a whole. But as discussed in the introduction, it is not immunization that pushes the protection of life over into its potential negation, but the autoimmune disease. The more pressing question here, however, is why *only* Nazism? For Esposito, Nazism was the complete realization of biopolitics, a 'truly fatal leap' when biopolitical immunization intersects with nationalism and racism. 'As in so-called autoimmune diseases, here too [in Nazism] the immune

88 Roberto Esposito, *Bíos: Biopolitics and Philosophy* (2004), trans. Timothy Campbell (Minneapolis: University of Minnesota Press, 2008), 116, emphasis added.

system is strengthened to the point of fighting the very body that it should be saving, but it is now causing that body's decomposition'. 'Auto-immunity' becomes 'auto-genocide'. Yet there is something a little too easy about treating Nazism in this way, much as it is a little too easy to point to Nazism as the suicidal state. The easiness resides in the implicit assumption that 'Nazism lies entirely outside not only modernity but the philosophical tradition of modernity'.[89] Yet, if there is one thing that can be said for sure about Nazism, it is that it lies entirely *inside* modernity and its philosophical tradition.[90]

There is a sense in which Esposito himself knows this. His examples, the standard ones offered by many other writers when discussing Nazism, are, in fact, hardly peculiar to Nazism: 'The specificity of Nazism is demonstrated . . . by the particularity of the disease against which it intended to defend the German people. We aren't dealing with any ordinary sort of disease, but with an infective one.' But any analysis of liberal democracies in the last century will find the idea of an 'infective disease' as a justification for violent security measures, as we have seen. In the same manner, is not the Parasite that we discussed in chapter 4 also an endemic feature of Western politics per se, and not just Nazism? Indeed, this is the line Esposito eventually concedes in *Bíos* and elsewhere, where liberalism's preventive war on terror threatens an 'autoimmunitary turn' in biopolitics, in the sense that 'the immunitary machine demands an outbreak of effective violence on the part of all contenders'. In this context, Esposito invokes autoimmunity:

> The idea – and the practice – of preventive war constitutes the most acute point of this autoimmunitary turn of contemporary Biopolitics, in the sense that here, in the self-confuting figure of a war fought precisely to avoid war, the negative of the immunitary procedure doubles back on itself until it covers the entire frame.

He goes on:

---

89 Roberto Esposito, *Terms of the Political: Community, Immunity, Biopolitics* (2008), trans. Rhiannon Noel Welch (New York: Fordham University Press, 2013), 73, 86.

90 See my *Fascism* (1997).

> Just as in the most serious autoimmune illnesses, so too in the planetary conflict under way: it is excessive defense that ruinously turns on the same body that continues to activate and strengthen it. The result is an absolute identification of opposites: between peace and war, defense and attack, and life and death, they consume themselves without any kind of differential remainder.[91]

What Esposito fails to mention here is that it is security that is the real issue. He makes a slight gesture towards this, in his suggestion that 'immunity progressively transfers its own semantic center of gravity *from the sense of "privilege" to that of "security"*',[92] but it remains only a gesture, because to follow through on it would force it into an argument about security rather than genocide. It is in 'social systems . . . *neurotically haunted* by a continuously growing need for security' that one finds 'the risk from which the protection is meant to defend is actually created by the protection itself'.[93] Precisely, which is the very reason we should be talking about (liberal) security and not (Nazi) genocide. The 'neurotic' search for security describes the modern state in all its forms, including, and perhaps even most of all, its liberal democratic forms, which is perhaps why liberal democracy so often seems intent on a suicidal self-destruction. 'A society does not ever die "from natural causes"', but always dies from suicide or murder', observes Arnold Toynbee, who attributed such suicide to militarism.[94] In fact, it is the suicidal statecraft of security that is the real issue.

This is where Derrida's political reading of autoimmunity becomes useful. Derrida's argument was originally flagged in *Rogues* (2002), where he offers the Algerian elections of 1992 as an example of a tendency towards autoimmune suicide, on the grounds that the governing party believed the electoral process was about to 'lead democratically to the end of democracy' and so 'decided in a sovereign fashion to

91 Esposito, *Bíos*, 147–48.

92 Esposito, *Bíos*, 72, emphasis added.

93 Roberto Esposito, *Immunitas: The Protection and Negation of Life* (2002), trans. Zakiya Hanafi (Cambridge: Polity, 2011), 141, emphasis added.

94 Arnold J. Toynbee, *A Study of History*, abridgement of vols. I–VI (1946), ed. D. C. Somervell (Oxford: Oxford University Press, 1987), 273, 336, 345, 582. Also Jean Baudrillard, *The Spirit of Terrorism* (2002), trans. Chris Turner (London: Verso, 2003), 7.

suspend, at least provisionally, democracy for its own good'. Typical of 'assaults on democracy in the name of democracy', the decision exemplifies democracy's tendency towards an autoimmune suicide.[95] This example is both strange and strained, since it is far from clear how elections are part of the body politic's system of immunity-security. In contrast, Derrida's argument becomes more focused in an interview conducted a few weeks after the 'absolute terror' of 11 September 2001. As we discussed in chapter 2, part of the terror of an autoimmune disease is that the threat 'comes from within', meaning that one's vulnerability is without limit.

> Whence the terror. Terror is always, or always becomes, at least, in part, 'interior'. And terrorism always has something 'domestic', if not national, about it. The worst, most effective 'terrorism', even if it seems external and 'international', is the one that installs or recalls an interior threat, *at home* – and recalls that the enemy is *also always* lodged on the inside of the system it violates and terrorizes.[96]

The fact that the 9/11 hijackers came from 'within' opens the way for Derrida to treat the attacks as 'the first symptom of suicidal autoimmunity'.

> The aggression . . . comes, *as from the inside*, from forces that are apparently without any force of their own but that are able to find the means, through ruse and the implementation of *high-tech* knowledge, to get hold of an American weapon in an American city on the ground of an American airport. Immigrated, trained, prepared for their act in the United States by the United States, these *hijackers* incorporate, so to speak, two suicides in one: their own . . . but also the suicide of those who welcomed, armed, and trained them. For let us not forget that the United States had in effect paved the way for and consolidated the forces of the 'adversary' . . . by first of all creating the

95 Jacques Derrida, *Rogues: Two Essays on Reason* (2002), trans. Pascale-Anne Brault and Michael Naas (Stanford, CA: Stanford University Press, 2005), 33.

96 Jacques Derrida, 'Autoimmunity: Real and Symbolic Suicides – A Dialogue with Jacques Derrida', in Giovanna Borradori, *Philosophy in a Time of Terror: Dialogues with Jürgen Habermas and Jacques Derrida* (Chicago: University of Chicago Press, 2003), 187.

> politico-military circumstances that would favour their emergence and their shifts in allegiance.[97]

Now, picking up again on comments we made in the introduction, we need to note that Derrida, like Esposito, jumbles things a little. He comments that 'an autoimmunitary process is that strange behavior where a living being, in quasi-suicidal fashion, "itself" works to destroy its own protection, to immunize itself against its "own" immunity'.[98] This is a misleading account of autoimmunity and autoimmune disease, frequently repeated in the secondary literature aiming to deconstruct the war on terror and international politics. Although Derrida is right to point to the fact that the attackers did come from within, to think of this through the 'the general law of this autoimmune process', or a 'general logic of auto-immunization', makes little sense, since the attackers were hardly a part of the body politic's system of security-immunity. The attacks may have been an 'immune failure' in the sense that they failed to detect an enemy, but not all immune failures are autoimmune diseases; as previously noted, Derrida tends to conflate and confuse autoimmune disorders with immunodeficiency. The point, however, is that although the terror of the enemy attackers was real, the absolute terror of the autoimmune disease stems from the fact that an enemy *is imagined* and is imagined as *within*. The outcome is, as we have seen, that the security system starts destroying its own body politic, allies treated as enemies and friends as foe in a suicidal immune reaction against the body.

To sum up, and to connect back to discussion in earlier chapters, what is terrifying about the body's overreaction in the name of immunity turns out to be what is terrifying about the sovereign body's overreaction in the name of security: under the banner of security-immunity it seeks at all costs to protect itself against any danger and secure itself against any insecurity, even if the ultimate cost is the life of the very body it thinks it is defending. The principle that 'Society Must Be Defended!' spirals out of control, and (Self-)defence turns into (Self-) destruction. The security system declares war on the enemy; thinks that the enemy is within; seeks out the enemy; announces and denounces the suspected enemy cells within the body without realizing that the cells

97 Derrida, 'Autoimmunity', 95.
98 Derrida, 'Autoimmunity', 94.

are not enemy cells at all; starts to eliminate them; and continues to do so until there is nothing left to eliminate. In the name of security-immunity, the body politic destroys itself.

What we can now add to this picture is an understanding of the nervous state that underpins the process. The nervous state ratchets up the hunt for yet more enemies within. Esposito suggests that 'the immune system is revealed as the nerve center through which the political governance of life runs',[99] and we have already cited him to the effect that we find ourselves in a social system 'neurotically' haunted by the ever-expanding need for security, in the same way that he can casually throw out the idea that the war on terror is the 'madness', of immunization.[100] Nothing is made of this relation to 'nerves', 'neurosis' and even 'madness', despite the fact that he is trying to develop an immunitarian biopolitics. In fact, a great deal hangs on it, as the nerves of state and its immunitary processes work in tandem. It is the nervousness of the nervous state that generates an autoimmune disorder, the sovereign power's own security practices generating a nervous political condition which in turn generates more security practices. The nervous state finds itself unable to extricate itself from this condition. Through the lens of a political psychoneuroimmunology working with a unified notion of nerves and immunity, we might say that the immune system of the nervous state turns against the very thing it is meant to protect, but also, and at the same time, the nervous system of the immune state generates more and more immunity-security operations; the outcome is the same.

In *The Nervous System*, Taussig segues from the opening essay 'Why the Nervous System?' to an essay on Walter Benjamin and the state of emergency, in which 'exceptional' and 'extreme' measures are taken in the name of security, those times when it becomes abundantly clear that security expects nothing less than for us to accept the violence meted out in its name. Taussig does not make much of the connection between the two essays in *The Nervous System*, but in a later book, he writes a diary of his experience of a *limpieza* in Colombia.[101] A *limpieza* is the Colombian name for a state of emergency characterized by 'military

99 Esposito, *Immunitas*, 150.

100 Esposito, *Terms*, 62, emphasis added.

101 Michael Taussig, *Law in a Lawless Land: Diary of a* Limpieza *in Colombia* (Chicago: University of Chicago Press, 2003), 17–18.

rule' and the attempt to impose 'law and order' through paramilitary violence and political assassinations. The word '*limpieza*' also connotes a cleaning of the body, a purge, a clean-up operation. In other words, a state of emergency, a period in which security's takeover of the body politic becomes clear, most evident in the way it ultimately destroys that body in the name of security. What if the only system is a nervous system? asks Taussig. Better still, and to rephrase a question from previous chapters: What really is this System that turns on itself through its own nervousness?

This is the protean nervousness through which the political must now be imagined, the destructive side of the plasticity of contemporary bodies of security. The nervous state reveals itself in hyper-security operations that are normal, not an emergency. What is normal is the nervous terror and the terrifying nervousness that flows through the body and manifests itself in an organized if unevenly executed violence against that very body. The nervous state has a vivid imagination: it dreams of monstrous mutating forces within its own body, of parasites and demons seeking mastery over the host, of foreign enemies masquerading as friends, of non-Self cells masquerading as Self, and myriad other diseases, viruses and mutating cells. The nervous state cannot not be nervous while there are enemies to imagine fighting, and there are always enemies to imagine. Perhaps the secret of bourgeois modernity is its death drive, a capacity for self-destruction carried out in the name of security. Perhaps this is why the security wars of the body politic look like reducing all of us to collateral damage.

6

Immunity's Fiction

In the city there's a thousand men in uniforms.
And I've heard they now have the right to kill a man.
The Jam, 'In the City' (Polydor Records, 1977)

Of course I'm dangerous. I'm the police. I can do terrible things to people.
With impunity.
Rustin Cohle, *True Detective*, season 1 (HBO, 2014)

We live in an age of collateral damage. The British Library catalogue lists over seven hundred books with 'collateral damage' in the title; over four hundred of them have been published in the last decade. The US state now operates a 'Collateral Damage Estimation' system, a 'Collateral Damage Methodology', and a 'Fast Assessment Strike Tool – Collateral Damage'. In terms of numbers, 'collateral damage' is said to account for between 5 and 10 percent of deaths in World War I, for over half the deaths in World War II and for around 90 percent of the deaths in wars fought by the end of the twentieth century. And, as if the late twentieth century was not a dangerous enough time to be a civilian, the twenty-first century was already being described as 'the heyday of collateral damage' before its first decade was even over.[1] The extent of this

1 Stephen J. Rockel, 'Collateral Damage: A Comparative History', in Stephen J. Rockel and Rick Halpern (eds.), *Inventing Collateral Damage: Civilian Casualties, War, and Empire* (Toronto: Between the Lines, 2009), 4.

damage is more than a little strange, given that we are told time and again that civilians are protected as far as possible from the violence of war; that states seek to protect noncombatants (the terms 'civilian' and 'noncombatant' are more often than not used interchangeably in this literature); that any state claiming to be built on ethical principles recognizes this need for protection; and that this protection has been codified in international law. In other words, we are said to live in an age of noncombatant immunity. 'Noncombatant immunity has served as a fundamental limit on political violence', is how one of the hundreds of books called *Collateral Damage* begins. It goes on to say that so powerful is this limit that noncombatant immunity is now 'a basic human right codified in international law'.[2] This is the gist of much of the literature on collateral damage.

So: we live in an age of noncombatant immunity designed to provide civilians with some semblance of security, that is also an age of collateral damage in which civilians appear to lack any meaningful security; an age in which civilians are announced as being formally and legally immune from the ravages of state violence and yet are killed in their thousands. How can this be? How is it possible that immunity is a human right ascribed to noncombatants, codified in international law and adhered to by liberal democratic states, and yet so many noncombatants are being killed by those same states? 'These two developments are mutually contradictory', say some.[3] Others say that such 'accidents' are part of the 'fog of war' (the 'shit happens' explanation for dead bodies). Such deaths, we hear, are the price we pay for security. But 'the price we pay for security' is the very phrase used to describe autoimmune disease. So, perhaps there is some mileage to be gained in exploring what is at stake in all this destruction taking place through and around the idea of immunity.

One approach to this is to insist that there is a principle of distinction between two 'zones' or 'spheres' – military/civil, combatant/

2 Sahr Conway-Lanz, *Collateral Damage: Americans, Noncombatant Immunity, and Atrocity after World War II* (Abingdon, Oxon: Routledge, 2006), xi. Likewise, Neta Crawford, *Accountability for Killing: Moral Responsibility for Collateral Damage in America's Post-9/11 Wars* (Oxford: Oxford University Press, 2013), 47–53; Bruce Cronin, *Bugsplat: The Politics of Collateral Damage in Western Armed Conflicts* (Oxford: Oxford University Press, 2018), 1.

3 Thomas Hippler, *Governing from the Skies: A Global History of Aerial Bombing* (2014), trans. David Fernbach (London: Verso, 2017), 12. Also Alex J. Bellamy, *Massacres and Morality: Mass Atrocities in an Age of Civilian Immunity* (Oxford: Oxford University Press, 2012).

noncombatant, soldier/civilian, protector/protected – and that the distinction is breaking down. This principle of distinction is said to be found in the legal and international codes and agreements. For example, the Lieber Code of 1863, article 22, states that as a principle 'the unarmed citizen is to be spared in person, property, and honor', and the same or similar phrases appear in the Geneva Convention of 1864, the Brussels Conference of 1874, the Hague Regulations of 1899 and 1907, the Universal Declaration of Human Rights in 1948, and the Geneva Conventions of 1949. Article 51 of the Additional Protocols to the Geneva Conventions, agreed in 1977, reiterates the point: 'The civilian population as such, as well as individual civilians, shall not be the object of attack.' This principle of distinction also figures large in just war theory, which has in turn reinforced international law's portrait of itself as just. In *Just and Unjust Wars*, for example, a book which did much to put the principle of distinction centre stage, Michael Walzer points to the importance of noncombatant immunity to the distinction and hence to just war theory in general. Tucked away in a footnote is the suggestion that 'we are all immune to start with' and thus have a 'right not to be attacked'. This right is 'lost by those who bear arms' but 'retained by those who don't bear arms at all'.[4] Yet Walzer does little with this idea of immunity. Neither, for that matter, do most of the writers who articulate the same idea.[5] Immunity is simply taken as read, as somehow obvious or even, dare we say, natural.

This final chapter returns to immunity's roots in law. I aim to address the idea of noncombatant immunity without simply imagining two worlds on either side of a line (the principle of distinction) and then stating the obvious about the line being breached. I want to think instead about noncombatant immunity as a *sort of fiction*. But I also want to suggest that exploring the nature and limits of immunity as a sort of fiction allows us to reveal immunity's inner truth, touched on by Thomas Gregory when he suggests that 'far from protecting civilians from the death and destruction of war, the principle of noncombatant immunity works to legitimize the killing of civilians providing that certain

4 Michael Walzer, *Just and Unjust Wars: A Moral Argument with Historical Illustrations*, 4th edn. (New York: Perseus Books, 2006), 136, 228, 145–46, 236–37.

5 For example, Jean Bethke Elshtain, *Just War against Terror: The Burden of American Power in a Violent World* (New York: Basic Books, 2003), 53, 62, 190.

conditions are met'.[6] The truth of immunity's fiction is state violence exercised in the name of security. Immunity works to legitimize the killing of civilians because it is the state and its security system that *claims immunity for itself and its agents*. In pursuing the idea that immunity is integral to the political management of violence, killing, and death, we will find that immunity's fiction has a narrative twist. The twist is not that the state claims to treat civilians as immune but then has few qualms about destroying those same civilians; that is a blindingly obvious part of the narrative. Rather, the twist is that the state ultimately claims immunity for itself and for its agents of violence. The fiction of civilian immunity then, reveals the truth of immunity's violence.

The Civilian: Born to Die

Just war theory and Christianity alike claim the moral high ground when it comes to questions of war. One way in which they do so is by articulating a principle of noncombatant immunity. In *Just War Tradition and the Restraint of War*, for example, James Turner Johnson suggests that the early laws of war were defined principally by the chivalric code of the knights which limited the use of violence towards particular persons, and that this was reinforced by 'a rudimentary definition of noncombatant immunity' produced by the Church.[7]

Yet early Christianity had neither the concept of 'civilian' nor that of 'noncombatant immunity'. Thinkers such as Augustine, Aquinas, and Vitoria were far more interested in 'innocence'. They were interested in 'innocent' persons not to declare them immune, however, but to identify when they might rightfully be slaughtered.[8] Augustine

6 Thomas Gregory, 'Targeted Killings: Drones, Noncombatant Immunity, and the Politics of Killing', *Contemporary Security Policy*, vol. 38, no. 2 (2017), 212–36 (218).

7 James Turner Johnson, *Just War Tradition and the Restraint of War: A Moral and Historical Inquiry* (Princeton, NJ: Princeton University Press, 1981), 47.

8 The interest in innocence is retained by the liberal tradition, which likes to play a game of distinguishing liberal and terrorist killing: terrorists kill people because they do not respect innocent lives, whereas liberal states claim to respect 'innocent' lives. 'Innocence' gets attached to 'immunity', and the killing of any 'innocent' person by a liberal state must by definition be some kind of mistake. The strongest critique of the concept of innocence has come from within feminism – see R. Charli Carpenter, *'Innocent Women and Children': Gender, Norms and the Protection of Civilians* (Aldershot:

wrote little in defence of the innocent in war; permitted the slaying of the innocent among the enemy population if the necessities of just war demand it; believed that it was not very likely that there would be many 'innocents' among the enemy anyway; was more interested in the 'innocence' of the Christian soldier than the innocence of 'civilians'; and had nothing to say about what now goes by the name of 'immunity'. For Augustine, the just war was equated with righteousness, and any violation of righteousness warranted unlimited punishment. Moreover,

> the subjective *culpa* or guilt of the enemy merited punishment of the enemy population without regard to the distinction between soldiers and civilians. Motivated by a righteous wrath, the Just warriors could kill with impunity even those who were morally innocent.[9]

Note that term 'impunity', because it is a key term in debates about sovereign violence, and we will return to it below.

It is undoubtedly the case that particular people and sites were understood to have a certain kind of protection, but this was connected to ecclesiastical property and the power of the Church. For example, it is sometimes said that the 'Peace of God' movement 'forms one of the roots of the doctrine of noncombatant immunity'.[10] The 'Peace of God' refers to those decrees and pronouncements of medieval Church councils protecting certain classes of the population from legitimate military attack, usually starting from the one held at the monastery of Charroux in Poitou in 988. Those involved in religious duty were to be protected, and this included priests, monks, and clerics but also sometimes women, children, pilgrims, and merchants. Similar injunctions were declared in council after council such that, by the time of the Third Lateran Council in 1179, peasants and their animals were also protected from attack. Now, in just war theory those deemed immune

---

Ashgate, 2006); Helen M. Kinsella, *The Image before the Weapon: A Critical History of the Distinction between Combatant and Civilian* (Ithaca, NY: Cornell University Press, 2011); Laura Sjoberg, *Gendering Global Conflict: Toward a Feminist Theory of War* (New York: Columbia University Press, 2013).

9 Frederick H. Russell, *The Just War in the Middle Ages* (Cambridge: Cambridge University Press, 1975), 19.

10 Johnson, *Just War*, 127.

from violence 'are all types of person who, because of their social function, have nothing to do with warmaking; thus they are not to have war made against them'.[11] But the supposed 'immunity' of the people identified in the council decrees does not rest on a principle of distinction between those who have nothing to do with the conduct of war and those who do. The idea that some people were not to have war made against them had nothing to do with the idea that they themselves had nothing to do with war. The peasant, for example, was integral to war, since the peasant tilled the land, produced much of the wealth over which war was fought, and provided sustenance for those doing the fighting. When the peasantry did have 'protection' it was because of their role in tilling the land and providing resources for the Church. Their 'immunity', such as it was, came not through being understood as a 'noncombatant' or 'civilian' but through the exemption granted to the Church on whose land they worked. Moreover, the idea that one can find in such immunities some kind of universal right pertaining to the noncombatant is the very opposite of what was meant by immunity at the time which, if anything, implied an exemption from anything pertaining to what might be understood as 'universal'.[12] All of which might explain why the Crusades resulted in the death of a rather large number of people who might otherwise be understood as 'noncombatants': they were people who could be killed because they were not 'innocent'. Like Augustine, none of the other early Christian thinkers had very much to say much about civilian or noncombatant immunity in the context of war, not least because these concepts had not been invented. Aquinas, for example, generally follows Augustine's train of thought, resting his argument on the question of the subjective guilt or innocence of those who might be killed, and even the innocent might legitimately be punished for sins they had committed in other connections, making it difficult if not impossible to find in Aquinas's work anything that might count as an argument for noncombatant immunity. By the early colonial period a thinker such as Vitoria would still be organizing his arguments around the idea of innocence, albeit with a slightly broader sense as to who might count as 'innocent'. More

11 Johnson, *Just War*, 127.

12 Kinsella, *Image*, 40–41; Richard Shelly Hartigan, *The Forgotten Victim: A History of the Civilian* (Chicago: Precedent Publishing, 1982), 75.

pointedly, when Vitoria does talk about immunity it is solely the exemption of *ambassadors* that is the issue.[13]

This mention of the figure of the Ambassador gives us an important insight into the ways in which immunity was being imagined at that time, but also the ways it is used now too, and so we need to say more about this figure. To get there, however, we need to take a little historical detour, which will in turn enable us to make more sense of the absence of the concept of noncombatant immunity in this period.

There is a simple way of approaching the idea of immunity which says that 'immunity' was a term used by the Romans to offer a certain set of *protections*, which is how we now have the idea of a civilian immunity protecting those not engaged in war. Unfortunately, the conceptual history of 'immunity' is not that simple. As we touched on in the introduction, immunity in Roman law conferred *exemption* from various kinds of state obligations. A compound of *in-* (not), *munus* (a gift as well as a service, but also the root of our term 'municipal'), and *-tas* (denoting an abstraction), the Latin word *immunitas* was concerned with freedom from municipal obligations or burdens such as taxes, duties, services, functions, tributes, and participation.[14] The entries for *immunes* and *immunitas* in Adolf Berger's *Encyclopedic Dictionary of Roman Law* run as follows:

> *Immunes.* Persons permanently exempt from military service (e.g., priests, persons over forty-six years of age, those who served ten years in cavalry or sixteen – later twenty-five – years in infantry). Temporarily relieved from service were the furnishers of the army, persons employed in lower official service . . . *Immunes* were also those who for any reason were exempt from public charges, taxes, and the like. See IMMUNITAS . . .
>
> *Immunitas.* Exemption from taxes or public charges (MUNERA). It was granted as a personal privilege to individuals, as a privilege of a social group (public officials, soldiers) or of a community in Italy or in a province. The extension of *immunitas* was different; it varied

13 Vitoria, 'On Civil Power' (c. 1528), in Vitoria, *Political Writings*, eds. Anthony Pagden and Jeremy Lawrance (Cambridge: Cambridge University Press, 1991), 40.

14 Alexander Callander Murray, 'Immunity, Nobility, and the *Edict of Paris*', *Speculum*, vol. 69, no. 1 (1994), 18–39 (18).

> according to the kind of the charges or the profession of the persons exempted (physicians, teachers, clergymen, etc.). *Immunitas* was granted by the senate through a decree (*senatus-consultum*) and under the Empire by the emperor through a general enactment (*edictum*) or a special personal privilege. Of particular importance were the exemptions in the domain of municipal administration.[15]

Thus, what was at issue was the idea of *privilege*, in the sense of a law that applied only to certain classes of persons or individuals. There also existed municipalities that were *civitates liberae et immunes* – 'cities free and immune' – by virtue of an exemption from payments of tribute and by being able to make their own laws. In this period, then, an immunity was conceived of as an exemption, and an exemption was conceived of as a liberty.

This notion of exemption continued to play a role in our understanding of immunity through the centuries and points to the fact that, when it comes to law, immunity operates by defining exemptions *to* the law granted *by* the law. Such an act does not lift the weight or power of the law but *confirms* it, since only the law can exempt anyone or anything from laws. Immunity in its origins thus appears to involve the *law setting aside the law* by exempting (that is, liberating) certain legal subjects from some of the legal responsibilities usually applicable to all citizens.

Yet the modern world was not born directly from the Roman, and what happens in the period between is important for the history of immunity. (Those 'biopolitical' accounts of immunity that point to the use of 'immunity' by the Romans as the basis of contemporary immunity are wide of the mark, ignoring several hundred years of historical developments in the meaning of the term before it entered the language of biology.) Frankish immunities occupy an important development between the Roman and early modern notions of immunity, but one often ignored in the simpler accounts. The immunity charters of the Merovingian period (481–751) tended to be grants to ecclesiastical establishments referring to the spiritual benefits accruing to the king in return for the grant, and they required that the financial benefits of the grant be applied directly to the upkeep of the establishment or, perhaps,

15 Adolf Berger, *Encyclopedic Dictionary of Roman Law*, *Transactions of the American Philosophical Society*, vol. 43, no. 2 (1953), 333–809 (492).

to help provide food for the monks and alms to the poor. Hence the sources attest to grants of immunity from taxes and other public charges, which also then meant that royal officials were banned from collecting taxes or from enjoying rights of hospitality. That is, again, an *exemption* and hence a *liberty*: a freedom from taxation or the right to collect taxes from dependents and to prohibit tax collectors and other public officials from entering the immune property. This required the existence and sanction of a superior public authority with the power to grant the right associated with the exemption. Hence immunities in Merovingian history were *exemptions* granted by kings to their subjects, usually the Church.[16] At the same time, however, immunity was in this period beginning to take on new meanings. In his account of feudal society, Marc Bloch argues that immunity designated the privilege of exemption from fiscal burdens but also, at the same time, the immunity of a territory from visitation by royal officials, and this involved delegation to the lord of certain judicial powers over the population.[17] What we have with immunities as they develop through the Frankish period are the beginnings of a connection to territorial jurisdiction. To put it bluntly, immunity was gradually coming to sit at the crossroads of power, property, and territory. To put it even more bluntly, and to momentarily get ahead of ourselves, it is this constellation of power, property, and territory inherent in the idea of immunity, rather than some Christian-liberal just war notion of protecting civilians, that underpins so much of the violence that eventually takes place in the name of the modern state.

In this regard, the term 'immunities' gradually but increasingly came to be thought of through the language of *defence* and *protection*. A key shift would appear to be the Carolingian coup of 751, which brought about a profound transformation in the conception of government. Barbara Rosenwein's research suggests that the Carolingians had a 'hands-off' policy vis-à-vis churches and monasteries but, gradually, became part and parcel of a well-run kingdom and so could not simply be jettisoned. Instead, they were *transformed*. The key to this transformation was the idea of *tuition* (protection), which came to be paired

16 Murray, 'Immunity', 20–24; Alan Harding, *Medieval Law and the Foundations of the State* (Oxford: Oxford University Press, 2002), 14.

17 Marc Bloch, *Feudal Society*, vol. II (1940), trans. L. A. Manyon (London: Routledge, 1965), 362.

with 'immunity', such as in the phrase *sub nostra defensione et inmunitatis tuitione*: 'under our defense and the protection of immunity'. By the ninth century, the increasing alignment of *tuitio* as *protection* with the privileges of immunity betokened nothing less than a sea change from exemption to protection and, concomitantly, from prohibition to control. For example, in 757 the monastery at Gorze was granted a privilege by Chrodegang which introduced new terms such as 'subjection' and 'protection'. The privilege granted to Gorze was less an *exemption* and more a *protection* in the form of *immunity as defence*. Likewise, the privilege issued to Saint-Calais in 760 by Pippin III gave Pippin responsibility for its *tuitio* and *immunitas*; Rosenwein suggests this as the earliest pairing of these two words.[18]

By explicitly associating *immunitas* with *tuitio* or *defensio*, immunities from roughly the time of Louis the Pious onwards became guarantees of protection. Bishops became the 'shepherds' – the *protectors* – of church property, protection went hand in hand with control, and the standard formula increasingly became *immunitas atque tuitio*: immunity *and* protection. In Cluny, for example, monks through the tenth and eleventh centuries used their contacts and prestige to gain a papal immunity, granting it protection, and further privileges were to follow. This culminated in the autumn of 1095 with a visit from Pope Urban II. After consecrating Cluny's major and second altar, and overseeing the consecration of three other altars, Urban II marked off the several boundaries of Cluny's area of jurisdiction, a sacred 'ban'. Turning to the people, he called on them not to violate the ban, threatening excommunication if they did. What he was assigning with the area of jurisdiction, he declared, were 'certain clear limits of immunity and security around the monastery'.[19]

The addition of ideas about 'protection' and then 'security' essentially opened new political possibilities and changed the rules of the game, not least as the fundamental presuppositions of immunity shifted from the religious to the legal and extended into the 'international'. Rosenwein shows the impact of this on ideas about political space but also the extent to which the *tuitio* in question became increasingly active as an

18 Barbara H. Rosenwein, *Negotiating Space: Power, Restraint, and Privileges of Immunity in Early Medieval Europe* (Ithaca, NY: Cornell University Press, 1999).

19 Cited in Rosenwein, *Negotiating Space*, 3.

idea. *Immunitas* was taken less and less to mean that public power is restrained and more and more to mean that, combined with *defensio* and *tuitio*, public power was and could be extended in order to guarantee special privileges; Rosenwein goes so far as to call it a 'new-style immunity'. So active a principle was this new-style immunity that it became the justification for fortifications, castles, moats, and walls, such that the perimeter of an immunity was increasingly defined by the castle walls. The work of scholars on other countries bears this out: Wendy Davies's account of immunities in Wales, for example, shows that the Welsh idea of *nawdd* (protection) intensified and increasingly territorialized from the eleventh century onwards.[20] At the same time, what this also points to is the fact that immunities came to be increasingly granted to lords who were gradually claiming more of the rights associated with independent lordship, most notably the right of territorial exclusion.

All these ideas can be seen clearly in, and hence take us back to, the figure of the Ambassador, a person charged with a diplomatic mission to a foreign sovereign or country and who highlights what is at stake in the early modern discussions of immunity but also, as we shall see, modern violence. The two key thinkers of the laws of war and peace from the early seventeenth century, Francisco Suarez and Hugo Grotius, reveal what is at stake.

Asking the question as to who on the enemy's side might be liable to punishment, Suarez comments that some persons are guilty and others innocent. 'It is implicit in natural law that the innocent include children, women, and all unable to bear arms; by the *ius gentium*, ambassadors, and among Christians, by positive [canon] law . . . religious persons, priests.' All other persons are considered guilty.[21] Three figures rolled together: those unable to bear arms (women and children), ambassadors, and priests. Commenting on this, Richard Shelly Hartigan suggests that the three sources used by Suarez to establish innocence are natural law, *jus gentium*, and canon law but that only the first of these could

20 Rosenwein, *Negotiating Space*, 115, 131–38, 142–44; Wendy Davies 'Adding Insult to Injury: Power, Property and Immunities in Early Medieval Wales', in Wendy Davies and Paul Fouracre (eds.), *Property and Power in the Early Middle Ages* (Cambridge: Cambridge University Press, 1995), 137, 144–45.

21 Francisco Suarez, *A Work on the Three Theological Virtues: Faith, Hope and Charity* (1621), in Suarez, *Selections from Three Works*, ed. Thomas Pink (Indianapolis: Liberty Fund, 2015), 962.

really advocate immunity for women and children.[22] Yet there is a problem with such a claim: Suarez mentions immunity but *only* for ecclesiastical persons and ambassadors.[23] Likewise Grotius, who follows the line that war can be carried out against 'not only those who are actually in Arms . . . but also all those who reside within his Territories', has next to nothing to say about immunity other than to cite those immunities enjoyed by ambassadors and churchmen.[24] Suarez and Grotius were merely recognizing that in political discourse and the language of diplomacy the word 'Ambassador' connoted a person, especially a minister of high rank, formally commissioned to convey a message, usually 'diplomatic', from one political authority to another. The part of its etymology in the Old Spanish and Latin *ambactus*, meaning 'servant' or 'vassal', points to the fact that such persons were 'servants' of the state. By the early sixteenth century, the word developed stronger connotations of a diplomat of the highest rank, appointed and accredited as the resident representative of a monarch or state in a foreign state. This combined with the idea of the 'Embassy', initially referring to a diplomatic mission but eventually coming to connote an actual physical space.

The point is that the figure of the Ambassador and the space of the Embassy were created with immunity because they were understood to be standing in for the sovereign power which they represented: the Ambassador as a servant and representative of the Sovereign, and the Embassy as a protected zone of extraterritorial sovereignty within the territory of another power. This was supported by the myth that the Ambassador is, by definition, an ambassador for 'peace'. Despite the fact that, as Garrett Mattingly points out, everyone knew full well that the ambassadors being sent across Europe were far from working in the service of universal peace, the myth was important to the grant of immunity within the state within which ambassadors were operating.[25]

All of which required a certain leap of imagination, one so great that Grotius calls the Ambassador and the Embassy 'a sort of fiction'.

22 Hartigan, *Forgotten Victim*, 94–95.

23 Francisco Suarez, *A Treatise on Laws and God the Lawgiver* (1612), in Suarez, *Selections*, 325, 390, 394, 675.

24 Hugo Grotius, *The Rights of War and Peace* (1625), ed. Richard Tuck (Indianapolis: Liberty Fund, 2005), 900, 1280.

25 Garrett Mattingly, *Renaissance Diplomacy* (Harmondsworth: Penguin, 1973), 45, 256, 496–97.

> I am fully persuaded, that tho' it has prevailed as a common custom every where, that all People that reside in Foreign Countries, should be subject to the Laws of the Countries; yet that an Exception should be made in Favour of Embassadors [sic], who, as they are, by a Sort of Fiction, taken for the very Persons whom they represent . . . so may they by the same kind of Fiction be imagined to be out of the Territories of the Potentate, to whom they are sent.[26]

We are beginning to get a sense of immunity's fiction. As an 'actor on a European stage', as Linda Frey and Marsha Frey dub the figure of the Ambassador, the Ambassador was understood to be 'standing in' for the sovereign they were representing. This was paralleled by the 'curious fiction of exterritoriality', as Mattingly and Timothy Hampton call the Embassy, which rendered the Embassy a kind of sacred space of sovereign action and facilitated the idea that the Ambassador was to be immune for the period of their diplomatic mission in the Embassy; the sacredness was important to the political immunity that continued to exist when diplomacy became secularized.[27]

We have yet to see how much violence this sort of fiction can involve. But note first what is manifestly *not* at stake in this history: noncombatant immunity. The individuals who appear in the texts and treaties defined as possessing immunity do not possess it by virtue of being a civilian or noncombatant. Far from it. But then from where do these ideas come?

One of the narratives of the history of warfare concerns a major shift in the late eighteenth and early nineteenth centuries. Compared to the mid-eighteenth century, when Rousseau was defining war as a relation between states, and writers such as Vattel were suggesting that 'war is carried on by regular troops' and 'the people, the peasants, the citizens, take no part in it', the narrative in question argues that the late eighteenth century witnessed the rise of a new unlimited war, as major-general and military historian J. F. C. Fuller puts it. This involved what Fuller calls 'the armed horde . . . on a national footing', but which others

26 Grotius, *Rights*, 912.

27 Linda S. Frey and Marsha L. Frey, *The History of Diplomatic Immunity* (Columbus: Ohio State University Press, 1999), 207; Mattingly, *Renaissance Diplomacy*, 42, 259, 266; Timothy Hampton, *Fictions of Embassy: Literature and Diplomacy in Early Modern Europe* (Ithaca, NY: Cornell University Press, 2009), 3, 75.

more politely describe as 'the wars of peoples'. The terms are used to describe a period of total war and mass conscription. Witness the stark figures for the revolutionary French army given by Hans Delbrück: in November 1792 the French army tackled the Austrians at Jemappes, near Mons, with a force of just over forty thousand men, mostly volunteers; by March 1793, having moved from volunteers towards a mandatory levy, they had three hundred thousand men; by the end of 1793, following the *levée en masse* instituted in August that year, the French troops numbered 770,000. War became 'democratized', 'the people' mobilized, armed, and imbued with ideological fervour. 'Democracy made all men equal in theory, but it was conscription which did so in fact', Fuller comments.[28] At which point, we get a major development in the way immunity is imagined, because it is only now that we witness the invention of noncombatant immunity. And this, it turns out, is the *invention of another sort of fiction*.

The word 'noncombatant' came into use in English in the first decade of the nineteenth century; the *OED* gives 1811 as the first use of the word. Likewise, although the term 'civilian' as applied to a practitioner of Roman civil law has a long history, the use of 'civilian' to describe a non-military person dates from the same period as 'noncombatant': the *OED* gives 1794 as the first use. In other words, what we see emerging in the late eighteenth and early nineteenth centuries is what appears initially to be a rather bizarre historical parallel: the rise of the idea of the noncombatant civilian came into being with the rise of total wars of the Napoleonic era. Yet total war has, at its core, the entire population and the entire social and economic system, and hence involves everyone as a combatant in some sense. Moreover, the conduct of war involves more and more weapons that could destroy more and more people in more and more anonymous ways, from the machine gun to the drone, and these weapons have time and again been shown to completely undermine the concept of noncombatant. These ideological, organizational, and technical components of 'total' war conjoined to make the concept of the 'civilian' obsolete from the very moment of its birth. We

28 Emer De Vattel, *The Law of Nations* (1758), trans. Thomas Nugent (Indianapolis: Liberty Fund, 2008), 550; Hans Delbrück, *History of the Art of War*, vol. IV (1920), trans. Walter J. Renfroe (Lincoln, NE: Bison Books, 1990), 395–96; Major-General J. F. C. Fuller, *The Conduct of War 1789–1961* (London: Methuen, 1961), 30–33.

have the invention of a class of people called noncombatant at the very moment when every member of the population is defined as a combatant. 'Noncombatant' is invented as an empty class. A sort of fiction. The civilian: born to die.

'The immunity of the civilian . . . has been shattered', wrote George Orwell in 1944.[29] He was far from alone, as critical commentary at the end of the war focused on the massive strategic bombing programmes destroying millions of lives, and then the use of nuclear bombs to continue the job. The language that had been used about air power in the previous decades had pointed to a new reality: that the battlefield now extended into every corner of a state and that air power allowed the 'strategic bombing' of 'vital centres', 'economic infrastructures', 'commercial hubs', and 'communications networks'. As we might say in the light of chapter 3, *the enemy was the system*. From the outset, the literature on air power held that the possibility of attacking cities in order to disable the mass war effort meant 'there will be no distinction any longer between soldiers and civilians' (Douhet in *The Command of the Air*, 1921), or that 'the civilian and the fighting man are now merged in one' (Charlton in *The Menace of the Clouds*, 1937).[30] The same point was made time and again in the 1920s, from J. M. Spaight's *Air Power and War Rights* (1924), B. H. Liddell Hart's, *Paris; or The Future of War* (1925), and William Mitchell's, *Winged Defense: The Development and Possibilities of Modern Air Power* (1925). By the end of World War II, the massive bombing campaigns had made it abundantly clear that modern war is a war on cities and therefore a war on those who lived in them, from which only one position emerged, stated in capital letters in an official US intelligence review of the use of strategic air power in Japan, published in July 1945: 'THERE ARE NO CIVILIANS IN JAPAN.'[31] In effect, the strategic bombing campaigns of World War II explicitly removed whatever sanctity might once have been attached to the

29 George Orwell, 'As I Please', 19 May 1944, in *Collected Essays, Journalism and Letters*, vol. III (London: Secker and Warburg, 1968), 152.

30 Giulio Douhet, *The Command of the Air* (1921), ed. Brijesh Dhar Jayal (Dehradun, India: Natraj Publishers, 2003), 14; L. E. O. Charlton, *The Menace of the Clouds* (London: William Hodge and Co., 1937), 23.

31 *Fifth Air Force Weekly Intelligence Review*, no. 86 (15–21 July 1945), in W. F. Craven and J. L. Cate (eds.), *The Army Forces in World War II*, vol. V (Chicago: University of Chicago Press, 1948), 696.

noncombatant, making clear that there really is no such thing as a civilian and hence can be no such thing as civilian immunity.

Orwell was far from alone in seeing it this way, as the following examples show. At the beginning of the war, J. M. Spaight, a leading international and military lawyer, had in the light of the developments in air power shifted his position about the supposed 'two great classes' (that is, the principle of distinction discussed above). 'The technique of war-waging has changed enormously since the principle of the immunity of non-combatants from direct violence was accepted as a basis of the law of war', Spaight argues, suggesting that because war is now waged in the factories as well as the battlefields, 'the time-honoured distinction between members of the armed forces and civilians' has been abrogated. There are people who might be considered 'non-combatants' but 'who cannot be held to be immune from direct attack'. As a second example we can take the seventh edition of *Wheaton's Elements of International Law* published in 1944. The text still presented from earlier editions the principle of distinction and the idea of immunity, but the seventh edition held that although 'the separation of armies and peaceful inhabitants into two distinct classes is perhaps the greatest triumph of international law', it is nonetheless also the case that 'the progress of events has nullified the triumph, and that, probably, it is just as well to abolish a distinction in itself, illusory and immoral'. A year later the judge advocate general of the US War Department, Lester Nurick, wrote in the *American Journal of International Law* about subjecting cities to bombing: 'Where does this leave the "fundamental" doctrine that a noncombatant is relatively immune from attack?' The principle of distinction between combatants and the civilian population may be 'one of the fundamental principles of international law', states Nurick, but it has 'been so whittled down by the demands of military necessity that it has become more apparent than real'. As a fourth example we can take a comment by eminent international lawyer H. Lauterpacht in 1952, reflecting like the others on what aerial bombing meant for 'the time-honoured principle of immunity of non-combatants': the reality now is that the civilian population 'enjoys no such immunity'.[32]

32 J. M. Spaight, 'Non-Combatants and Air Attack', *Air Law Review*, vol. 9, no. 4 (1938), 372–76 (372–74); A. Berriedale Keith, *Wheaton's International Law*, 7th edn., vol. II, *War* (London: Stevens and Sons, 1944), 169–71; Lester Nurick, 'The Distinction between Combatant and Noncombatant in the Law of War', *American Journal of International Law*, vol. 39, no. 4 (1945), 680–97 (680, 696); H. Lauterpacht, 'The Problem of the Revision of the Law of War', *British Year Book of International Law*, vol. 29 (1952), 360–82 (374).

These examples go some way to showing why it is that the Geneva Conventions of 1949 have *absolutely nothing to say about noncombatant immunity*. Article 23 of the Third Convention holds that 'no prisoner of war may at any time be sent to, or detained in areas where he may be exposed to the fire of the combat zone', but this is not to protect them. Rather, the article holds that the prisoner's presence in the zone may not be used 'to render certain points or areas immune from military operations'. Article 28 of the Fourth Convention, dealing with the protection of civilians, has just one sentence: 'The presence of a protected person may not be used to render certain points or areas immune from military operations.' In other words, it is against the Geneva Conventions to use a human body to render an area immune from violence. 'Immunity' here has nothing to do with protecting human beings from violence and everything to do with protecting the state's right to exercise that violence.

It is abundantly clear that even if the idea of a noncombatant civilian possessing something called 'immunity' did once make sense (though as we have seen, it never did), it was mortally injured with the rise of early total war, killed off by the invention of air power, and buried at the end of World War II. But given that noncombatant immunity was only really born with the era of total war, what this really means is that it was *obsolete from the very beginning*. Strictly speaking, it was only in 1977 that the principle of noncombatant immunity was finally, fully, and formally agreed upon as a foundational norm in the international law of armed conflict, by which time it was so obviously a political fiction that no one in their right mind could ever believe that states would adhere to it.

What I want to now suggest is that the situation is in fact even worse than this. I want to suggest that the fiction in question performs a crucial role in *facilitating the very violence from which civilians are meant to be protected*. Why? Because if the concept of immunity protects anything, it is the state and its powers of violence. Absolute power and legitimate violence operate most forcefully and directly through bodies and their pain, a stark reminder of sovereign power.[33] Just as feudal immunities could be breached by the monarchical powers that granted them, so

33 For the fiction surrounding violence, see Elaine Scarry, *The Body in Pain: The Making and Unmaking of the World* (Oxford: Oxford University Press, 1985); Peter Gratton, *The State of Sovereignty: Lessons from the Political Fictions of Modernity* (New York: SUNY Press, 2012); Sonja Schillings, *Enemies of All Mankind: Fictions of Legitimate Violence* (Hanover, NH: Dartmouth College Press, 2017).

modern immunities can be breached by the powers that have donned the kingly robes. Far from protecting civilians, the principle of immunity turns out to facilitate their very destruction. Insofar as the state's immune system appears to delimit a special status for the individual and collective bodies of the civilian, so that same system can and will turn against those very bodies. Turning against its own bodies? Its own body? This is beginning to sound like the autoimmune disease. Helen Kinsella suggests that

> the recurrence of metaphors of disease in the justifications for killing those we might otherwise exempt not only from (military) service but also from violence demonstrates that the debt the concept of protection owes to immunity is greater than simple exemptions.[34]

Whenever one says 'immunity', the idea of disease rears its head, and so too does the idea of violence. What, then, if legal immunity is a cloak for sovereign violence? What if instead of immunity-security from violence, the civilian becomes subject to the violence of immunity-security?

The State: Born to Kill

'Assassination' as a term for killing has gone out of fashion in military circles. The fashion now is for 'targeted killing'. But as a fashion item, targeted killing's key accessory is collateral damage.

Early in the invasion of Iraq during the 'war on terror', the US selected fifty 'high value individuals' to be targeted. According to one Defense Intelligence Agency analyst, 'all survived'. 'Not so lucky were the "couple of hundred civilians, at least."' The 'targeted killing' of Saddam Hussein, for example, included attempts such as the one in 2003 in which a restaurant that he was thought to be in was destroyed. According to a Pentagon analyst, 'We thought that Saddam was there. He wasn't, but we did kill a bunch of civilians.'[35] As the number of targets expanded, so too

34 Kinsella, *Image*, 29.

35 DIA analyst Marc Galasco, speaking on CBS, 25 Oct. 2007, and Pentagon analyst interviewed by Cockburn, both cited in Andrew Cockburn, *Kill Chain: Drones and the Rise of High-Tech Assassins* (London: Verso, 2016), 138.

did the number of deaths, to the extent that during one four-and-a-half-month period in the Afghan campaign approximately 90 percent of the people killed in targeted strikes were not the actual target.[36] The number of people killed in 'targeted killings' in which they were not the target is exacerbated by the wider attacks that go by the name 'signature strikes', a term used when striking groups of people based on 'patterns of life'. 'Patterns of life' plays directly on the notion of biological systems that we discussed in earlier chapters, treating the enemy as an *organic* product – 'we are killing these sons of bitches faster than they can grow them now',[37] commented one CIA head – but also a *form of life* that *cannot be permitted any kind of immunity*. Hence the signature strikes essentially target people whose identity cannot be known, whose immunity is unclear, and whose security is considered irrelevant. This has led to attacks on village meetings, wedding parties, funeral gatherings, groups of people coming to the aid of those injured in strikes (known as 'double tap attacks'), Red Cross warehouses, and even, in one instance in Afghanistan in February 2002, three men scavenging for scrap metal. Following the infamous attack on the wedding party, the US defense secretary at the time, Donald Rumsfeld, stated, 'Let's not call them "innocents". We don't know quite what they were.'[38] 'We have shot an amazing number of people, but to my knowledge, none has ever proven to be a threat', said General Stanley A. McChrystal about his time as senior American and NATO commander in Afghanistan.[39]

Narrowing the picture down generates reports about 'targeted killing' and 'signature strikes' such as this from the Open Society Justice Initiative report *Death by Drone: Civilian Harm Caused by U.S. Targeted Killings in Yemen* (2015): 19 April 2014, strike in al-Sawma'ah District, 4 civilians killed and 5 injured; 7 August 2013, strike in al-Mil, 2 civilians, including 1 child; 1 August 2013, strike in Wadi Sir, 1 civilian killed; 9 June 2013, strike in al-Sabir area, 1 child killed and 1 child injured; 17

36 Jeremy Scahill, 'The Drone Legacy', in Jeremy Scahill and staff of the *Intercept* (eds.), *The Assassination Complex: Inside the Government's Secret Drone Warfare Program* (London: Serpent's Tail, 2016), 9.

37 Cited in Daniel Bates, 'Post 9/11 CIA Has Become a Killing Machine Focused on Hunting Down Terrorists "Faster Than They Can Grow Them"', *Daily Mail*, 2 Sept. 2011.

38 Donald Rumsfeld cited in Thom Shanker, 'A Nation Challenged: The Military', *New York Times*, 22 Feb. 2002.

39 Cited in Richard A. Poppel, 'Tighter Rules Fail to Stem Deaths of Innocent Afghans at Checkpoints', *New York Times*, 26 March 2010.

April 2013, strike in Wesab al-Aali District, 3 civilians killed; 23 January 2013, strike near al-Masna'ah village, 2 civilians killed; 2 September 2012, strike in Walad Rabei' District, 12 civilians killed, including 3 children and a pregnant woman, and 2 civilians injured; 17 May 2012, strike in al-Maseelah village, 1 civilian killed. Individual cases include that of Momina Bibi, a sixty-seven-year-old grandmother who was picking okra in a field in Ghundi Kala village in Pakistan when she was assassinated by a Predator drone on 24 October 2012. To get a sense of the wider picture, it is estimated that approximately one thousand civilians were killed by coalition attacks in Afghanistan in October and November of 2001, that more than eleven thousand civilians were killed in attacks in Iraq between 2003 and 2008, and approximately one thousand were killed in Pakistan between 2004 and 2017. Estimates include the one from David Kilcullen to the US House of Representatives in April 2009, that between 2006 and 2009 'we have killed 14 senior Al Qaeda leaders using drone strikes. In the same time period, we have killed 700 Pakistani civilians in the same area.' A study by the Washington-based Brookings Institution just a few months later found that 'for every militant killed, 10 or so civilians also died'. The Bureau of Investigative Journalism's tally of drone strikes between 2010 and 2020 has counted 14,040 US drone strikes across Pakistan, Afghanistan, Yemen, and Somalia, and estimates these have killed between 910 and 2,200 civilians, and between 283 and 454 children.[40]

Unfortunately, it often turns out that the knowledge that the dead were not a threat could come only after they were dead. Concerning the case of the three men killed while scavenging for scrap metal, US state officials stated that it appeared to them to be a 'meeting on a hillside'. 'We're convinced that it was an appropriate target', they said, even though 'we do not know yet exactly who it was.'[41] The only logic that can underpin the idea that even though the state does not know who a person is they are still an appropriate target is one which presupposes that *civilian immunity does not exist for anyone.* One can be formally recognized as a

40 Testimony from David Kilcullen at a Hearing before the Full Committee on Armed Services, House of Representatives, 23 April 2009; Daniel L. Byman, 'Do Targeted Killings Work?', Brookings Institution, 14 July 2009; the Bureau of Investigative Journalism drone warfare statistics are at www.thebureauinvestigates.com.

41 Pentagon spokeswoman Victoria Clarke, cited in John F. Burns, 'A Nation Challenged: The Manhunt', *New York Times*, 17 Feb. 2002.

bona fide 'innocent civilian' only after being killed. The message: you may not be a threat, but we can only be sure of this after we have killed you. Your immunity from violence by virtue of you being a civilian can be claimed and confirmed only posthumously, so to speak.

The truth is that for all the talk of civilian immunity, to declare that assassination in the form of targeted killing is now a legitimate 'act of war', that signature strikes against groups of barely identifiable people going about their daily business is permitted, and that whoever else gets killed during such attacks is simply collateral damage, is to announce that immunity is a fiction.

This is also the reason for the emergence of a conceptual and legal discourse discussing the conditions under which civilian immunity can be ignored. The very phrase 'targeted killing' implies a target and suggests that the attack complies with the laws of armed conflict in accordance with the Geneva Conventions. As a 2010 report to the UN Human Rights Council makes clear, with a number of choice phrases, 'targeted killing is only lawful when . . .'; 'a State killing is legal only if . . .'. Highlighting such comments, Fleur Johns observes that the capacity for targeted killing is approached by the law with the possibility that such acts might be illegal but are unlikely to ultimately be so; the law's task becomes one of tethering the violence to firm legal ground.[42] Indeed, to openly discuss such targets *as targets* is to offer up the very transparency that is taken to somehow be understood as proof of the legality of the act. This grounding in legality is crucial since it reveals that a security system produced to protect the body politic becomes a system for the annihilation of bodies.

Yet there is an important extension to this, which reverses the logic of immunity in such acts of violence: if anything possesses immunity, it is the state and its representatives. And this immunity might even apply to the state killing its own civilians. In his introduction to the volume of once secret and now released 'drone memos,' Jameel Jaffer comments that 'if drone strikes are a cure, they are also part of the disease'.[43] What

42 Philip Alston, *Report of the Special Rapporteur on Extrajudicial, Summary or Arbitrary Executions*, UN General Assembly, Human Rights Council (28 May 2010), paras. 30 and 32; Fleur Johns, *Non-Legality in International Law: Unruly Law* (Cambridge: Cambridge University Press, 2013), 2.

43 Jameel Jaffer, 'Introduction', in Jameel Jaffer (ed.), *The Drone Memos: Targeted Killing, Secrecy and the Law* (New York: New Press, 2016), 20.

disease is this? Might it be a 'disease' in which the state's security system destroys part of its own social body? Following through on this question will allow us to engage with a final twist in the fiction of immunity. Let us get at this in a roundabout way.

On 30 September 2011, the US state used drone weaponry to kill US citizens Anwar al-Awlaki and Samir Khan. Anwar al-Awlaki had been on the US kill list, Khan had not. (Two weeks later, al-Awlaki's sixteen-year-old son, Abdulrahman al-Awlaki, another US citizen, was also killed in a US drone strike.) Anwar al-Awlaki tends to now be regarded as the first American citizen designated by the US security system to be a target to be killed on sight, with no arrest or trial necessary. The fact that al-Awlaki's death sentence was planned and announced over a year in advance meant there was time for the sentence to be challenged in court by the American Civil Liberties Union (ACLU) and the Center for Constitutional Rights (CCR), acting on behalf of his father. In assessing the case, the court came to what it described as a 'somewhat unsettling' conclusion: namely that 'there are circumstances in which the Executive's unilateral decision to kill a U.S. citizen overseas is . . . judicially unreviewable.'[44] Leading up to the killing, legal papers had been prepared within the Department of Justice asking whether the fact that al-Awlaki was a US citizen meant that he had constitutional protection from such a death sentence. The heavily redacted documents that have since been made public are sufficiently readable to see that what was at stake was a citizen's claim to immunity from a lethal operation conducted upon them by the security system of their own state. One memo to the attorney general held that 'being a U.S. person does not give a member of al-Qa'ida a constitutional immunity from attack.'[45] This point was repeated time and again in various memos, and every time it was the logic of a citizen's 'immunity' that was challenged: 'The fact that a central figure in al-Qaida or its associated forces is a U.S. citizen . . . does not give that person constitutional immunity from attack.'[46] When Attorney General Eric Holder addressed the issue in a speech in March 2012, he reiterated the point:

44 United States District Court for the District of Columbia, *Civil Action No. 10-1469 (JDB)*, 30 Oct. 2010, 78.

45 US Department of Justice, 'Memorandum for the Attorney General', 19 Feb. 2010, in Jaffer (ed.), *Drone Memos*, 69.

46 Justice Department White Paper, 25 May 2011, in Jaffer (ed.), *Drone Memos*, 156.

> Based on generations-old legal principles and Supreme Court decisions handed down during World War II, as well as during this current conflict, it's clear that United States *citizenship alone does not make such individuals immune from being targeted* . . . Some have argued that the President is required to get permission from a federal court before taking action against a United States citizen who is a senior operational leader of al Qaeda or associated forces. This is simply not accurate. 'Due process' and 'judicial process' are not one and the same, particularly when it comes to national security. The Constitution guarantees due process, not judicial process.[47]

As the 2010 court observed, judicial approval is required when the state seeks to target a citizen for electronic surveillance, but not when the state targets that same person for death.[48] Such targeting appears to be a death-penalty case in which there is no indictment, no presentation and assessment of evidence, and no accused present. What there is instead is a pronouncement of a death sentence, in advance, by the prosecutors.[49] Commentators have thus assessed al-Awlaki's sentence and subsequent execution as confirmation that a civilian can be killed by their own state without that state resorting to a court of law. More importantly, from our point of view, it confirms that a state's security system can and will be turned against its own citizens regardless of any supposed immunity that they might be thought to possess. Arguments that see the problem as simply one of 'American immunity' and hence an issue for the US Constitution somewhat miss the point.[50] The issue is not American immunity, but *immunity*. For a start, we find the same act carried out by other states. In September 2015, the British state killed Reyaad Khan and Ruhul Amin, two of its own citizens that it believed to be fighting in Syria for the Islamic State of Iraq and the Levant (ISIL). Other states such as Israel, Nigeria, Pakistan, and Saudi Arabia are well known for such acts.

47 Attorney General Eric Holder, speech at Northwestern University School of Law Chicago, 5 March 2012, emphasis added. Points repeated in Eric H. Holder, letter to Patrick J. Leahy, chair of the Committee on the Judiciary, US Senate, 22 May 2013.

48 District Court for the District of Columbia, *Civil Action No. 10-1469*, 2.

49 Jaffer, 'Introduction', 4–7.

50 Patrick Hagopian, *American Immunity: War Crimes and the Limits of International Law* (Amherst: University of Massachusetts Press, 2013); H. Jefferson Powell, *Targeting Americans: The Constitutionality of the U.S. Drone War* (Oxford: Oxford University Press, 2016).

Yet, as well as revealing immunity's fiction, the cases also reveal another aspect of the violence, which is that the state in question rebuts any legal challenges by *claiming immunity for itself and its officers*. On 2 October 2018, Jamal Khashoggi, a *Washington Post* columnist, walked into Saudi Arabia's embassy in Turkey to collect some divorce papers. He was, instead, questioned, tortured, strangled, and dismembered. A report by Helena Kennedy, a leading human rights lawyer, and Agnès Callamard, a UN special rapporteur on extrajudicial executions (*A Perverse and Ominous Enterprise*, 2019), found that Khashoggi's murder was premeditated, involving high-level Saudi officials up to and including Crown Prince Mohammed bin Salman, and that ultimately the Saudi state, which has been torturing and executing somewhere in the region of one hundred members of its own population a year (many of them political activists), bore full responsibility. Yet little could be done. Why? Because of the *privileges and immunities* granted to the Saudi embassy in Turkey. The *privilege* of killing was accompanied by *immunity* from prosecution. 'How could this happen in an embassy?', Khashoggi can be heard asking his murderers on the tape before they murdered him. The answer is straightforward: it could happen because an Embassy is a space of sovereign immunity.

Such claims of immunity are rooted in the long tradition of immunity for the Ambassador and the Embassy and act as a reminder of the *absolute nature of state power*. They are also a reminder of the absolute uselessness of law in challenging this power. William Blackstone pointed out many years ago that the supposition built into law is contained in a simple maxim: *The King can do no wrong*.[51] Not much has changed now that the state has donned the robes of the king. Modern conceptions of sovereignty have never really been able to shake off the feudal notion that 'all earthly Lordship is a limited representation of the divine Lordship of the World'.[52] 'Off with their heads', declares the state executive about the next person on the kill list, and the machine gets to work. 'Turns out I'm very good at killing people', observed one modern Head of State, who also happened to be a winner of the Nobel Peace Prize, a

51 William Blackstone, *Commentaries on the Laws of England*, vol. I (1765–1769) (Chicago: University of Chicago Press, 1979), 237.

52 Otto Gierke, *Political Theories of the Middle Age* (1881), trans. F. W. Maitland (Cambridge: Cambridge University Press, 1900), 30.

trained lawyer, and a student of the writings of Augustine and Aquinas.[53] He was writing about his own state's killing of one of its own citizens – namely, Anwar al-Awlaki.

In conducting the research for his book on the killing of al-Awlaki, Scott Shane conducted a survey among constitutional lawyers to find out what they thought about the legality of the killing. This is a group of people who should know a bit about the supposed immunities possessed by a citizen. Broadly, two themes emerged from the survey. One was a belief that since al-Awlaki had joined the enemy in a war, he could expect no immunity based on his citizenship, analogous to a German American who had fought for the Nazis in World War II. The second was an equally common belief based on a different analogy, one that Shane describes as rather 'bracing'. The analogy is that killing al-Awlaki was *like a police shooting.*

> The police shooting parallel was . . . raised repeatedly by government officials who supported targeting Awlaki. 'My view was that Anwar al-Awlaki was actively plotting to kill American citizens', said Gerald Feierstein, who was the American ambassador to Yemen during the hunt. 'To me, he was like a guy walking down an American street carrying an M-16. The police would take him out.'[54]

On this view, targeted killing comes within the ambit of police action, reinforcing once more the myth of law underpinning the operation. In other words, people will be more likely to accept it if it is presented in terms of a 'law enforcement paradigm' rather than a 'laws of war paradigm'. This is a standard ploy in the deployment of violence: call it a police action. This is one reason why thousands of civilians could be killed in colonial wars: because these were designated 'police actions' and hence not subject to the laws of war (not that such laws would have made any difference). It is also why just war theory falls back time and again on the same notion.

The bracing analogy merely reinforces the point I have been making about the immunity claimed by the state for itself and its officers, which

53 President Barack Obama, cited in Mark Halperin and John Heilemann, *Double Down: The Inside Account of the 2012 Presidential Election* (London: W. H. Allen, 2014), 55.

54 Scott Shane, *Objective Troy: A Terrorist, a President, and the Rise of the Drone* (New York: Random House, 2015), 225–26.

becomes clear when we recall the long history in which thousands of the state's own civilians have been killed by police. When it comes to our killing at the hands of the state to which we are said to belong as citizens, the part of the body politic's security apparatus that goes by the name 'police' is most likely to be the one that kills us. As the progeny of sovereign power, the police must possess immunity from prosecution for acts of violence. Again, the figures concerning such killings are always uncertain and subject to dispute, but if we take some reputable research such as the *Guardian*'s 'Counted' webpage and the 'Killed by Police' database, we get some fairly consistent figures for the US: 1,111 people killed in 2014; 1,146 killed in 2015; 1,093 in 2016; 1,147 in 2017; 1,166 in 2018. For the UK, the pressure group INQUEST has been counting the deaths in police custody or through contact with police since 1990, and at the time of writing the figure is 1,784. Number of convictions? One. The *Guardian*'s 'Deaths Inside' database also counts the number of Aboriginal deaths in police custody in Australia since the 1991 Royal Commission into the issue. Number of deaths since 1991: 432. Number of convictions: zero. In some countries such as Argentina, there were so many deaths at the hands of police in periods such as the 'Dirty War' that the courts have had recourse to the language of genocide. In Syria, the state has at times been executing so many people that a 2017 report on the country by Amnesty International was called *Human Slaughterhouse*. Such practices are exacerbated by the proliferation of broader everyday modalities of war such as the war on drugs, since, in a flip of the bracing analogy, being declared a 'war' helps legitimize the police killings.

'Hitler massacred three million Jews. Now, there are three million drug addicts. I'd be happy to slaughter them', said the Philippine president Rodrigo Duterte in October 2016. He was of course unable to literally slaughter them himself, so what did he do? He granted the state's representatives the privilege of immunity from prosecution for any extrajudicial killings they committed. He granted them, in other words, exactly what the state in liberal democracies offers its own police officers: *immunity from prosecution* for acts of violence conducted in their role as officers of the state, up to and including the torture and killing of the state's own subjects. Once again, the war power and the police power reveal their unity.

Despite all the talk down the centuries of making civilians immune from state violence, most states, and the international state system as a whole, appear far more interested in *exempting* themselves, their servants and officials, for those acts of violence conducted against civilians. States, for centuries, also granted the same 'privileges and immunities' to corporations, a body politic created by the body politic and a legal form allowing the corporation to act as a sovereign power, and therefore with a legally sanctioned immunity for plunder, murder, and torture; the corporation as a police power.[55] Herein lies the power of 'these privileges and immunities': the *privilege* of killing to be accompanied by the *immunity* from prosecution. It is the killers working for the state that are always protected, because they act in the name of the state. In the case brought by the ACLU and the CCR concerning al-Awlaki's planned assassination, the court held that because the claim was brought 'against the President, the Secretary of Defense, and the Director of the CIA in their official capacities' it was, in effect, 'a suit against the United States itself'. Citing case law, the court went on to hold that 'it is axiomatic that the United States may not be sued without its consent and that the existence of such consent is a prerequisite for jurisdiction'. Moreover, it added that 'all purported waivers of sovereign immunity will be strictly construed . . . in favor of the sovereign'.[56] What was at stake in the case was not al-Awlaki's protection rooted in his immunity but, rather, the immunity of the sovereign power from the case brought by the plaintiff. It was the state and its security – *the System* – that was thought to need protecting, not the citizen. Only sovereign power can really be imagined as possessing immunity, revealing immunity as a function of the *body of the state*.

The immunity of the state's operatives – the soldier, police officer, security official, and diplomat – lies in the fact that they are part of the body of the state, represent the state, and act with its authority, as is made clear in countless court decisions. Article 2 of the UN Convention on Jurisdictional Immunities of States and Their Property (2004) also grants immunity to 'representatives of the State acting in that capacity'. The operatives are the state's *ambassadors*. Ambassadors are persons who stand in for the *person of the state*, bodies that represent the *body of*

55 Mark Neocleous, *Imagining the State* (Maidenhead: Open University Press, 2003), .72–97; Joshua Barkan, *Corporate Sovereignty: Law and Government under Capitalism* (Minneapolis: University of Minnesota Press, 2013), 4–5, 16.

56 District of Columbia, *Civil Action*, 59.

the state. They carry the authority to enact violence and are therefore protected by the state from prosecution by a series of 'privileges and immunities' granted to those carrying out acts of violence in its name. Here we can recall the point made earlier: immunity concerns *exemption*, and exemption implies *proximity to the centre of power*. The farther one is from the centre, the less one's life means. Civilians, with their distance, mean nothing. Migrants, less than nothing. Soldiers and police officers have the privilege of proximity which renders them immune, exempted from certain responsibilities in the same way as the original Ambassadors or Diplomats. All of which might help explain why, for example, US troops in Vietnam were regarded as members of a diplomatic mission and the twenty thousand US troops stationed in Iraq in 2018 were there on diplomatic visas. The soldiers had been granted diplomatic 'privileges and immunities'.[57] By a sort of fiction, they can be taken to be the very Person of the State. They carry the authority to enact violence and are therefore protected by the state from prosecution by a series of 'privileges and immunities' granted to those carrying out acts of violence in its name. Such acts are always 'to be connived at', as Grotius put it.[58]

We might benefit here from noting some comments made by Chief Justice Marshall in the case of *The Schooner Exchange v. McFaddon*, a famous case heard in 1812 by the US Supreme Court and often treated as one of the first definitive statements of the doctrine of state immunity. Briefly, a ship called *The Schooner Exchange* owned by John McFaddon and William Greetham sailed for Spain from Baltimore in October 1809. A year later, it was seized by the French and then commissioned as a French warship. When the vessel was later forced to dock in Philadelphia because of a storm, McFaddon and Greetham filed an action in the district court to seize the vessel. The case eventually reached the Supreme Court, where Marshall delivered the ruling. Marshall noted that a state has absolute and exclusive sovereign jurisdiction within its own territory, but also that it could waive jurisdiction. He also noted that international custom presumed that jurisdiction was waived in a number of situations, most obviously when diplomatic representatives

---

57 Major General George S. Prugh, *Law at War: Vietnam, 1964–1973* (Washington, DC: Department of the Army, 1975), 87–88, 149–50.

58 Grotius, *Rights*, 912.

are treated as exempt from the jurisdiction of domestic courts. In the court's judgement, customary international law held that a friendly warship that enters an open port is exempt from the sovereign jurisdiction of the state which claims legal authority over the port. The point of all this is that what was revealed was that the ship possessed 'an immunity from the ordinary jurisdiction'. The comparison made was that the immunity granted was 'as extensive as that of an ambassador, or of the Sovereign himself'. To imagine the immunity held by figures such as foreign diplomats and ambassadors, Marshall suggested, we must either 'consider him as in the place of the sovereign he represents' or, alternatively, we must 'by a political fiction suppose him to be extra-territorial, and, therefore, in point of law, not within the jurisdiction of the sovereign at whose Court he resides'.

In announcing this decision for the court, Marshall did not call on any previous decisions concerning state immunity. Neither could he find any specific legal rules on the matter.[59] Rather, the powers inherent in the immunity appeared to stem simply and straightforwardly from the *very nature of sovereignty itself*. In the Supreme Court's words, 'We are asked, whence we infer the immunity of the public armed vessel of a sovereign. We answer *from the nature of sovereignty*.' Sovereign power enjoys immunity by virtue of the nature of sovereignty; the state enjoys immunity because it is the state. Immunity is a *political fiction* that *can be written only by the state itself*.

To make sense of this, then, we might backtrack into the work of Thomas Hobbes. Hobbes has been described by Roberto Esposito as the person 'responsible for inaugurating modernity's most celebrated immune scenario'.[60] By treating immunization as a negative form of the protection of life, Esposito suggests Hobbes not only places the protection of life at the centre of his thought but also subordinates such protection to the sovereign power. With such a view, 'the immunitarian principle has already been founded'.[61] This plays on the classic Hobbesian imagery of the body politic and the politics of security for

59 Xiaodong Yang, *State Immunity in International Law* (Cambridge: Cambridge University Press, 2012), 45.

60 Roberto Esposito, *Immunitas: The Protection and Negation of Life* (2002), trans. Zakiya Hanafi (Cambridge: Polity, 2011), 86.

61 Roberto Esposito, *Bíos: Biopolitics and Philosophy* (2004), trans. Timothy Campbell (Minneapolis: University of Minnesota Press, 2008), 46.

which he is so well known and with which we began this book. And yet this does not really do justice to Hobbes's actual argument, especially the fact that when Hobbes describes the state of nature it is in terms of its immunities as much as its liberties. Hobbes presses home the point that the social contract constitutes a pledge to sacrifice one's Self. On this account, the ultimate 'Libertie, or Immunitie from the service of the Commonwealth' comes in the state of nature. Beyond that, the idea of liberty or immunity applied only to the sovereign power. For Hobbes, the liberty of which we speak when discussing the ancient Greeks and Romans 'is not the liberty of *particular men*, but the *liberty of the Commonwealth*'. When we say 'the Athenians and Romans were free', we mean that they were 'free Commonwealths', not that individuals were free to resist the sovereign power. And hence when we speak of immunity, we should be speaking too of the Commonwealth rather than the individual.

> There is written on the turrets of the city of Luca in great characters at this day, the word *Libertas*; yet no man can thence infer that a particular man has more liberty or immunity from the service of the Commonwealth there than in Constantinople. Whether a Commonwealth be monarchical or popular, the freedom is still the same.

Thus *no one is immune* from service to the Commonwealth; immunity from the service to the Commonwealth exists in the state of nature, but that is what one has given up with the creation of the Commonwealth.[62] In *Behemoth* (1668), Hobbes also claims that immunity lies with the Commonwealth. What does this mean? It means that it is the sovereign that is granted immunity by virtue of being authorized to act as is deemed necessary by the sovereign power itself. Acting in the name of security, law, and order, whatever the sovereign does involves 'no injury to any of his subjects'. In Hobbes's view, the body politic is constituted by the bodies of the subjects. Because every Subject is by the institution of sovereignty 'Author of all the Actions, and Judgments of the Soveraigne Instituted', it follows that whatsoever the sovereign does 'can be no injury to any of his Subjects' and hence neither 'ought he to

62 Thomas Hobbes, *Leviathan* (1651), ed. Richard Tuck (Cambridge: Cambridge University Press, 1991), 149.

be by any of them accused of Injustice'. By instituting the Commonwealth, 'every particular man is Author of all the Soveraigne doth; and consequently he that complaineth of injury from his Soveraigne, complaineth of that whereof he himselfe is Author'.[63] Every subject authorizes the actions and judgements of the sovereign, so it follows that whatever the sovereign does, whatever punishment or act of violence the sovereign carries out, cannot be unjust. To complain about what the sovereign does is to thus complain about the very thing that one has authorized. It is to accuse no one other than oneself. In that sense, the sovereign is insulated from complaint because the sovereign's actions are the subject's own, meaning that the Hobbesian sovereign is *authorized without qualification*.

> Nothing the Soveraign Representative can doe to a Subject, on what pretence soever, can properly be called Injustice, or Injury; because every Subject is Author of every act the Soveraign doth . . . And therefore it may, and doth often happen in Common-wealths, that a Subject may be put to death, by the command of the Soveraign Power; and yet neither doe the other wrong.[64]

This applies even if the Sovereign *puts to death an 'Innocent Subject'*. The fact that the Sovereign's actions are authorized by the subjects – that the body politic is constituted by the bodies of the citizens, that the civilian Self is always already part of the System, that Subject and Security have been forced to coincide – means that *there is no civilian Self that can be immunized against the violence of the System*. But it also means that the Sovereign has and must have nothing less than *complete immunity*. This Sovereign power to kill any of the Sovereign's subjects is by no means contradicted by the subject's right of self-defence that we discussed in chapter 2.

What emerges from accounts of the absolute power of the state such as Hobbes's is a claim for the immunity of the sovereign power, and it is a claim now made by the modern liberal state, which is one reason why Hobbes's account of sovereignty is still so compelling. The sovereign power is absolute and authorized to be absolute, and so too is its immunity. Contrary to the claims by Luhmann, Esposito, Sloterdijk,

63 Hobbes, *Leviathan*, 124.
64 Hobbes, *Leviathan*, 148.

and many others that community cannot exist without some form of immunitary apparatus, *it turns out to be modern sovereignty that is unable to exist without it.*

This also explains Hobbes's reference to immunity's close ideological cousin: impunity. Possessing immunity means that the sovereign power possesses impunity.[65] Hobbes's view was common at the time, and sovereign powers retain a hold of it. 'Impunity', from the Latin *impūnis* ('unpunished'), granted exemption from punishment or penalty. This became important to the political discourse of early modernity and the justification of sovereign power. It was used to try to circumvent the complaint that permission has not been granted to the sovereign to conduct such acts and atrocities. Grotius explicitly states that when classical authors write about killing 'innocent' people such as prisoners of war, 'they do not mean a Permission . . . but an Impunity'.[66] Certain things, he tells us in the chapter on spoiling an enemy's territory, 'by the Law of Nations may be damaged or destroyed with Impunity'.[67] Dressed in the garb of security, designated servants of the state *can kill with impunity*. Despite all the rhetoric of 'self-defence', despite the idea of legitimate defence grounded in law, and despite natural law assumptions about the defence of the organism, the only unchallengeable 'self-defence' is that carried out by sovereignty and its operatives. The rationale for the violence that the state uses in its 'self-defence' is nothing other than life itself and the life of its own Self as the System; that is, the life and security of its own body politic. It is this political self-defence claimed by the body politic that becomes evident in the practices surrounding immunity of state operatives from criminal liability for killing civilians in the line of duty. *Immunity* reveals *impunity*. Of course, the police are dangerous, as Rustin Cohle says in *True Detective*, in a passage cited as an epigraph to this chapter: they are dangerous because they can do terrible things to people *with impunity*. And the reason why immunities exist for certain groups of people lies in the need for their *security as a class of persons representing sovereignty*.

65 Thomas Hobbes, *The Elements of Law: Natural and Politic* (1640), ed. Ferdinand Tönnies (London: Cass, 1889), 113.

66 Grotius, *Rights*, 1279.

67 Grotius, *Rights*, 1467.

Kenneth Culp Davis once suggested that 'one can read a hundred judicial opinions about sovereign immunity without ever encountering a reason in favour of it. Decisions based on sovereign immunity customarily rest on authority, the authority rests on history, and the history rests on medievalisms about monarchs.' Other lawyers follow suit, arguing against the 'medievalisms' and the 'anachronistic principles' upon which immunity has been founded, especially the idea that the king can do no wrong. 'Sovereign immunity must go', they cry in unison.[68] But the medieval anachronisms remain integral to modern sovereign power. Most states find that the king's robes really do fit rather well, even if they also feel obliged to dress them up with some liberal democratic accessories. As case law often finds, the doctrine of sovereign immunity stems from the ancient common law monarchical tenet that the king can do no wrong. The privilege of not being dragged through his own courts is 'the grandest of his immunities', as Frederick Pollock and Frederic William Maitland comment about Henry III, since without such immunity the sovereign power would spend its life in court.[69] But more telling, for us now, is that as well as thinking that it can do no wrong, the state has assumed this part of the king's power to kill with impunity. In the middle of the twelfth century, John of Salisbury suggested that the Prince, as an absolute law unto himself and bearer of the public persona, can *shed blood blamelessly*.[70] The modern state likes to think the same of itself. The state: born to kill. Born to kill in the name of security.

68 Kenneth Culp Davis, 'Sovereign Immunity Must Go', *Administrative Law Review*, vol. 22, no. 3 (1970), 383–40; John E. H. Sherry, 'The Myth That the King Can Do No Wrong: A Comparative Study of the Sovereign Immunity Doctrine in the United States and New York Court of Claims', *Administrative Law Review*, vol. 22, no. 1 (1969), 39–58; Noel P. Fox, 'The King Must Do No Wrong: A Critique of the Current Status of Sovereign and Official Immunity', *Wayne Law Review*, vol. 25, no. 2 (1979), 177–98; Erwin Chemerinsky, 'Against Sovereign Immunity', *Stanford Law Review*, vol. 53, no. 5 (2001), 1201–24.

69 Sir Frederick Pollock and Frederic William Maitland, *The History of English Law before the Time of Edward I*, vol. I (1898) (Indianapolis: Liberty Fund, 1968), 544–45.

70 John of Salisbury, *Policraticus* (Cambridge: Cambridge University Press, 1990), 30.

Coda: Fiction

In the introduction, we noted Blackstone's observation, in his *Commentaries* on English law, that the discovery of legal fictions tends to startle students of the law. Nonetheless, he claims, the student will upon further consideration find out that legal fictions are in fact 'highly beneficial and useful'. At the Council of Lyon in 1245, Pope Innocent IV argued that the *universitas* was a person without a body and, as such, was a 'fictitious person' (*persona ficta*), and he considered the corporation to be the same. Hobbes pointed out that 'there are few things, that are uncapable of being represented by Fiction', giving as examples churches and hospitals.[71] Universities, Churches, Hospitals. As Maitland once quipped, sometimes the most interesting persons we ever know are legal fictions.[72]

Many liberals baulk at the idea of legal fictions. Bentham considered them to be plain falsehoods and swindles that run like syphilis through the law – note the metaphor of disease – and undermine the province of legislation. Yet, as Lon Fuller writes, the legal fiction is not intended to function as a straightforward lie. 'A fiction is distinguished from a lie by the fact that it is not intended to deceive.' For Blackstone, legal fictions are 'troublesome, but not dangerous', and he sought to console the student, and probably to console himself too, by reminding us that legal fictions can be traced back to the Romans and that their purpose is to simply 'prevent a mischief, or remedy an inconvenience'. In the end, he writes, legal fictions turn out to be rather 'expedient', in being rooted in established traditions of law and thus better serving the purpose of justice. Legal fictions are necessary, we are told, because the law has retained feudal relics which cannot be removed without doing severe damage to the law, but which also cannot keep pace with changes in contemporary society. 'We inherit an old Gothic castle, erected in the days of chivalry, but fitted up for a modern inhabitant.' Or as Fuller puts it, if all goes well in social life, there is no need for legal fictions; such fictions 'represent the pathology of the law'. The fiction is thus a little like

71 Hobbes, *Leviathan*, 113.

72 F. W. Maitland, 'Moral Personality and Legal Personality' (1903), in *The Collected Papers of Frederic William Maitland*, vol. III (Cambridge: Cambridge University Press, 1911), 309.

an illness in the body: 'Only in illness, we are told, does the body reveal its complexity.'[73]

Yet might there not also be other reasons for the fictions, at least for us today, and more intricately connected to power? We might not like to be told that we are being peddled a fiction, but the fiction might in fact be useful in helping us grasp the truth. Once we realize, for example, that the corporation as a legal fiction is the gift of the sovereign, then the truth of capitalist power, colonial violence, imperial appropriation, and the creation of the modern state come more sharply into view. Blackstone's stress on a need for legal fictions indicates that the law is relying on something for which it does not have an adequate theoretical basis, and so the fiction has a utility.[74] Many of the discussions of legal fictions point to those pertaining to monarchical authority, captured in some of the maxims through which we still talk about sovereign power, such as 'The King is dead, long live the King'. One such maxim has already appeared in this chapter, 'The King Can Do No Wrong', cited from Blackstone's *Commentaries* but gratefully inherited by the impersonal sovereignty of the modern state. The reason the fiction is not a lie rests on the fact that the author of the fiction – that is, the one who *authorizes* it – does not intend to deceive and fully expects that we know that what is in place is a fiction. The fiction is not outside truth but is in fact expected to be accepted as true; herein lies its power. A fiction such as 'The King Can Do No Wrong' must be understood as working within truth. Such fictions emerge from and serve to undergird the very forms of power that render them true.[75] Indeed, we can push this further and suggest that the fiction can undergird forms of power because it also helps sustain the very existence of such power. The derivation of 'fiction' from *fingere* plays less on the modern sense of *feigning* and more on the original sense of *creating*, or *fashioning*, or *shaping*. In other words, legal fictions help *fabricate* the very power that does not yet exist but which the state needs to be brought into being. This is the reason the fiction is

73 Lon L. Fuller, *Legal Fictions* (Stanford, CA: Stanford University Press, 1967), viii, 6; Blackstone, *Commentaries*, vol. III, 43, 107, 267–68.

74 John H. Langbein, 'Introduction to Book III', in Blackstone, *Commentaries*, vol. III, vi.

75 Michel Foucault, 'Interview with Lucette Finas' (1977), trans. Paul Foss and Meaghan Morris, in Meaghan Morris and Paul Patton (eds.), *Michel Foucault: Power, Truth, Strategy* (Sydney: Feral Publications, 1979), 74.

so important to law, because the power that lies within the fiction gets grounded in and legitimized through law. This is also why we need to take fiction seriously. It is also why critical theory can be understood as the unmasking of fictions.

The fiction is the fabrication and deployment of power. The essence of political violence, and the law of this violence, is carried in the fiction, but a fiction always elided by the sovereign power itself. The fiction of noncombatant immunity combined with the fiction of the immunity of the state's ambassadors has a political reality: the dead body produced through security. Perhaps all immunity is a fiction in this way. Perhaps all security is too. Perhaps that is why security and immunity live together, conjoined within the sovereign power, and why they kill together too, even the very bodies they are said to be protecting.

Index